The twofold purpose of this book is to serve as an introductory text for direct energy conversion courses as well as fill the gap between the introductory texts and the specialized papers intended for those who are already versed in one method or another.

The first chapter places the field of direct energy conversion in relationship to energy and development. The second chapter treats briefly the physical principles. The following nine chapters deal with nine separate methods of converting energy directly, and the last chapter treats in less detail more than fifteen additional methods of converting energy directly into electricity. In this text information has been gathered from published and unpublished works and in some instances original material has been added by the author. The MKSC system of units has been used throughout the text unless otherwise stated.

This book can be used for both senior and graduate courses in direct energy conversion. Since the ten chapters dealing with the different ways of converting energy are only slightly interdependent, it is possible to choose several chapters for one- or two-semester courses without the necessity of covering the entire book or following it in its given order. The problems given at the end of each chapter are of different levels of sophistication. A large list of references has been added at the end of each chapter to give the teacher more freedom in extending the text toward areas of his interest depending on the engineering department in which he is lecturing.

This book may also be of interest to those engineers and students who are working with one method of direct conversion and who wish to acquire a picture of the entire discipline. It may be used as a reference book for those engineers and scholars practicing in one of the fields of direct energy conversion.

M. ALI KETTANI
University of Pittsburgh

DIRECT ENERGY
CONVERSION

ADDISON-WESLEY PUBLISHING COMPANY
Reading, Massachusetts · Menlo Park, California · London · Don Mills, Ontario

This book is in the
ADDISON-WESLEY SERIES IN ELECTRICAL ENGINEERING

Consulting Editors:

DAVID K. CHENG
LEONARD A. GOULD
FRED K. MANASSE

To my parents

PREFACE

Direct energy conversion is a new interdisciplinary field of engineering which is presently developing from the research work of engineers and scientists. When the author was asked in 1966 to help develop a graduate course on direct energy conversion at the University of Pittsburgh, he realized that a textbook was necessary. This book has been written in an attempt to fulfill this need.

The books then in existence treated only five methods of energy conversion and were of two types: (1) books compiled from the work of different authors and (2) books written as an introduction to the most advanced methods in the field and consequently intended for undergraduate students only. The purpose of this book is to fill the gap between the introductory works and the specialized papers intended for those who are already versed in one method or another.

The first chapter places the field of direct energy conversion with respect to energy and development. The second chapter treats briefly the physical principles. The following nine chapters deal with nine separate methods of converting energy directly, and the last chapter treats in less detail more than fifteen additional methods of converting energy directly into electricity. The different chapters are written, whenever possible, in the following sequence: introduction, physical bases, other devices based on the same physical principles, types of generators, generator analysis, materials, related topics, systems, and conclusions. The mksc system of units has been used throughout the text unless otherwise stated. In this text, information has been gathered from published and unpublished works and in some instances original material has been added by the author.

This book can be used for both senior and graduate courses in direct energy conversion. In a senior course, Chapters 1 and 2 serve as a basic introduction to the field. Chapter 2 does not pretend to really develop the needed physics in any detail; it gives only the background necessary for the understanding of the following chapters. Since the ten chapters dealing with the different ways of converting energy are only slightly interdependent, it is possible to choose several chapters for a one- or two-semester course without the necessity of covering the entire book or following it in its given order. The problems at the end of each chapter are of different levels of sophistication. About half of them have been chosen to match the level of a senior class and to give the student confidence in developing mathematical solutions and in understanding the text. A complete list of references has been added at the end of each chapter to give the teacher more freedom in extending the text toward areas of his interest, depending on the engineering department in which he is lecturing. Indeed, this textbook may be used in many of the engineering branches, especially electrical and mechanical engineering, but also chemical, nuclear, and aeronautical engineering. Chapter 9 would be useful as part of a bioengineering course. For all these reasons, the density of information has been kept as high as possible, without straining the reader who has a satisfactory background in undergraduate physics. Repetitions have been avoided except where absolutely necessary for the understanding of the text.

For a graduate course, Chapter 2 can be omitted and the student referred to the references given at the end of the chapter. The complete text will give full benefit if it is used for a two-semester course. Here again, a shorter or more specialized course can be designed by a proper selection of chapters. The student who wishes to study in greater detail specific problems of interest to him should consult frequently the list of references at the end of each chapter. About half of the problems given are devised so as to suit a graduate level of sophistication and to give the student enough training to extend the theory to special cases. It is hoped that at the end of the two-semester course, the student will have a "direct energy" thinking, enabling him to solve original problems and to participate in research efforts.

This book may also be of interest to those engineers and students who are working with one or the other method of direct conversion but wish to acquire a picture of the *entire* discipline. It may be used as a reference book for those engineers and scholars practicing in one of the fields of direct energy conversion.

The author wishes to thank Dr. Howard B. Hamilton, Head of the Electrical Engineering Department at the University of Pittsburgh, for his continuous encouragement during the writing of this work. The author is grateful to the reviewers advising his publisher for their invaluable suggestions and comments. Special thanks are extended to Mrs. Barbie Arnold for proofreading the manuscript, to Miss Margaret Kisso, Miss Patricia Wesesky, and Miss Carol Johnston for typing the several versions of this text, and to Mr. Varouj Najarian for drawing the graphs.

M. A. K.

Riyadh, Saudi Arabia
July 1969

GLOSSARY

a = acceleration (m/sec^2)

a^* = MHD geometric parameter

α = activity

A = atomic weight

$\mathbf{A}$ = vector potential (W/m)

$\mathscr{A}$, $\mathscr{A}_1$, $\mathscr{A}_2$, $\mathscr{A}_3$, $\mathscr{A}_4$, $\mathscr{A}_5$ = constant values

$\mathscr{A}_c$ = photovoltaic curve factor

$\mathscr{A}_r$ = reflectance

$\mathscr{A}_v$ = photovoltaic voltage factor

b = coefficient of expansion (m^3/sec)

b^* = coefficient of expansion (m/$^\circ$K)

ℓ = mobility (m^2/V-sec)

$\mathbf{B}$ = magnetic field (W/m^2 = Tesla)

$\mathscr{B}_1$ = Stephan-Boltzmann constant (W/m$^2\cdot{}^\circ$K^4)

$\mathscr{B}_2$ = electron emission constant (A/m$^2\cdot{}^\circ$K^2)

c = speed of light in vacuum (m/sec)

c_f = fuel cost per unit heat input ($/J)

c_t = total cost per unit energy ($/J)

c_τ = fixed cost per unit energy produced by topper or tailer ($/J)

c_1 = fixed cost per unit energy produced by conventional plant ($/J)

c_m = specific heat per unit mass (J/$^\circ$K·kg)

c_p = specific heat per unit mass at constant pressure (J/$^\circ$K·kg)

c_v = specific heat per unit mass at constant volume (J/$^\circ$K·kg)

C = heat capacity (J/$^\circ$K)

C_E = Ettinghausen coefficient (m$^3\cdot{}^\circ$K/W·A)

C_F = capacitance (F)

C_N = Nernst coefficient (V·m^2/W·$^\circ$K)

C_p = heat capacity at constant pressure

C_t = total cost ($)

C_v = heat capacity at constant volume
$\mathscr{C}$ = plasma factor
$\mathscr{C}_E$ = equilibrium constant of a reaction
$\mathscr{C}_f$ = friction coefficient
$\mathscr{C}_F$ = ferroelectric Curie constant
$\mathscr{C}_M$ = magnetic Curie constant
$\mathscr{C}_r$ = mean reflection coefficient
$\mathscr{C}_\alpha$ = slippage coefficient
$\mathscr{C}_\alpha'$ = EHD coefficient
$\mathscr{C}_\beta$ = constant value
$\mathscr{C}_\gamma$ = constant value
$\mathscr{C}_\lambda$ = constant value
$\mathscr{C}_v$ = EHD coefficient
$\mathscr{C}_1$ = ferroelectric coefficient
$\mathscr{C}_2$ = ferroelectric coefficient
$\mathscr{C}'$ = performance characteristic of a refrigerator
d = distance (m)
d = piezoelectric coefficient
$\mathbf{D}$ = electric displacement (C/m)
D_1 = EHD velocity parameter
$\mathscr{D}$ = diffusion coefficient (m^2/sec)
$\mathscr{D}_p$ = diffusion coefficient for holes (m^2/sec)
e = electronic charge (C)
e_s = emissivity coefficient
e = piezoelectric coefficient
$\mathbf{E}$ = electric field (V/m)
E_{br} = breakdown strength (V/m)
$\mathscr{E}$ = energy (J)
$\mathscr{E}_a$ = energy level of acceptors (J)
$\mathscr{E}_c$ = lowest energy level in the conduction band (J)
$\mathscr{E}_{\mathrm{cap}}$ = energy stored in a capacitor (J)
$\mathscr{E}_d$ = energy level of donors (J)
$\mathscr{E}_F$ = Fermi level (J)
$\mathscr{E}_g$ = energy gap of the forbidden band in a semiconductor (J)
$\mathscr{E}_{\mathrm{ind}}$ = energy stored in an inductor (J)
$\mathscr{E}_{\mathrm{kin}}$ = kinetic energy (J)
$\mathscr{E}_n$ = energy of an electron in the nth quantum state (J)
$\mathscr{E}_{\mathrm{pot}}$ = potential energy (J)
$\mathscr{E}_v$ = highest energy level in the valence band (J)
$\mathscr{E}_v$ = energy loss per unit volume due to viscosity (J/m^3)
f = frequency (sec^{-1})
f_B = Fermi factor for bosons
f_e = distribution function of the electronic velocities
f_F = Fermi factor for fermions

f_p = plasma frequency (sec^{-1})

f = fugacity (N/m^2)

F = force (N)

F_n = Faraday number (C/mole)

F' = force per unit volume (N/m^3)

$\mathscr{F}$ = free energy (J)

g = generation rate of particles (m^3/sec)

g_r = radiative generation rate (sec^{-1})

g = piezoelectric generator coefficient (V·m/N)

G = power gain

$\mathscr{G}$ = Gibbs free energy (J)

h = Planck's constant (J·sec)

h = piezoelectric stress coefficient (N/m·V)

$\mathbf{H}$ = magnetic flux density (A/m)

$\mathscr{H}$ = enthalpy (J)

i = ratio of the partial pressure of the seed vapor to the partial pressure of the carrier gas in MHD

$\bar{i}$ = complex current density ratio

ι = degree of ionization

I = electric current (A)

I_0 = dark or saturation current (A)

$j = \sqrt{-1}$

J = current density (A/m^2)

J_s = saturation current density (A/m^2)

k = Boltzmann constant (J/molecule·°K)

k_p = constant of propagation of a wave (rad/m)

k = piezoelectric electromechanical coupling coefficient

K = thermal conductance (W/°K)

K_1 = charge density parameter

$\mathscr{K}_F$ = Faraday loading factor

$\mathscr{K}_H$ = Hall loading factor

ℓ = thickness, length (m)

L = inductance (H)

$\mathscr{L}$ = Lorentz number (V/°K^2)

m = mass flow (kg·m^{-2}·sec^{-1})

m_A = ratio of the illuminated area over the effective area presented to the sun

m_b = specific weight of fuel cell battery (kg/W)

m_e = photovoltaic energy ratio

m_l = load ratio

m_m = mirror ratio

m_p = pinch ratio

m_r = ratio of reflected radiation

m_t = pinch time ratio

m = magnetic moment (A·m^2)

M = mass (kg)

M_c = total mass of a gas in a container (kg)

M_t = total weight of fuel cell system (kg)

M_s^0 = specific weight of fuel cell system (kg/W·sec)

$\dot{M}$ = mass flow (kg·m^{-2}·sec^{-1})

$\mathcal{M}$ = Mach number

$\mathcal{M}$ = magnetic dipole per unit volume (A/m)

n = number of particles per unit volume (m^{-3})

n_e = negatively charged particle density (m^{-3})

n_p = positively charged particle density (m^{-3})

n' = number of moles in a gas

$\dot{n}$ = number of interactions per unit time per unit volume (sec^{-1}·m^{-3})

$\dot{n}_{Cs}$ = number of cesium particles arriving at the surface of the cathode per unit time (m^{-2}·sec^{-1})

n = integral number

n_f = number of degrees of freedom

n_g = number of degenerate states

n_z = atomic number

N = total number of molecules in a container

N_0 = Avogadro's number (molecules/mole)

N_1 = EHD interaction parameter

$\mathcal{N}$ = number of moles in a gas

p = nonequilibrium concentration of holes (m^{-3})

p_b = probability function

p_n = concentration of holes in the n-region (m^{-3})

p_p = concentration of holes in the p-region (m^{-3})

p_r = pressure (N/m^2)

$\dot{p}$ = momentum (kg·m/sec)

$\dot{p}_x$ = momentum in the x-direction (kg·m/sec)

$\not{p}$ = pyroelectric constant (C/m·°K)

$\not{p}$ = dipole moment (C·m)

P = power (W)

P_i = ion internal partition function

P_h = reversible heat rate (W)

P_n = neutral internal partition function

$\mathcal{P}$ = polarization (C/m)

q = total electric charge (C)

$\mathbf{q}$ = heat flow vector (C/m^2)

$\not{q}$ = Lagrangian coordinate

Q = cross section (m^{-2})

Q_t = total cross section (m^{-2})

$\mathcal{Q}$ = heat (J)

$\mathcal{Q}_c$ = heat from cold region (J)

$\mathcal{Q}'_c$ = heat from cold region in a refrigerator (J)

$\mathcal{Q}_h$ = heat from hot region (J)

$\mathcal{Q}'_h$ = heat from hot region in a refrigerator (J)

r = distance (m)

r_D = Debye radius (m)

r_n = electron recombination (sec^{-1})

r_p = hole recombination (sec^{-1})

r_r = radiative recombination rate (sec^{-1})

$\imath$ = constant of a gas (J/kg)

R = resistance (ohm)

R_l = load resistance (ohm)

R_p = parallel resistance (ohm)

R_s = series resistance (ohm)

$\mathcal{R}$ = universal constant of gases (J/mole·°K)

$\dot{\mathcal{R}}$ = gas constant per mole (J/mole·°K)

s = distance (m)

$\mathcal{s}$ = compliance coefficient (m^3/N)

$\mathcal{s}^*$ = elastic stiffness coefficient (N/m^3)

S = surface area (m^2)

S_p = streaming potential (m^3/C)

$\mathcal{S}$ = entropy (J/°K)

t = time (sec)

$\mathcal{t}$ = relaxation time (sec)

T = temperature (°K)

T_c = temperature of cold region (°K)

T_e = temperature of electrons (°K)

T_h = temperature of hot region (°K)

T_i = temperature of ions (°K)

$\mathcal{T}$ = ferroelectric ratio

u = velocity in the x-direction (m/sec)

U = internal energy (J)

v = velocity in the y-direction (m/sec)

$\mathcal{v}$ = velocity vector (m/sec)

$\mathcal{v}_d$ = drift velocity (m/sec)

$\mathcal{v}_s$ = speed of sound (m/sec)

V = voltage (V)

V_T = voltage at surface of tube (V)

$\mathcal{V}$ = volume (m^3)

$\mathcal{V}_f$ = final volume (m^3)

$\mathcal{V}_i$ = initial volume (m^3)

$\dot{\mathcal{V}}$ = reaction rate (m^3/sec)

w = velocity in the z-direction (m/sec)

$\mathcal{w}$ = random velocity (m/sec)

W = work (J)

W' = work done by a refrigerator (J)

x = strain, distance (m)

x = real part of the current ratio in MHD

x_b = impact parameter (m)

x_w = amount of precipitable water in the atmosphere (m)

X = stress (N/m^2)

y = coordinate (m)

Y = Young modulus (N/m^2)

$\mathscr{Y}$ = nonlinearity factor (m/V)

z = coordinate (m)

$\varkappa$ = ferroelectric coefficient (F·m^2)

Z = thermoelectric figure of merit (°K^{-1})

Z_E = Ettinghausen figure of merit (°K^{-1})

Z_N = Nernst figure of merit (°K^{-1})

α = ratio of the electric field in the y- and x-directions in MHD

α = attenuation constant (m^{-1})

α_λ = photon absorption coefficient (m^{-1})

β = imaginary part of the electric field ratio in MHD

β_c = droplet charge coefficient in EHD

β_e = electron parameter

β_I = ion parameter

β_v = photovoltaic voltage ratio

γ = ratio of specific heats

γ_m = ratio of maximum efficiency over Carnot efficiency

Γ_{dif} = diffusion current density (sec^{-1}·m^{-2})

Γ_r = random current density (sec^{-1}·m^{-2})

δ = piezoelectric strain coefficient (C/N)

δ_c = cross section coefficient (m^2/V)

ε = permittivity (F/m)

ε_r = relative permittivity

ε_0 = permittivity of free space (F/m)

ζ = second thermodynamic potential (kg/sec^2)

η = efficiency

η_c = Carnot efficiency

η_i = ideal efficiency

η_p = pressure ratio

η_Q = collection efficiency of a photovoltaic cell

η_τ = efficiency of a topper

η_1 = efficiency of a conventional converter

θ = angle (degrees)

Θ = emissivity

κ = thermal conductivity (W·°K/m)

λ = mean free path, wavelength (m)

Λ = activity coefficient

μ = permeability (H/m)

μ_r = relative permeability
μ_0 = permeability of free space (H/m)
ν = light frequency (sec^{-1})
ν_c = angular collision frequency (rad/sec)
ν_R = refractive index
ξ = first thermodynamic potential (kg/sec^2)
ξ_0 = thermionic dimensionless variable
Ξ = Seebeck coefficient (V/°K)
Π = Peltier coefficient (J/A)
Π_m = pyromagnetic coefficient
Π_p = hydrostatic pressure (N/C)
ρ = mass density (kg/m^3)
ρ_e = charge density (C/m^3)
ρ_p = probability density
σ = conductivity (mho/m)
Σ_s = thermionic parameter
τ = Thomson coefficient (V/°K)
τ_p = lifetime of holes (sec)
ϕ = angle (degree)
ϕ_r = rotational transform angle (degree)
Φ = flux of particles (m^{-2}·sec^{-1})
Φ_s = solar intensity (J/m^3)
χ = dielectric susceptibility
χ_M = magnetic susceptibility
ψ = wave function
ψ_0 = thermionic dimensionless variable
$\boldsymbol{\psi}$ = vectorial pressure (N/m^3)
Ψ = volumetric loading ratio
ω = angular frequency (rad/sec)
ω_c = cyclotron angular frequency (rad/sec)
ω_p = plasma angular frequency (rad/sec)
υ = viscosity coefficient (N/sec·m^2)
Υ = fuel cell transfer coefficient

CONTENTS

10   PIEZOELECTRIC POWER GENERATION

11   FERROELECTRIC POWER GENERATION

12   MISCELLANEOUS METHODS OF POWER GENERATION

A   FERROMAGNETIC POWER GENERATION

B   NERNST POWER GENERATION

1 ENERGY CONVERSION

1.1 Introduction

Since his appearance on earth, man has tried, with relative success, to master the forces of nature for his own benefit. Our primitive ancestors first applied energy conversion in the dawn of prehistory, when they tried to replace or amplify the limited forces of their muscles by the use of such primitive weapons and tools as stones and animal bones. Later, the discovery of fire and of the wheel were real revolutions in energy conversion, and, consequently, in the way of life of mankind. Man's continuing search for new sources of energy eventually led to the discovery of electricity, in the eighteenth century.

Electrical energy is the most convenient form of energy to which all other forms of energy may be converted. It is easy to transport, easy to control, and easy to transform into any form of work desired by the consumer. Most of the time, however, energy can be converted into electricity only through many intermediate transformations, which leads to limitations in efficiency, reliability, and compactness. With increasing understanding of the physical behavior of matter and with progress in material technology, it is becoming possible to convert energy more and more directly into electricity.

Direct energy conversion methods are now receiving increased attention and are undergoing concentrated development on both national and international levels, especially in defense, space, transportation, utilities, fuels, and mining. Direct, subsidiary, and derived uses and benefits of direct energy conversion are numerous. Required to fulfill the demands of a large variety of modern energy applications, several direct energy conversion methods are destined to play a basic role in the world's economy and technology.

The chain of energy handling can be divided into three stages: fuels, conversion methods, and usage. The vital middle link, energy conversion, is undergoing a revolution. Giant strides are being made on several continents, particularly spurred by breakthroughs that are an outgrowth of the space race. The developments in this middle link (i.e., conversion methods) can make or break the role of the other links. In the competitions among countries, regions, and industries, energy conversion may well be decisive. An entire spectrum of possibilities exists and careful, sensible, utilitarian matches depending on the particular application must be made.

As early as 1795 [93],* direct conversion devices such as Volta's battery had been developed, but direct energy conversion as an organized discipline is a very new field. Today, the problems of direct energy conversion are newer, more challenging, and more essential to the economy than ever. In spite of the need for it, this field has lagged in dissemination and implementation for a variety of reasons. In addition, the workers in each particular method have been so involved in their own specialities that they have little time for, or interest in, the other methods. Fortunately, these barriers are gradually breaking down.

Once the fusion of technologies has been mastered, one can realistically meet the ever-increasing needs of mankind. New energy sources, as well as ways to use the old sources, are at hand. New conversion methods are needed, new problems are waiting to be solved, and new opportunities are continually opening for engineers and scientists of all disciplines.

To supply power for use in outer space, on land, and on and in the seas, power sources with a variety of characteristics are needed. Some applications make particular demands and need energy supplies that are reliable, light in weight, capable of unattended operation for long periods of time, and silent (to avoid detection), while other applications have other needs and specialized requirements.

A succession of discoveries in direct energy conversion followed Volta's work on batteries. In 1802, Sir Humphrey Davy [27, Chap. VII] suggested converting chemical energy to electricity, leading to the fuel cell principle. Thermoelectricity originated in Seebeck's [84, Chap. V] discovery, in 1821, that electrical currents are generated at the junction of dissimilar metals which are at different temperatures. Faraday's [20, Chap. IX] discovery, in 1831, that when mercury contained in a glass tube was moved between the poles of a magnet, a voltage was generated perpendicular to the directions of the magnetic field and the mercury's motion, led to the idea of generating electricity by using the magnetohydrodynamic principle. In 1834, Peltier [94, Chap. V] discovered that the passage of an electric current through a junction of two different conductors creates a heating or cooling at the junction, depending on the direction of the current. The photovoltaic effect was first discovered by Becquerel [2, Chap. VIII] in 1839, when he noted that voltage could be generated when light is directed toward one of the electrodes in an electrolytic solution. In 1856, William Thomson (Lord Kelvin) [93, Chap. V] related Seebeck's effect to Peltier's by noting a lateral transfer of heat in a conductor carrying an electric current in a temperature gradient. Seventeen years later, Smith found that light can reduce the electrical resistance of a circuit element made of selenium. This effect is related to the photovoltaic effect and is known

---

* Numbers are keyed to references at the end of the chapter.

as photoconductivity. Work on semiconductors was started by Bell in 1880, and on thermionic emission by Edison [25, Chap. VI] several years later. Einstein's [29] proposition, in 1909, that matter is nothing but a form of energy was important in the development of the nuclear and fusion fields.

After their discovery, however, most of these effects remained scientific curiosities, and only a few of them were used in some secondary applications. With the advent of World War II and Sputnik I, there has been a continuing explosion of interest and application of many of the direct energy conversion devices based on these nineteenth century discoveries. This is due to the need to exploit new energy sources and to the progress in all the other fields of science and technology leading to higher efficiencies for direct energy conversion devices.

1.2 Energy Conversion and Energy Sources [2, 12, 51]

Energy is a fundamental engineering concept involving the capacity of doing work. Kinetic energy is the energy acquired by a material body of mass M moving with a speed v:

$$\mathscr{E}_{\text{kin}} = \tfrac{1}{2}Mv^2.$$

Potential energy is the energy available for extraction, such as the energy stored by water in a reservoir at some height above the ground level. This stored energy is potentially available and is transformed into kinetic energy by the force of gravity, which causes the water to fall and be set in motion. In this case, the amount of stored potential energy is

$$\mathscr{E}_{\text{pot}} = Fd,$$

where F is the force of gravity and d is the height of the fall. Potential energy is stored in many ways, as, for example, in chemical and nuclear substances. A material such as coal, oil, or uranium which stores potential energy and can release it readily is called a fuel.

Power is the amount of energy delivered per unit time. For instance, a trickle of water delivers a very small amount of energy every second, and therefore has very little power. A waterfall delivers a great amount of kinetic energy, i.e., mechanical energy, every second, and consequently provides a great amount of power.

Energy can exist in several forms, such as thermal, kinetic, electromagnetic, nuclear, chemical, or electrical. Energy can be converted from one form to another in many ways. For instance, when thermal energy is converted into mechanical energy in an automobile engine, the stored energy is the chemical energy available in the gasoline, which itself originated from the electromagnetic energy of the sun through photosynthesis in living matter. By combustion of the gasoline in a cylinder, thermal energy is released. The hot gases press against the pistons and by expansion convert the heat energy into mechanical energy to turn the shaft of the engine. This sequence has been summarized in Fig. 1.1.

In a conventional power plant, coal, oil, or natural gas is the available material in which energy is stored. The thermal energy released by combustion serves to heat water, turning it into steam. The steam expands against the blades of the turbines, thus transforming the heat energy into mechanical energy to the shaft. The shaft is coupled to an electric generator which finally transforms the transmitted mechanical energy into electricity.

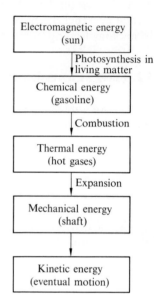

Figure 1.1. Conversion path leading to the mechanical energy of motion of an automobile. This energy also originates from the sun.

Today, most electrical energy is obtained from conventional fuels such as coal and gasoline, or nuclear fuel such as uranium and thorium, through the services of heat engines. Although the heat engine is only one stage of many in a conversion process, most of the losses occur there. For instance, in a typical central power station, the efficiency of the boiler and the thermodynamic efficiency of the steam turbine are about 90% each, whereas the electrical efficiency of the generator is often around 98%. The efficiency of the heat engine, around 40% at best, compares poorly with these figures. This efficiency obeys Carnot's limitation, as will be explained in Chapter 2. The Carnot efficiency is the maximum efficiency of an ideal heat engine which takes heat from a hot source at an absolute temperature T_h and releases it to a cold sink at a temperature T_c after doing work. This efficiency, given by

$$\eta_c = 1 - T_c/T_h,$$

cannot be very large. Indeed T_h is limited for obvious mechanical and metallurgical reasons to about 1200°K, and T_c is limited by the compression ratio to a value no lower than about 700°K.

To improve the efficiency of a power plant, it seems obvious that the number of converting stages should be minimum. This can be achieved through direct energy conversion. Furthermore, if Carnot's limitation can be eliminated by avoiding the heat engine stage, there will exist possibilities for much higher efficiencies. Two important devices for converting energy seem to achieve this goal—fuel cells and photovoltaic converters—since they are isothermal (i.e., at constant temperature) in their operation.

The sources of energy in the world can be divided into two main groups. The first is the *energy capital* due to the storage of solar energy in the form of chemical energy by the action of living organisms millions of years ago. The fuels in which this energy is stored, such as coal, gas, and oil, are called *fossil fuels*. These fuels are exhaustible if their use is continued indefinitely. Another type of energy capital is supplied by matter itself, since matter is a special form of energy which is practically inexhaustible. The second group is the *energy income*, the resources of which are continuously renewed and replenished, either instantaneously by the direct use of solar energy or after a short period of storage in the form of winds, rivers, waves, natural steam, or other forms (wood, animal waste).

Note that except for matter, all these energies originate from the sun. Whether a specific energy is recognized as an income or a capital depends on its storage time scale. This is represented in Fig. 1.2. For instance, in the case of hydraulic

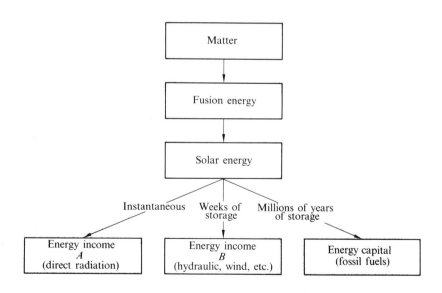

Figure 1.2. Energy income and energy capital, all originating from the sun and from matter. The difference is in the duration of the storage period.

energy, the heat of the sun vaporizes the seas and carries the produced vapor up into clouds; then the winds, which are also due to the temperature gradients in the atmosphere, carry the clouds to higher altitudes; precipitation results when the clouds cool; then water is collected in basins and falls down through rivers and eventually through hydraulic turbines to produce electric power.

The most used primary source of energy today is fossil fuel [6, 55, 56], which supplies 85% of the world's present need. This fuel powers most of the existing power stations using steam turbines, gas turbines, or internal combustion engines. Since fossil fuels are energy capital, the existing reserves are not replenished, and continuous use may lead to their exhaustion, thus creating a serious energy problem if mankind is not prepared to tap other sources. Indeed, world-wide energy consumption is increasing at an accelerated rate. From the time of Christ until the middle of the last century, it is estimated that mankind used around 8 Q of energy.* In the last century 4 Q were consumed, and the need in the next century will be between 100 and 400 Q, due to the increase in population and higher standard of living. The United States, with only 6% of the world's population, consumes 36% of the world's energy. The known recoverable reserves of fossil fuels are very limited and cannot satisfy this large need for long. As shown by Table 1.1, prepared by D. C. Duncan and V. E. McKelvey [15] of the U.S. Geological Survey, the known recoverable reserves of the United States are not larger than 5.5 Q, whereas those of the entire world are smaller than 23 Q.

Even if the undiscovered marginal and submarginal resources which are presently not economically recoverable were considered, the problem of diversifying the energy sources would not be alleviated. Indeed, these resources are estimated at about 81 Q for the U.S. and 452 Q for the entire planet.

Nuclear energy [17, 35, 54] is now being put into commercial service in many countries, especially Great Britain. According to present theories, mass is nothing but a particular form of energy. This means that energy can be neither destroyed nor created, and therefore the sum of mass and energy is conserved in every

Table 1.1

Reserves in fossil fuels

Fuel	U.S., Q	World, Q
Coal	4.600	18.00
Natural gas	0.310	2.11
Petroleum	0.278	1.70
Oil	0.298	1.70
Total	5.486	22.91

---

*   1 Q = 10^{18} Btu = 2.93×10^{14} kWh.

transformation. This is, in fact, the generalized first law of thermodynamics, which will be discussed later.

If mass is a particular form of energy, it should be possible to convert it. In the early 1910's Einstein [29] gave the now familiar relation for equivalence of mass and energy,

$$\mathscr{E} = Mc^2,$$

where c is the speed of light in vacuum. Energy cannot appear without destruction of mass. This is dramatized in the fission and fusion reactions. In fission, a heavy nucleus is broken down into two or more smaller nuclei, with the release of a huge amount of nuclear energy. It is noted that the total mass of the particles of fission is smaller than the mass of the fissured particle which is the fuel; mass has been converted into energy. If only one 1 kg of mass were entirely transformed, the total amount of energy would be equal to the present daily consumption of energy of the entire world. The two potentially usable fuels are uranium and thorium. In the case of uranium, only isotope U^{235} is fissionable. This isotope makes up 0.7% of natural uranium, and the rest is mainly U^{238}. The known deposits of uranium and thorium are shown in Q units for the United States and the entire world in Table 1.2 as reported by A. Cambel [15].

Table 1.2

Reserves in nuclear fuels

Fuel	U.S., Q	World, Q
Uranium	22	80
Thorium	7	48

Hydropower [27, 41], the major source of energy income, is in a well-advanced stage of development. A large portion of the hydropower resources in North America and in Europe are being exploited. For instance, in 1963, hydroelectric plants in the United States were producing 172 billion kWh annually, whereas the total reserve capability was estimated to be about four times this amount. However, larger potentialities are still waiting to be used in the underdeveloped countries of Asia, South America, and Africa.

The solar [23, 80, 98] radiation received by the earth is practically limitless and yet virtually unused. Solar radiation received only in the United States is about one thousand times the country's consumption of all forms of energy. With the advances in direct energy converters, great hopes are now seen for mastering this huge amount of energy. Indeed, solar energy can fuel photovoltaic, fuel cell, thermionic, thermoelectric, and other types of energy converters by the use of concentrators. This problem of concentration is the main handicap in solar cell progress. Solar energy can also be tapped through the use of sea water in hydroelectric plants and evaporation of the used water by solar energy in those depressions which are below sea level, such as Djerid (Tunisia), Kattara (Egypt), the Caspian Sea, and the Dead Sea.

Other sources of energy do not offer great promise, but can be considered in some instances and for special applications. Among these, wind power [4] has the advantage of being a power income, free and lacking any kind of end products. The most important aerogenerator, built in the 1940's in Vermont, produced 1250 kW [15, p. 334]. The other sources considered are wood, animal wastes, the tidal power of the oceans [33] (in France), the temperature gradients in the oceans [3] and the earth (geothermal energy), natural steam, and other sources of lesser importance.

1.3 Energy Conversion and Heat Engines [22, 46, 48]

A multitude of devices and schemes has been conceived to transform energy from one form to another. Six of these forms—electromagnetic, thermal, kinetic, nuclear, electrical, and chemical—are shown in Fig. 1.3. The arrows show the process of transformation from one form to another. Any path can be taken, although it seems obvious that the most direct ones would be the most efficient. For instance, the electromagnetic energy of the sun can be transformed into electricity by following a lengthy path. It can be first transformed into chemical energy by photosynthesis, then to thermal energy by combustion, then to kinetic energy by thermal expansion, and finally to electricity by the use of a conventional generator. It is, however, highly improbable that a system using such a circuitous path would be efficient. Indeed, it would seem much more convenient to convert the solar energy directly into electricity by the use of the photovoltaic effect.

It was seen that the efficiency of any conversion process is limited by Carnot's efficiency as obtained from the ideal Carnot cycle (explained in the next chapter) whenever there is a heat engine in the system. In direct energy conversion, the heat engine is not always avoided, except for the two schemes mentioned above (photovoltaic and fuel cells).

In heat engines, thermal energy is converted into mechanical energy, which in turn is converted into electricity by a conventional electrical generator. Thermodynamic descriptions of their operation are given by power cycles such as the theoretical Carnot cycle, which cannot exist in practice.

For a steam power plant, the Rankine power cycle is used. In this plant, the working fluid is evaporated in a boiler. The vapor is then adiabatically expanded through a steam turbine to do work. It is then condensed into liquid and pumped back to the boiler. The cycle of this operation is illustrated in Fig. 1.4. Line a-b represents the compression of the liquid at constant volume in the boiler by the addition of heat. Then vaporization of the liquid takes place at constant pressure (line b-d). Line d-e represents a reversible adiabatic expansion in the steam turbine, and line e-a represents the condensation of the vapor-liquid mixture at

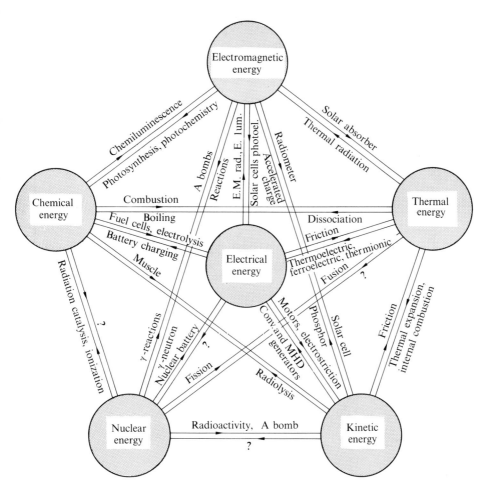

Figure 1.3. Energy conversion chart. The circles represent the different forms of energy and the arrows the ways of converting energy from one form to another.

constant pressure. The efficiency of this cycle is less than Carnot's. Also note that Rankine's cycle should be corrected for any new elements introduced in the circuit, such as superheat, reheat, or regenerative heaters.

Reciprocating gas power plants use Brayton power cycles. These cycles are also used in many other applications, such as aircraft propulsion. These heat engines use piston compressors and expanders with heat exchangers for heating and cooling the working gas. Since the fuel is burned directly in compressed air, the temperature limitations of the heat exchanger are avoided. The Brayton cycle is illustrated in Fig. 1.5. Line *a-b* represents the heating of the gas at constant entropy (see Chap. 2), which is then heated at constant pressure (line *b-c*). The compressed hot gas then expands in the gas turbine to deliver mechanical work

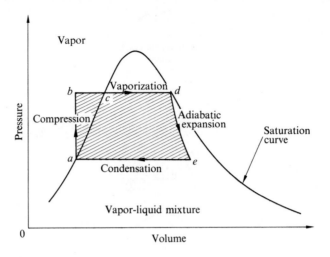

Figure 1.4. The Rankine p_r-$\mathcal{V}$ cycle. This cycle is delineated by a pair of isobaric lines, an adiabatic line, and a line at constant volume.

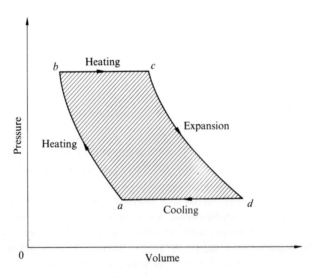

Figure 1.5. The Brayton p_r-$\mathcal{V}$ cycle. This cycle is delineated by a pair of isentropics and a pair of isobaric lines.

(line c-d) at constant entropy, and it is finally cooled at constant pressure, as shown by line d-a. For a closed cycle, the gas is heated and cooled in heat exchangers, whereas for an open cycle, air is drawn from the atmosphere at a and rejected to the open air from the turbine at d.

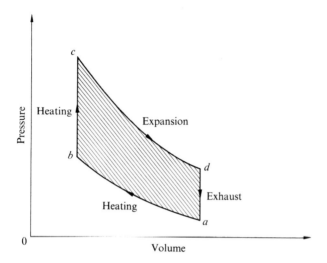

Figure 1.6. The Otto p_r-$\mathscr{V}$ cycle. This cycle is delineated by a pair of isentropics and a pair of lines at constant volume.

The operation of spark-ignition engines using gasoline can be described by the Otto power cycle. This cycle is shown in Fig. 1.6. The air fuel mixture is compressed by a piston along line a-b at constant entropy. The fuel is then ignited by a spark when the piston is near the top of its stroke, leading to a constant-volume compression and heating by combustion (line b-c). Then the gas expands to deliver mechanical work when the piston moves back (line c-d), and finally the combustion products are released as shown by line d-a. In the four-cycle engine, the combustion gases are exhausted to the atmosphere after compression, and fuel and air are admitted during the fourth stroke. In the two-cycle engine, combustion products are rejected and fuel and air are admitted at the second stroke.

The theoretical Diesel cycle consists of an isentropic compression followed by a constant-pressure heating and then expansion for the production of work, followed by the rejection of the combustion products to the atmosphere in a constant-volume process. In Diesel engines, ignition of the fuel is caused by the use of a much higher compression ratio than in the Otto cycle type of engine.

The Stirling power cycle has a theoretical efficiency comparable to that of Carnot's. Small power plants using this type of cycle were built for farms and, after World War II, for the U.S. Marines. The working gas is heated by external combustion, and thermal energy is stored during the cycle by the use of a regenerator. The Stirling cycle consists of an isothermal compression followed by an addition of heat at constant volume; then the gas is expanded first at constant temperature and then at constant volume to close the cycle.

Other types of cycles are used in several other heat engines. For instance, the Walter cycle is used in some submarine propulsion power systems, where the fuel is petroleum and the oxidizer is an oxygen-rich hydrogen peroxide product. In some space power generation systems, liquid hydrogen is used as the working fluid and the heat is provided by the losses from the other power systems in the spaceship. In this type of "tailer," a cryogenic power cycle is used.

1.4 Direct Energy Conversion Schemes [5, 16, 31]

To eliminate a complicated series of stages in energy conversion, it seems attractive to follow the most direct and consequently most efficient process. In some applications, it is necessary to obtain energy from a certain source in a certain location, and only certain methods of direct energy conversion are applicable to that kind of conversion. Often, such characteristics as ruggedness, silent operation, ease of maintenance, and simplicity are important factors for the choice of a specific converter. Here are outlined the most important direct energy schemes, starting with those which are not subject to Carnot's limitation.

Fuel cells [63] and photovoltaic cells. The hydrogen-oxygen cell is typical of all fuel cells and is the most efficient of all the types. Usually a fuel cell [40] converts the chemical energy of a fuel directly into electrical energy; it is not subject to the limitations of the Carnot cycle. Fuel cell research and engineering work are underway in the United States, the U.S.S.R., and Western Europe. The potential applications of fuel cells are manifold. For example, fuel cells found a primary use in the Apollo project as space elements. Other important contributions to this field have been made in a host of terrestrial applications ranging from defense needs to electrical vehicles.

The need for energy is increasing dramatically, year after year, and sources which were abundant yesterday may very well disappear tomorrow. The time will come when every reasonable source of energy will be tapped for the benefit of the human race. The sun is the principal source of energy for the earth and for man, and yet solar energy suffers a terrible waste. For instance, the 30,000 square miles of northern Chile, with continuous bright sunshine and practically no rainfall can theoretically yield as much heat per year in the form of solar radiation as is now produced annually with the whole world's fossil fuel supply. Since the beginning of recorded history, the energy of the sun has been used by concentrating it with lenses or mirrors and then converting it into heat. With the advent of semiconductor technology, direct generation of electricity became possible by using the photovoltaic effect in a *pn* junction.

Fusion power [10, 34, 73]. Nuclear fusion uses deuterium, which is a practically unlimited source of fuel. Two deuterium nuclei fuse into a heavier one, whose mass is smaller than the total mass of the original particles. The rest of the mass is converted into power—a much greater amount than in the case of fission. This phenomenon is at the origin of all energy delivered in the universe by the stars, including our own sun. To create the same phenomenon on an artificial industrial scale, it is necessary to satisfy many conditions, the most important of which is to reach the fusion ignition point, which may be at a temperature as high as several hundreds of millions of degrees Kelvin. Electrical power can be extracted directly in several ways, such as by converting the heat generated in a conventional power plant or by using the magnetic field generated by the hot fusion plasma to directly induce electrical currents in a circuit. Controlled thermonuclear research is underway in many parts of the world.

Magnetohydrodynamic and thermionic power generation. Magnetohydrodynamic (MHD) generators use high-velocity, electrically conducting gases interacting with magnetic fields to produce electrical power. To be conducting, the gas must be ionized; an ionized gas is called a *plasma*. The MHD electrical generation principles are similar to those resulting from the study of the induced electric fields that arise when conductors move through magnetic fields, so long utilized in rotating generators where solid copper wires are the conductors. In MHD generators, the moving conductor is a gas, with the attendant benefit of not depending on solid moving parts. Not until the 1930's [46, Chap. IV], however, were MHD phenomena seriously considered for electrical power generation, because of the stimulation of theoretical as well as experimental work on ionized gases and high-velocity gas flow in shock tubes, and the tremendous progress in material technology. MHD energy conversion provides an exciting possibility for improving the efficiency of central station power generation, as well as for such special applications as silent propulsion of submarines.

The thermionic converter is a heat engine based on thermionic emission, i.e., the boiling off of electrons from a hot metal surface. A great deal of research on different types of thermionic energy conversion schemes is presently conducted in many leading laboratories in the world. A thermionic converter is essentially a metal-vapor filled device with a hot cathode acting as an electron emitter and a cold anode acting as a collector. The emitter and collector are thermally insulated and provisions are made to heat the emitter and cool the collector. The thermionic converter is a low-voltage, high-current device with very important applications—compact power plants for stationary or naval applications, portable electrical generating systems, and electrical power sources for space vehicles, among others.

Thermoelectric power generation [14, 28]. The oldest direct conversion heat engine is the thermocouple, discovered by Seebeck. However, this effect was only recently considered for energy production, as a direct result of the advance in semiconductor theory. To use thermoelectricity for power production necessitated the improvement of thermoelectric materials. This was accomplished in the early 1950's when the first nuclear-heated thermoelectric generator was built in the AEC Mound Laboratory in Miamisburg, Ohio, followed by the development of the SNAP-3 (systems for nuclear auxiliary power) generator. Nuclear thermoelectric generators have been launched into space aboard satellites and installed on remote land stations and at the bottom of the oceans. Thermoelectricity has already entered the commercial markets where propane-fueled thermoelectric generators for use in camping equipment are available. Higher efficiencies are expected in the near future, with the result that thermoelectric power will replace dynamic conversion in many applications and locations.

Electrohydrodynamic (EHD) power generation. Here, electrical energy is converted into mechanical energy by exploiting the kinetic energy of a gas stream to push molecular ions from a region of low electrical potential to one of high electrical potential, where a collector can gather them. A corona electrode and an attractor electrode, acting as a corona discharge unit, serve to insert molecular ions at the point of low electrical potential. The ions leave the corona and head toward the attractor, but a working gas stream is sent into the tube to carry them away toward a collector electrode against an opposing electric field. Hence the gas does work and the heat lost by the gas is directly converted into electricity. An EHD generator is a cheap high-voltage source which has many uses, for example, electrostatic precipitation for smokestacks, salt precipitation for desalination, power supplies for high-voltage electron devices, and smoke eliminators for automobile exhaust pipes.

Dielectric power generation. When an electric field acts on an insulating material, the material becomes polarized and acts as an insulating dipole. Some materials are polarized under pressure and are said to be *piezoelectric*. The piezoelectric effect permits the direct transformation of mechanical energy into electrical energy. It is an example of the more general electrostriction phenomena in dielectrics.

In some piezoelectric materials, a polarization exists even in the absence of an electric field. The surface charges may be enhanced by heating or cooling the material, thus allowing the direct conversion of heat into electricity. Such materials are called *pyroelectric*.

When the spontaneous polarization in a pyroelectric material can be effected by the application of an external electric field, the material is called *ferroelectric*. Ferroelectricity allows the direct conversion of heat into electrical energy, and the converter is an electrical capacitor whose capacitance is changed by temperature. If heat is added, the net capacitance will decrease and the voltage will rise; the capacitor is then made to discharge through the load, extracting a net gain in electrical energy from the originally added heat.

Other direct energy conversion methods. Thermal energy can be converted into electricity by utilizing the variation of the permeability of a *ferromagnetic* material with temperature. A ferromagnetic material, such as cobalt or nickel, is a material in which magnetic effects are very strong. A varying voltage may be induced in a ferromagnetic-cored coil which is submitted to a thermal cycle.

Phase transition can be used for generating electricity. For instance, an electrical potential is produced by the freezing of water. The voltage produced between the ice and the solution from which it originated can be as high as 200 V for some special inorganic solutions. The voltage is determined by the freezing rate of the solution.

When a fluid flows on a solid surface, *electrokinetic* potential may be generated, thus converting mechanical energy into electricity. An electronic transducer used to measure pressure fluctuations in oil pipelines is based on this phenomenon.

The *photoemission effect* can be used to convert radiant energy directly into electrical energy. When the photoemissive layer is illuminated, photons collide with electrons and release some energy to them in the same manner as in a thermionic diode. The electron is then emitted with a kinetic energy equal to the excess of the energy received from the photon over the work function of the material.

The *photoelectromagnetic effect*, discovered in 1934 in the Soviet Union [90, Chap. XII], can also be used for the conversion of radiant energy. Here the illumination is applied on a semiconductor material in the presence of a magnetic field. Charge carriers are then formed near the surface of the material and tend to diffuse under the effect of a particle concentration gradient.

Another example is the nuclear battery. It depends upon the emission of charged particles from a surface coated with a radioisotope. The particles are collected on another surface, thus creating a voltage directly from radioactive decay.

These and other potential energy conversion schemes are being vigorously investigated by scientists and engineers. There may be many other methods that probably have not yet been thought of, and the field is wide open for research.

1.5 New Sources and Economical Development [1, 39, 63]

Civilization was born as a response of man to the natural challenge of a harsh life in a difficult environment, as was pointed out by the British historian Arnold Toynbee [88]. Thus, civilization is a direct result of the success of a human group in converting the natural sources of energy to its own benefit, as is attested by the earliest civilizations of Egypt, Mesopotamia, or the eastern Mediterranean islands. This success can be visualized as a material one, the progress of which depended on successive inventions of new tools and machines or, in broader words, of new "energy converters" for the use of the energy capital and energy income to supplement human and animal labor. Progress was rather slow until the industrial revolution in Europe, followed in the nineteenth century by what can be called a "domestic revolution," when energy conversion on an efficient industrial scale reached the individual and raised his standards of health and living.

This progress did not have the same rate of advance in all the countries of the world. On the contrary, great disparities kept widening, reaching grave levels,

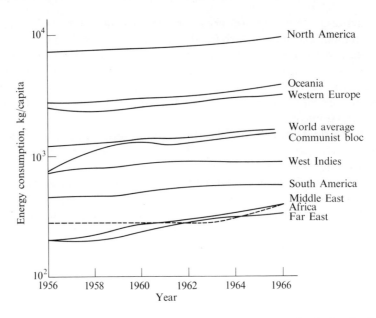

Figure 1.7. Energy consumption in kilograms of coal equivalent per capita during the decade 1956–1966 for several regions of the world. The consumption of a North American is more than forty times higher than that of an average Afro-Asian. This ratio does not seem to decrease with time.

Reference: United Nations Statistical Yearbook, 1966.

as shown in Fig. 1.7. From this figure, it can be seen that the energy consumption of an average North American is equal to the energy consumed by about 25 average Africans, for instance. The relationship between the standards of living and energy consumption for different countries is obvious, as clearly shown in Fig. 1.8. The individual output is closely related to the energy available for the individual, which in turn is related to the capital investment.

Closing the gap between the most advanced countries and those less developed necessitates investment and good planning in two important areas: first, electrification, which means that the available sources of energy should be transformed into electricity; and second, consideration of the different aspects of energy consumption which respond to and amplify local needs. In most of the underdeveloped countries, power networks are either embryonic or limited to zones of development around the large cities. Most of the areas outside the large cities have isolated power plants based on Diesel engines, small hydropower turbines, and perhaps gas turbines. In these areas, the cost of electricity is extremely high due to a combination of such factors as the small scale of the generating plant, low load factor, and high fuel cost. In such areas, the use of fuel-saving direct energy schemes using solar or wind energy or any combination of local sources can be extremely advantageous. A third type of area is the most widespread in the underdeveloped countries. These are the "remote" areas where there is no generation of electricity. Often the bulk of the population of the underdeveloped countries lives in such areas and it may be a long time before the central governments can afford to build power transmission lines or central stations in them. Direct energy conversion can be of great help in these areas.

As for the consumption of energy, its primary function is to raise the agricultural productivity of the land. Lack of water is often the most limiting factor for any rural development. Indeed, in the tropics, to raise 1 ton of wheat necessitates the use of about 8000 tons of water, which is lost mainly by transpiration and evaporation. To avoid the fluctuations of weather, it is therefore necessary to establish irrigation schemes. These necessitate pumps which should be powered by electricity. Electrification of rural areas will also facilitate the spread of workshops and village industries based on or related to agricultural production.

Investment in energy is a necessary prelude to any industrial development of a country. It was found [26] that on the average, an increase of industrial production of 1% involves an increase of 0.67% of energy consumption. It is important to

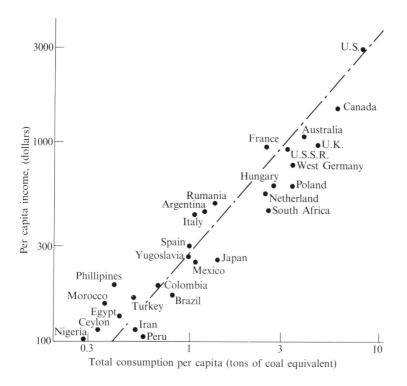

Figure 1.8. Per capita income and fuel consumption in 1958 for several nations. The per capita income is almost a linear function of the per capita consumption.
Reference: United Nations Statistical Yearbook, 1966.

note that labor and capital can be interchanged; thus labor can be considered a particular form of capital. If in the West there is a tendency to substitute capital for labor by a systematic mechanization, it seems wiser for the less developed countries with an abundance of labor to use less complex and smaller plants using more labor.

Any primary industry should be based on the availability of local raw materials. The emplacement of these materials will often dictate the location of the industry to save the cost of transport. However, in countries producing oil and gas, transport by pipelines is cheap, and consequently the industrial development of such countries is proportionally facilitated. Transport of energy by high-voltage transmission lines is relatively inexpensive. In cases where solar energy or wind power can be used, there will be practically no distribution costs involved. It is also interesting to note that for an industrial development, a good water supply is necessary. For instance, 200 tons of water are needed to make 1 ton of steel and 20 tons are needed to refine 1 ton of gasoline [40]. Water is therefore a necessary element for any agricultural or industrial development.

1.6 Transmission [9, 13] and Storage [45, 97] of Energy

In the early days of the electrical industry—as it is now in most of the under-developed countries—electrical power production, transmission, and distribution concerned only local authorities. This resulted in low overall efficiencies due to the low load factors. Transmission was only over small distances and consequently there was no need to use high transmission voltages.

To obtain the highest conversion efficiencies possible, it is today becoming more and more necessary to build central generating stations of the largest possible size. To maximize the load factor, it is becoming necessary to connect generating stations and load centers effectively over the largest possible area. These problems are now becoming of national interest and in many cases they are even an international concern.

To avoid Joulean (heat) losses in the transmission process, high transmission voltages are used. The voltage at which the power is generated is stepped up by transformer banks to the desired value of transmission voltage, as shown in Fig. 1.9. Transformers are also used near load centers to transform the high transmission voltage to the low consumption values. At the energy conversion center, electrical power is generated at 2.3 to 24 kV, and the transmission voltages range from a few hundred to about seven hundred kilovolts, whereas the nominal voltage of most loads is 115 or 230 V.

Most of the power transmitted is alternating current (ac), but direct current (dc) transmission is being developed in many countries, especially Canada. It is found to be less costly than ac transmission when the amount of power transmitted and the distances involved are both very large.

For long-distance transmission of energy, overhead transmission lines are used, whereas underground transmission cables serve relatively short distances

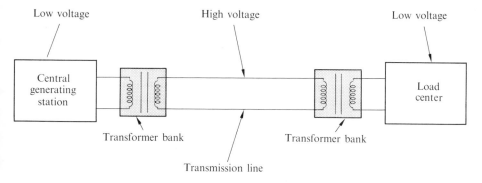

Low voltage High voltage Low voltage

Central generating station

Transformer bank

Transmission line

Transformer bank

Load center

Figure 1.9. A simplified electrical power system.

as in metropolitan areas. Both dc and ac transmission methods use overhead lines and underground cables.

Direct current transmission is found to have many advantages, leading to two important results. First, a given transmission facility can transport at a given high voltage more dc than ac energy with the same reliability and efficiency. Second, there is no "charging current" phenomenon in dc transmission and therefore there is no limitation on the permissible length of the dc transmission line or cable. Direct current transmission through a submarine cable becomes highly advantageous to regulate the flow of energy between two load centers. The most interesting example is the 160-MW cable connecting the French and English power systems across the English Channel.

Since power is usually consumed in the ac form, additional rectifiers and inverters are required at the terminals of a dc transmission line. A dc system will become economically competitive only when the gains in power by reduction of losses become at least equal to the additional costs of the rectifiers and inverters. This will require distances larger than at least 500 miles.

Transmission facilities are also used to interconnect separate power generating stations. For instance, all major electric systems in the United States are grouped into six interconnected systems. By interconnecting several systems, the fluctuations in instantaneous local consumption can be limited, thus reducing the cost of consumed power. Furthermore, if a unit is out of service for some reason, the problem of power breakdown is less acute when several other units can replace it without much delay.

Although this field of engineering is well developed, many further developments are being considered. The economical answer to the question of transport in the form of electricity or in the form of fuel is not always obvious. Furthermore, there is still room for reducing the Joulean losses in the process of transmission; for instance, if superconductors are used, Joulean heating could be eliminated altogether. Many problems of stability of large interconnected systems are currently being investigated to avoid the poor dynamic stability of these systems.

Industrial storage of energy can be accomplished in the form of fuel in reservoirs or as electricity. In the second case, there are three possibilities: capacitive storage, inductive storage, or chemical storage in batteries.

Capacitors store electrical energy and can theoretically be used to store large amounts of power for long periods. The total energy stored by a capacitor is given by

$$\mathcal{E}_{cap} = \tfrac{1}{2}\mathcal{V}\varepsilon E^2,$$

where $\mathcal{V}$ is the volume of the dielectric, ε its dielectric constant, and E is the electric field strength. The value of E, and consequently the stored energy, are limited by the breakdown strength E_{br} of the dielectric. Table 1.3 shows the maximum energy storable per unit volume for several dielectric materials. The dielectric should also have extremely low conductivity to avoid leakage losses. Very few research programs have been carried out in storing energy by using dielectrics for any appreciable length of time or any large amount of power.

Table 1.3

Dielectric materials

Material	ε_r	E_{br}, MV/m	$\left(\dfrac{\mathcal{E}_{cap}}{\mathcal{V}}\right)_{max}$, J/cm^3
Vacuum	1.0	∞	∞
Mica	6.0	200	106
Amber	2.7	90	9.67
Quartz	5.0	40	3.54
Teflon	2.1	60	3.35
Paper (impregnated)	3.0	50	3.32
Polyethylene	2.3	50	2.54
Glass	6.0	30	2.39
Titanium dioxide	100.0	6	1.60
Bakelite	5.0	25	1.38

Inductors store magnetic energy. The amount of energy stored can be much higher than that stored by capacitors. However, great losses are almost unavoidable because most of the materials which have high permeabilities also have high electrical resistivities and, since the energy stored is at a very high current, the Joulean losses may become prohibitively high. The energy stored is equal to

$$\mathcal{E}_{ind} = \tfrac{1}{2}\mathcal{V}\mu H^2,$$

where μ is the permeability of the material, and H is the magnetic flux density in amperes per meter. With advances in superconducting materials, the problems of Joulean losses might be eliminated. Large inductive storage is still impractical and only storage on a small scale and for special applications is common today.

The primary battery which stores energy in the form of chemical energy is the simplest storage device. It comprises two electrodes (anode and cathode) and an electrolyte, which is an ionic conductor. The energy stored by these batteries is, however, very low for a given volume and the storing price is rather expensive. Present research work is directed toward the improvement of the energy to weight ratio, compactness, and maintenance-free scaled batteries.

1.7 Direct Energy Conversion Systems and Applications

The design of an energy converter is often dictated by the type of energy to be converted, although it is the duty of the engineer to seek out new and more efficient ways of transforming the primary sources of energy into electricity.

There are many reasons for the use of new and direct conversion schemes. These can be grouped into three important areas: efficiency, reliability, and the use of new sources of energy. It is hoped that when a process occurs directly rather than passing through several steps, it is likely to be more efficient. This will lead to less expenditure of the primary energy reserve and a lower investment per installed unit power. Efficiencies are, however, still low at this stage of development of most direct energy conversion schemes. As for reliability, there are places where energy conversion equipment must run for years without breaking down and without maintenance. These are situations where the ultimate in reliability is required: scientific satellites, manned space flights, equipment in remote areas, underwater stations, etc. Finally, the possibility of using new sources of energy seems enhanced by the development of the new direct energy converters. Today, two important types of direct energy converters can be observed: converters for space applications and those for terrestrial applications which comprise an entire spectrum both in military and civilian fields.

Space systems and applications [21, 60, 79]. In space applications, specific weight of a power source (gross weight per net power available) plays a key role in the selection of a specific system. The two other important factors are reliability and the cost of the installation.

Propulsion is the key to space exploration and it is based on the rocket principle. A rocket system carries all the material required for its mission and has the advantage of working efficiently in the virtual vacuum of outer space. The rocket is a system which produces a stream of gas molecules at very high velocities, which, by reaction, propels the vehicle in the direction opposite the stream. Rocket engines are usually classified according to the nature of the energy sources, which may be chemical, nuclear, or electrical. Another classification is based on whether the gas particles are produced by chemical reactions, heat transfer, or acceleration of electrically charged particles. In the majority of chemical rockets, propulsion energy is produced by the combustion of a fuel with an oxidizer, both of which are called propellants. Liquid propellant rocket engines consist of pumps, feed systems, and a thrust chamber assembly made up of a propellant injector, an ignition system, a combustion chamber, and a nozzle, as shown in Fig. 1.10. Solid propellant rocket engines have a similar thrust chamber.

For manned exploration of space, launched vehicles should be able to provide a large burnout mass at a very high exhaust velocity. There are many practical limitations to this requirement in the case of a chemical rocket. In this respect, nuclear rockets have the advantage that the temperature is limited only by the strength of the materials forming the rocket structure. Any propellant can be used, since the rocket will work as a heat transfer engine transferring the heat from the reaction to the working fluid. The electrothermal thrustors (resisjet, arc-jet) form another type of heat transfer rocket which is under consideration for space propulsion.

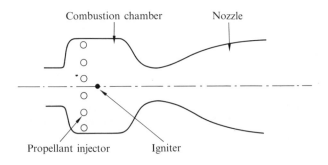

Figure 1.10. A thrust chamber of a liquid propellant rocket, made up of a propellant injector, an ignition system, a combustion chamber, and a nozzle.

Charged particle rocket motors can be classified in two categories: motors using electrostatic acceleration (ion rockets) and motors using electromagnetic acceleration (plasma rockets). In both cases, charged particles are accelerated directly by electric or magnetic fields at high velocities. Some advanced propulsion techniques include the use of cavity reactors, propulsion by nuclear explosions, thermonuclear propulsion, and photon propulsion, including so-called "solar sailing" in which the rocket would use the pressure exerted by the solar photons on its surface to move in the solar system.

Most of the new proposed propulsion methods listed above (electrothermal thrustors, ion rockets, plasma rockets, etc.) require relatively large amounts of electric power. Furthermore, electricity is necessary in both manned and unmanned space vehicles to operate the control and communication instruments as well as the space laboratories which will be built in the future. This will require the services of direct energy converters for the transformation of whatever type of energy is available into electricity.

One of the most important sources of energy in space is nuclear fission, and SNAP systems are based on this source. These fission reactors are most advantageous for power systems requiring more than 1 kW of energy, especially so far as the weight ratio is concerned. These reactors differ from rocket reactors only in some design details. SNAP-2 produced 3 kW and SNAP-8 produced 35 kW. For both, the fuel element was U^{235} and the moderator was the hydrogen atoms existing in a zirconium hydride mixed with the fuel. A liquid alloy of sodium and potassium served as a coolant to remove the heat produced by fission. This coolant was then passed through a boiler where liquid mercury was boiled to be used as a working fluid through a Rankine cycle, to convert heat to electricity by a converter of the conventional type. SNAP-50 uses the same principles with an alkali metal as a coolant.

Conventional generators involve moving parts and they are not simple. Direct conversion schemes are considered as a solution to many problems of power for space. SNAP-10A was devised to test the thermoelectric conversion concept by the utilization of fission heat. The core is similar to that in other SNAP systems and the coolant is a sodium-potassium alloy. The thermoelectric elements, made up of a germanium-silicon alloy, are connected in series around the circular reactor. The output power is about 500 W. The same type of reactor can be used as a source of heat for a thermionic converter. Thermionic emitters may be in contact with the reactor core on one side and with a cesium atmosphere on the other side separating them from the collectors. These collectors are cooled by a sodium-potassium alloy flow, which removes the rejected heat from the collectors to the radiators. It is hoped that this kind of power generator may yield, in the future, more than 1 MW of power for space propulsion.

Another source of energy in space is the radioactivity of radioisotope materials. The nuclei of a radioisotope material emit electrically charged particles, either positive alpha particles (Po^{210}, Cu^{242}, Pu^{238}, etc.) or negative beta particles* (Ce^{144}, Pr^{147}, Sr^{90}, Ce^{137}). The half-life of an individual isotope characterizes the rate of its radioactive decay. This life ranges from several microseconds to millions of years. When absorbed by solids, these radiations generate heat which can be converted into electricity by a thermoelectric or a thermionic converter. For instance, SNAP-1A was built to use cerium-144 to drive a turbogenerator, but the project failed due to many difficulties. SNAP-3 used polonium-210 to yield heat to a thermoelectric generator (Pb-Te) with an initial power output of 5 W and it operated successfully. Other thermoelectric generators which supplied electrical power to the satellite Transit IV-A (plutonium-238) and Transit IV-B in 1961 were similar to the larger SNAP-9A (25 W) which was built in 1963. SNAP-11 was proposed to supply power for the Surveyor spacecrafts, which land softly on the moon. It is a thermoelectric system using curium-242 as its heat

---

* A beta particle is a normal helium ion (proton).

source. Other alternatives are SNAP-13 using thermionic conversion, and SNAP-17 (strontium-90) and SNAP-19 (plutonium-238) using thermoelectric power generation schemes. Another variation in the use of radioisotope energy is given by the nuclear cell, where the alpha and beta particles would be directly collected to generate electrical power.

Solar energy has been the most common source of power for unmanned missions. Solar energy is converted directly into electricity by the use of the photovoltaic principle. Solar systems have the advantage of not having to carry their necessary fuel. The satellites which have used these facilities are Telstar and Relay, where the solar cells are arranged on the outer surfaces of the spacecrafts; in others, such as Mariner and Nimbus, the solar cells are laid out on flat panels. Solar energy can be used indirectly for power generation through a heat path. For instance, in the Sunflower system, the heat from the sun is collected by a mirror and stored by a lithium hydride, thus using mercury vapor through a Rankine cycle (with a turbogenerator) to produce about 3 kW of electrical power. The heat can also be transformed by fuel cells in a regenerative cycle into electricity. This has been the case for the Gemini and Apollo projects.

Terrestrial systems and applications [72, 85]. Terrestrial applications of direct energy conversion are many and diverse. Direct converters have been considered to generate electricity for meteorological, agricultural, medical, and military applications and in general in all the spectrum of human activity.

To accurately predict the weather, some stations have to be built in rather remote areas, such as in the Arctic, in Antarctica, or on remote mountains and the seas. In these areas, it is necessary to use equipment which can work satisfactorily without being attended. Thus, the use of direct energy converters becomes extremely attractive. A weather station of this type has been built using a SNAP-7 generator. In this generator, the radioactive energy of strontium-90 is transformed into electricity by a thermoelectric converter. The electricity powers the electronic system emitting information such as precipitation, wind velocity, temperature, and barometric pressure.

The waste heat of MHD generators has been considered for distillation of sea water. It has also been proposed to use this heat to create nitrogen compounds which will react with limestone to yield fertilizers for agriculture. Another interesting unit has been proposed for use in remote underdeveloped areas. It is a self-contained irrigation pumping unit powered by the sun, the heat of which is gathered by a collector to feed a thermoelectric generator. The generator produces 50 W and is able to drive a small pump. Larger units yielding 200 W and which meet the water needs of a small community [98] have been studied.

In medical engineering, work is going on to replace failing natural organs by artificial ones. These artificial organs will need a reliable source of electrical energy for their auxiliary equipment, such as Pacemakers for the artificial heart. This electrical energy can be supplied by thermoelectric generators using the human body itself as the heat source and the open air as the heat sink.

The U.S. Army is interested in the study of a unit that can operate in a terrain using any kind of fuel, such as charcoal, wood, grass, or even refuse. A unit based on thermoelectric power generation and yielding 150 W has been studied [5, 38]. Its advantage as a compact and independent source of electrical energy for communication equipment is obvious. Solar energy has also been considered to power a prototype telephone for remote areas through the use of the photovoltaic effect.

Another important application of direct energy converters is related to the development of the electric car [43]. Present cars have the disadvantage of exhausting gases which become a danger to public health in metropolitan areas. Fuel cells are considered for replacing the combustion engine by an electric motor. In this case, the products of combustion can be water and nitrogen, which are not harmful. This arrangement also has the advantage of limiting the level of noise in large cities, which is believed by many physicians to be an important source of hearing and mental troubles.

Many direct energy converters are already considered for generating electrical power in central power stations. However, the efficiency of most direct energy converters is not yet competitive with the conventional methods to be used solely for these applications in the immediate future. The possibility of associating conventional and direct conversion in power plants gives hopes of increasing greatly the overall efficiency of these plants. In most heat engines there is a range between the available flame temperature and the upper usable temperature, which is in general open to thermodynamic conversion. This range can be visualized as an additional energy which can be converted to electricity by the use of direct energy converters. This operation is called *topping* and the energy converter is referred to as a *topper*. As shown in Fig. 1.11, if η_τ is the efficiency of the topper and η_1 is the efficiency of the conventional converter, the total efficiency of the

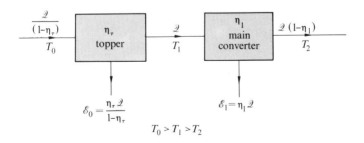

Figure 1.11. Power plant using a topper. $\mathcal{Q}$ = heat input to main converter, $\mathcal{E}_0$ = energy output of topper, T_0 = temperature at input of topper, T_1 = temperature at output of topper, T_2 = temperature at output of main converter and at input of tailer (see Fig. 1.12), $\mathcal{E}_1$ = energy output of main converter.

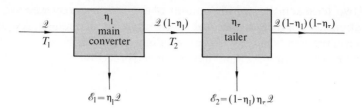

Figure 1.12. Power plant using a tailer. $\mathscr{E}_2 =$ energy output of tailer. (Definitions of the other quantities are given in Fig. 1.11.)

plant will be

$$\eta = \eta_\tau + \eta_1 - \eta_\tau\eta_1.$$

Similarly, the exhaust gases in most conventional converters are ejected at a temperature high enough to be used by a direct energy converter for the generation of more electrical power. This operation is called *tailing* and the direct converter is a *tailer*. As shown in Fig. 1.12, the overall efficiency of a power plant using a tailer of efficiency η_τ is

$$\eta = \eta_\tau + \eta_1 - \eta_\tau\eta_1.$$

There are many other applications which will be considered in the following chapters dealing with the individual types of direct energy conversion.

1.8 Economic Aspects of Power Generation [26, 65, 81]

The total cost of a power generation plant or system should be as low as possible. This cost depends on three important factors: the capital cost, the fuel cost, and the operating costs. The capital cost is the price of the kilowatt-hour installed, which depends on such factors as the expected lifetime of the station, the extent to which it is used, and its reliability. The fuel cost is directly related to the efficiency of the plant and the initial price of the raw fuel to be used. The operating and maintenance costs depend on the degree of simplicity and directness of the apparatus. Thus, it seems clear that efficiency alone is not always an overriding criterion. A device of low efficiency but with low capital costs can compete under some circumstances.

For instance, if the use of a topper in a power plant is to be defended, the total cost of the output kilowatt-hour for the conventional-topper plant should be larger than that for the plant alone. If c_τ is the fixed cost per unit energy produced by the topper of efficiency η_τ, c_1 that produced by the conventional plant of efficiency η_1 and c_f is the fuel cost per unit heat input $\mathscr{Q}$, the total cost C_t of the output power from the conventional-topper plant will be

$$C_t = \mathscr{Q}c_1\eta_1 + \frac{\mathscr{Q}c_\tau\eta_\tau}{1 - \eta_\tau} + \frac{\mathscr{Q}c_f}{1 - \eta_\tau}. \tag{1.1}$$

The first two terms on the right-hand side of this equation represent the capital and operation costs of the conventional plant and the topper, respectively, whereas the third term represents the cost of the fuel. From Fig. 1.11, it is seen that the total electric energy output is

$$\mathscr{E} = 2\eta_1 + \frac{2\eta_\tau}{1 - \eta_\tau}. \tag{1.2}$$

Therefore, the cost c_t per unit energy output will be equal to the ratio of Eq. (1.1) to Eq. (1.2):

$$c_t = \frac{c_1\eta_1(1 - \eta_\tau) + c_\tau\eta_\tau + c_f}{\eta_1(1 - \eta_\tau) + \eta_\tau}. \tag{1.3}$$

For the topping to be economical, c_t should be smaller than the cost of the conventional plant alone:

$$c_t < c_1 + c_f/\eta_1.$$

This condition becomes

$$\frac{c_\tau}{c_1} < 1 + \frac{c_f(1 - \eta_1)}{c_1\eta_1}. \tag{1.4}$$

For modern power plants this ratio is slightly larger than two. Also note that this ratio is independent of the topping efficiency and is only a function of the efficiency of the conventional power plant, the fuel cost per unit heat input, and the capital and operating costs of the conventional plant.

Similar calculations can be made for the tailer. In general, for any research and development policy, the economic factor has no immediate influence but comes after such factors as the economic growth and the possibility of increasing the levels of investment and employment in the future.

1.9 Future Trends [19, 49, 69]

The development of energy resources is a condition for any progress in the economy of a country. From many and varied predictions, it is certain that mankind's consumption of all forms of energy will continue to soar, as shown in Fig. 1.13. The future trends are multidirectional: to save capital on installation through direct conversion of energy, to save capital on maintenance by avoiding the moving parts in a system, and to save capital on the fuel by improving the overall efficiency.

The commercial attractiveness of fuel cells is as a source of electric energy for individuals and small communities rather than for central stations. It is hoped that in the future, electricity produced by gas-powered fuel cells will cost less for the consumer than electricity obtained from a central power plant. Solar energy research is enhanced especially as a potential source of development for the developing nations. It is also greatly related to space and defense needs. In both space and terrestrial applications, solar cells are promising in spite of the problems of weight ratio for space and economics for terrestrial systems.

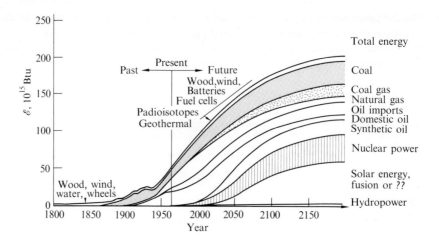

Figure 1.13. Energy consumption in the United States: past, present, and projected. From Gaucher [32].

In spite of the apparently unsuccessful attempts to obtain power from controlled thermonuclear reactions, basic research is continuing at an accelerated rate. This can be explained by the tremendous potentialities offered by fusion power. Fusion research is directly related to progress in MHD power generation. The possibility of obtaining electrical energy from an MHD power plant is most attractive, since a large variety of energy sources can be taken as the heat sources for the formation of the working plasma. Open-cycle MHD systems present large hopes as potential competitors for conventional power plants. Research and development payoffs can be evaluated only when high-temperature materials for electrodes are found, and after more advances in superconductivity and more progress in understanding plasma stability have been made.

Most of the thermoelectric and thermionic converters are related to space research. Basic research for civilian applications is underway.

A dynamic and coordinated research and development program is necessary for any national development. Most industrial countries of the world understand this and most of them have direct energy programs. Perhaps, in the future, international cooperation in this field will occur, leading to sensible progress in the way of living for all mankind.

PROBLEMS

1.1 (a) Calculate the kinetic energy spent by a stream of water flowing in a bed of 500 m^2 cross section through a denivelation of 100 m. Consider the force of gravity only. (b) What is its potential energy?

1.2 The temperature of a natural water reservoir is 150°C at a pressure of 5.5 atmospheres and a depth of 1000 m. A total of 12 identical wells have been drilled to bring the hot water at the surface. (a) Assuming that this water drives a Carnot engine, how much water should be produced per hour for the production of 500 MW of energy? (b) What is the radius of each well?

1.3 The average solar energy received in an American city is about 1 kW/m². What should be the surface area covered by solar cells of 16% efficiency to obtain an annual rate of energy of 5×10^7 MWh?

1.4 Using Einstein's relation, find the amount of energy released by the nuclear reaction

$$n_0^1 \rightarrow H_1^1 + e^- + \mathscr{E},$$

knowing that the atomic mass of the neutron is 1.008982 and that of the hydrogen atom is 1.008142.

1.5 The energy released by the burning of 1 kg of coal is about 770 Wh. How much coal is necessary to release as much energy as 1 g of U^{235}? The average energy produced by the fission of a U^{235} atom is 1.65×10^5 keV.

1.6 What is the efficiency of a Brayton engine?

1.7 A remote village is situated in a gorge through which wind blows constantly with an average velocity of 30 km/h. How much power could be generated by transforming the energy of the wind into mechanical energy through a cylindrical channel of 5-m radius?

1.8 It has been reported that it is possible to build a 100,000-kW sea thermal power plant using the temperature difference between the surface and the deep sea water. Describe such a plant and its feasibility knowing that at the Carribbean the surface temperature is 28°C and the deep sea water temperature is 6°C.

1.9 The Gulf Stream is an ocean "river" flowing from south to north in the Northern Atlantic Ocean. It has been estimated that this stream could generate 182 trillion kWh per year. (a) On what basis have those estimates been made? (b) Assuming that the speed of the stream is uniform along its cross section and equal to about 1 m/sec, estimate the cross sectional area of the Gulf Stream.

1.10 The utilization of tidal energy has been seriously considered in the United States and in France. The energy produced in a time t by a tide of height d from the equilibrium surface through a surface perpendicular to the velocity of the tide is

$$d\mathscr{E} = \rho(a_g d + \tfrac{1}{2}v^2)v \cdot dS,$$

where ρ is the density of the sea water, a_g is the gravitational acceleration, and dS is the converter cross section. Taking d as a sine function of time, find an expression for the average power generated by the tide.

1.11 What is the energy stored in the electric field produced by an electric point charge in a dielectric medium?

1.12 What is the amount of energy stored in the near field of an elementary conductor of length dz through which a current I flows? (*Hint:* This is an antenna-like problem.)

1.13 (a) What is the efficiency of a Carnot engine used to power an artificial heart and using the human body as a heat source and the open air as a heat sink? The patient is an Eskimo living in an ambient temperature of about -10°C? (b) How many extra calories should the patient supply to produce 40 W for his artificial heart? (c) How many calories should the

Eskimo produce when he visits Florida (ambient temperature 30°C)? It is then obvious that other sources of energy such as mechanical energy should be used in conjunction with the thermal energy mentioned above (see Chapter 10).

1.14 What is the efficiency of a plant using both a tailer and a topper?

1.15 What is the cost per unit power output of the device of Problem 1.14?

REFERENCES AND BIBLIOGRAPHY

1. Abdel-Rahman, I. H., *Social Aspects of the Sources of Energy*, Paper GEN/11, United Nations Conference on New Sources of Energy, Rome, Aug. 21–31, 1961.

2. Ailleret, R., *The Abundance of Natural Energy and the Choice of the Means of Harnessing It*, Paper GEN/12, United Nations Conference on New Sources of Energy, Rome, Aug. 21–31, 1961.

3. Anderson, T. H., and T. H. Anderson, Jr. "Thermal Power from Seawater," *Mechanical Engineering*, Vol. 88, No. 4, p. 41, 1966.

4. Angelini, A. M., *Reflections on the Economic Value of Geothermal Energy, Wind Power and Solar Energy, Especially after Conversion to Electrical Energy*, Paper GEN/1, United Nations Conference on New Sources of Energy, Rome, Aug. 21–31, 1961.

5. Angrist, S. W., *Direct Energy Conversion*, Allyn and Bacon, Boston, 1965.

6. Averitt, P., "Coal Reserves of the United States, A Progress Report, Jan. 1, 1960," *U.S. Geol. Surv. Bull. 1136*, 1961.

7. Benton, M., *Direct Energy Conversion—Literature Abstracts*, Office of Technical Services, 1961.

8. Berger, B., "Does the Production of Power Pollute Our Nation's Rivers?," *Power Eng.*, Vol. 65, No. 3, p. 60, 1961.

9. Bills, G. W., "Space Electric Power Transmission," *Electrical Engineering*, Vol. 78, No. 10, p. 1021, 1959.

10. Bishop, A. S., *Project Sherwood—The U.S. Program in Controlled Fusion*, Addison-Wesley, Reading, Mass., 1958.

11. Blomeke, Y. O., et al., *Estimated Costs for Management of High-Activity Power Reactor Processing Wastes*, Oak Ridge National Laboratory, ORNL-TM-599, 1963.

12. Breuvery, E. S., *New Sources of Energy and Energy Development*, Paper GR/1(GEN), United Nations Conference on New Sources of Energy, Rome, Aug. 21–31, 1961.

13. Bromberg, E., *An Annotated Bibliography of High Voltage Direct Transmission, 1932–1962*, Paper 63-388, General Meeting AIEE, New York, Jan. 27–Feb. 1, 1963.

14. Cadiff, I. B., and E. Miller (eds.), *Thermoelectric Materials and Devices*, Reinhold, New York, 1960.

15. Cambel, A. B., *Energy RD and National Progress*, U.S. Government Printing Office, Washington, D.C., 1964.

16. Chang, S. S., *Energy Conversion*, Prentice-Hall, Englewood Cliffs, N.J., 1963.

17. Clegg, T. W., and D. D. Foley (eds.), *Uranium Ore Processing*, Addison-Wesley, Reading, Mass., 1958.

18. Clerk, R. C., *The Utilization of Flywheel Energy*, SAE Paper 711A, June 1963.

19. Corliss, W. R., "Survey of Space Power Requirements 1962 to 1976," in *Power Systems for Space Flight*, M. A. Zipkin and R. N. Edwards (eds.), Academic Press, New York, 1963.

20. Corliss, W. R., *Direct Energy Conversion*, U.S. Atomic Energy Commission, Technical Information, 1964.

21. Corliss, W. R., and D. G. Harvey, *Radioisotopic Power Generation*, Prentice-Hall, Englewood Cliffs, N.J., 1964.

22. Coanady, G. T., *Theory of Turbomachines*, McGraw-Hill, New York, 1964.

23. Daniels, F., "Direct Use of Sun's Energy," *American Scientist*, Vol. 55, No. 1, p. 15, 1967.

24. Daniels, G. H., and E. A. Robinson, "Planning of Investment in Fuel and Power in Underdeveloped Countries," *World Power Conference Trans.*, Vol. 2, p. 221, Belgrade, Yugoslavia, 1957.

25. Early, H. C., and R. C. Walker, *Economics of Multimillion-Joulean Inductive Energy Storage*, AIEE Paper 57-79, July 1957.

26. Eckstein, O., "Investment Criteria for Economic Development and the Theory of Intertemporal Welfare Economics," *Quart. J. Econ.*, Vol. 71, No. 1, Feb. 1957.

27. Eckstein, O., *Water Resource Development*, Harvard University Press, Cambridge, Mass., 1958.

28. Egli, P. H. (ed.), *Thermoelectricity*, Wiley, New York, 1960.

29. Einstein, A., and L. Infeld, *The Evolution of Physics*, Simon and Schuster, New York, 1938.

30. Friedlander, G. D., "Science and the Salty Sea," *IEEE Spectrum*, Vol. 2, No. 8, p. 53, 1965.

31. Gardner, T. W., *Electricity Without Dynamos*, Penguin Books, Baltimore, 1963.

32. Gaucher, L. P., "Energy Sources of the Future for the United States," *Solar Energy*, Vol. 9, No. 3, p. 119, 1965.

33. Gibrat, R., *L'Energie des Marées*, Presse Universitaires de France, Paris, 1966.

34. Glasstone, S., and R. H. Lovberg., *Controlled Thermonuclear Reactions*, Van Nostrand, Princeton, N.J., 1960.

35. Glasstone, S., and A. Sesnonake, *Nuclear Reactor Engineering*, Van Nostrand, Princeton, N.J., 1963.

36. Glasstone, S., *Sourcebook on the Space Sciences*, Van Nostrand, Princeton, N.J., 1965.

37. Gratch, S., *Energy Sources and Devices for the Transportation Industry*, 7th Biennial Gas Dynamics Symposium, Northwestern University, Aug. 23–25, 1967.

38. Hamilton, R. C., *Recent Progress in Military Energy Conversion*, 7th Biennial Gas Dynamics Symposium, Northwestern University, Aug. 23–25, 1965.

39. Hartley, H., *Energy as a Factor in the Progress of Underdeveloped Countries*, Paper GEN/4, United Nations Conference on New Sources of Energy, Rome, Aug. 21–31, 1961.

40. Herbert, S. W., *Battery Research and Fuel Cells*, SAE Paper 5367, 1962.

41. Hirschleifer, T., et al., *Water Supply*, University of Chicago Press, Chicago, 1960.

42. Hodgkins, T. A., *Soviet Power, Energy Resources, Production and Potential*, Prentice-Hall, Englewood Cliffs, N.J., 1961.

43. Hoffman, G. A., *Battery-Operated Electric Automobiles*, The Rand Corp., Santa Monica, Calif., March 1963.

44. Horne, A., "Special Features," *EDN Magazine*, April 1967.

45. Inville, W. L. (ed.), *Proceedings of a Conference on Energy Conversion and Storage*, Oklahoma State University, Oct. 1963.

46. Jennings, B. H., and W. L. Rogers, *Gas Turbine Analysis and Practice*, McGraw-Hill, New York, 1953.

47. Kapur, T. C., *Socio-Economic Considerations in the Utilization of Solar Energy in Underdeveloped Areas*, Paper GEN/8, United Nations Conference on New Sources of Energy, Rome, Aug. 21–31, 1961.

48. Kovacik, V. P., *Dynamic Engines for Space Power Systems*, ARS, Vol. 32, Oct. 1962.

49. Landsberg, H. H., L. L. Fischman, and T. L. Fisher, *Resources in America's Future: Patterns of Requirements and Availabilities, 1960–2000*, Johns Hopkins Press, Baltimore, 1963.

50. Lauch, F. W., et al., *Portable Power from Nonportable Energy Sources*, SAE Paper, Nov. 1962.

51. Leach, G., *New Sources of Energy*, Phoenix House, London, 1965.

52. Lehninger, A. L., *Bioenergetics*, Benjamin, Amsterdam, 1965.

53. Levine, S. N. (ed.), *New Techniques for Energy Conversion*, Dover, New York, 1961.

54. Loftness, R. L., *Nuclear Power Plants—Design, Operating Experience and Economics*, Van Nostrand, Princeton, N.J., 1964.

55. Londown, T., *Petroleum Handbook*, Shell International Petroleum Co. Ltd., London, 1959.

56. Lowry, H. H., *Chemistry of Coal Utilization*, Wiley, New York, 1963.

57. Lynch, C. D., "Unconventional Power Sources," *Prod. Engr.*, Vol. 32, No. 28, p. 57, 1961.

58. Lynch, C. D., "Emerging Power Sources," *Science and Technology*, Oct. 1967.

59. Mackay, D. B., *Powerplant Heat Cycles for Space Vehicles*, Inst. Aeronautical Sci. Paper 59-104, 1959.

60. McNab, I. R., *Power Conversion in Space*, IEE Colloquium on Electrical Methods of Propulsion in Space, London, Feb. 13, 1964.

61. Mininger, R., *Minerals for Atomic Energy*, Van Nostrand, Princeton, N.J., 1956.

62. Mitchell, W., Jr., *Fuel Cells*, Academic Press, New York, 1963.

63. Mueller, H. F., *Problems of Energy Supply in Underdeveloped Countries with Special Regard to New Sources of Energy*, Paper GEN/7, United Nations Conference on New Sources of Energy, Rome, Aug. 21–31, 1961.

64. Murphy, G. W., R. C. Taber, and H. H. Stinhauser, *The Minimum Energy Requirement of Sea Water Conversion Process*, U.S. Office of Saline Water, R & D Progress Report 9, 1956.

65. Netschert, B. C., and G. O. Lof, *New Sources of Energy in the World Energy Economy*, Paper GEN/10, United Nations Conference on New Sources of Energy, Rome, Aug. 21–31, 1961.

66. Parker, A., *World Power Conference Survey of Energy Resources*, Central Office of World Power Conference, London, 1962.

67. Perry, H., et al., *Current Developments in the Conversion of Coal to Fluid Fuels in the U.S.*, Sixth World Power Conference, Melbourne, Australia, Oct. 22–27, 1962.

68. Perry, H., "Revolution in Coal Transport. Is it Near?," *Mechanization*, Vol. 27, No. 5, May 1963.

69. Putnam, V. C., *Energy in the Future*, Van Nostrand, Princeton, N.J., 1953.

70. Ramey, J. T., "Nuclear Energy = Potential for Desalting," *Mechanical Engineering*, Vol. 88, No. 4, p. 52, 1966.

71. Rich, G. E., and H. R. Schmidt, *Power for Deep Ocean Systems*, 7th Biennial Gas Dynamics Symposium, Northwestern University, Aug. 23–25, 1967.

72. Robba, W. A., *Status of Direct Conversion Programs in the United States with Special Emphasis on Civilian Nuclear Power*, Brookhaven National Laboratory, May 1960.

73. Rose, D. T., and M. Clark, Jr., *Plasmas and Controlled Fusion*, MIT Press, Cambridge, Mass. and Wiley, New York, 1961.

74. Russel, C. R., *Elements of Energy Conversion*, Pergamon Press, New York, 1967.

75. Sanders, R., *Project Plowshare. The Development of the Peaceful Uses of Nuclear Explosions*, Public Affair Press, Washington, D.C., 1962.

76. Schurr, S. H., and B. C. Netschert, *Two Statements on the Nations Energy Position*, Resources for the Future Inc., Washington, D.C., 1959.

77. Schurr, S. H., et al., *Energy in the American Economy. 1850–1975; an Economic Study of its History and Prospects*, Johns Hopkins Press, Baltimore, 1960.

78. Skrotzki, G. A., and W. A. Vopat, *Applied Energy Conversion*, McGraw-Hill, New York, 1961.

79. Snyder, N. W. (ed.), *Energy Conversion for Space Power*, Academic Press, New York, 1961.

80. Spanides, A. G. (ed.), *Solar and Eolian Energy*, Proceedings of the First NATO International Seminar, Sounion, Greece, Sept. 4–15, 1961; Plenum Press, New York, 1964.

81. Spring, K. H., "The Thermodynamic and Economic Basis," in *Direct Generation of Electricity*, K. H. Spring (ed.), Academic Press, New York, 1965.

82. Stern, A. C. (ed.), *Air Pollution*, Academic Press, New York, 1961.

83. Stodola, A., *Steam and Gas Turbines*, McGraw-Hill, New York, 1927.

84. Szasz, S. E., and V. D. Berry, Jr., *Oil Recovery by Thermal Methods*, 6th World Petroleum Congress, 1963.

85. Tabor, H. Z., "Power for Remote Areas," *International Science and Technology*, No. 65, p. 52, May 1967.

86. Taylor, C. F., *The Internal Combustion Engine in Theory and Practice*, Wiley, New York, 1960.

87. Thacker, M. S., *New Sources of Energy and Energy Development*, Paper GEN/11, United Nations Conference on New Sources of Energy, Rome, Aug. 21–31, 1961.

88. Toynbee, A. J., *A Study of History*, Oxford University Press, London, 1935.

89. Ubbelohde, A. R., *Man and Energy*, Brazillev, New York, 1955.

90. Urrows, G. M., *Nuclear Energy for Desalting*, U.S. Atomic Energy Commission, Technical Information, 1966.

91. Vincent, E. T., *The Theory and Design of Gas Turbines and Jet Engines*, McGraw-Hill, New York, 1950.

92. Vogely, W. A., and W. E. Morrison, "Patterns of U.S. Energy Consumption to 1980," *IEEE Spectrum*, Vol. 4, No. 9, p. 81, 1967.

93. Volta, A., "On the Electricity Excited by the Mere Contact of Conducting Substances of Different Kinds," *Phil. Trans. Roy. Soc.*, Vol. 90, p. 403, 1800 (French).

94. Warren, F. C., *Rocket Propellants*, Reinhold, New York, 1958.

95. Weeks, L. G., "The Next Hundred Years Energy Demand and Sources of Supply," *Geotimes*, Vol. 5, No. 1, 1960.

96. Weinberg, A., and G. Young, "Scale, Nuclear Economics and Salt Water," *Nuclear News*, Vol. 6, No. 5, May 1963.

97. Wiederhold, P. R., "Energy Storage for High-Power Discharges," *Astronautics and Aerospace Engineering*, Vol. 1, No. 4, p. 104, 1963.

98. Zarem, A. M., and D. D. Erway (eds.), *Introduction to the Utilization of Solar Energy*, McGraw-Hill, New York, 1963.

99. Zapp, A. D., "Future Petroleum Producing Capacity of the United States," *U.S. Geol. Surv. Bull.*, 1962.

100. Zapp, A. D., *Supplies, Costs, and Uses of the Fossil Fuels*, U.S. Department of the Interior, Energy Policy Staff, 1963.

2 PHYSICAL PRINCIPLES

Before treating the different types of energy converters separately, it will be helpful to review the important physical principles necessary for understanding the following chapters. This chapter is divided into four parts: thermodynamics, quantum mechanics, solid state, and plasma physics.

A. THERMODYNAMICS

2.1 Heat and Work

It is well known that when two systems at different temperatures are placed together, both will reach a final temperature which is between the initial temperatures of the two systems. It is known that heat has been transferred from one system to the other. It can be said that heat is what is transferred between a system and the world outside (of the system) as a result of temperature difference only.

As a result, heat can be measured, and a unit of heat can be defined as the heat necessary for the production of a standard change in temperature. The two important units of heat are the calorie and the British thermal unit (Btu). One calorie (cal) is the quantity of heat necessary to raise the temperature of 1 g of water from 14.5°C to 15.5°C. One Btu is the quantity of heat necessary to raise the temperature of 1 standard pound of water from 63°F to 64°F. It can easily be found that 1 Btu = 252 cal.

The heat capacity C is an important characteristic of a body. It is defined as the ratio of the quantity of heat $\Delta \mathcal{Q}$ supplied to a body to its corresponding temperature rise ΔT, or in the limit $C = \delta \mathcal{Q}/dT$.

The specific heat c_M is the heat capacity per unit mass M:

$$c_M \gamma = C/M.$$

The specific heat of a material has a unique value only when the conditions of

35

volume and pressure are specified. At constant pressure p_r the unique value of $c_M \gamma$ is

$$c_p = \frac{1}{M} \frac{\delta \mathcal{Q}}{dT}\bigg]_{p_r},$$ (2.1)

and at constant volume $\mathcal{V}$ it is

$$c_v = \frac{1}{M} \frac{\delta \mathcal{Q}}{dT}\bigg]_{\mathcal{V}}.$$ (2.2)

Heat can also be defined as energy flowing from one system to another because of a temperature difference between the two systems. The energy transmitted without affecting the temperatures is defined as work W. Both $\mathcal{Q}$ and W are associated with the interaction between the system and its environment.

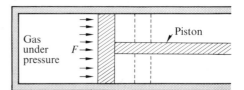

Figure 2.1. A gas under pressure works on a piston by exerting a force and producing a displacement of the piston.

If the system is a gas in a cylindrical container with a movable piston, as shown in Fig. 2.1, $\mathcal{Q}$ and W can be computed for a specific thermodynamic process. The gas at pressure p_r does work δW on the piston of surface area S by exerting a force $F = p_r S$ and producing a displacement ds. The amount of this work is $\delta W = F\, ds$, or

$$\delta W = p_r\, d\mathcal{V},$$ (2.3)

where $\mathcal{V}$ is the volume of the gas. The total work W done by the system on the piston is

$$W = \int_{\mathcal{V}_i}^{\mathcal{V}_f} p_r\, d\mathcal{V},$$

where $\mathcal{V}_i$ stands for initial volume and $\mathcal{V}_f$ stands for final volume as shown in Fig. 2.2. There are many ways in which the system can be taken from the initial state to the final state. It can be seen that the work done by a system depends on both the initial and final states and on the path between these states. The same can be said about the heat lost or gained by the system.*

---

*   The operator δ is chosen instead of the operator d to distinguish it as a variable which is independent of the path.

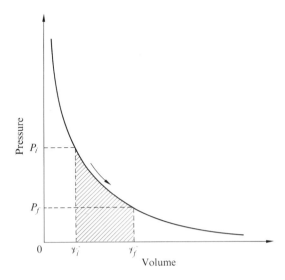

Figure 2.2. The p_r-$\mathscr{V}$ diagram. The work done by the system is represented by the shaded area.

2.2 The First Law of Thermodynamics

The first law of thermodynamics is nothing more than the principle of the conservation of energy: *the total energy of a closed system remains constant.* Energy may be converted, transformed, or transported, but it is indestructible and its total is invariant. A system that is in contact with other systems loses energy to, or gains energy from, the other systems. When a system has its state changed from S_i to S_f by following different paths, it is found that the difference $\mathscr{Q} - W$ is always the same, independent of the path followed. This difference depends only on the initial and final states and is defined as the internal energy function U. It must represent the internal energy change of the system. From the principle of conservation of energy,

$$dU = \delta\mathscr{Q} - \delta W, \tag{2.4}$$

which is another expression for the first law of thermodynamics.

A process is called *adiabatic* when the change of state is accomplished without any transfer of heat ($\delta\mathscr{Q} = 0$). An adiabatic process in which no work is performed on or by the system is called a *free expansion* process ($\delta\mathscr{Q} = \delta W = dU = 0$).

2.3 Kinetic Theory of Gases

It is found, experimentally, that all gases can be approximated by an ideal gas model if their densities are not too high. The equation of state of such gases can be written as $p_r \mathcal{V} = M_i T$, where i is the gas constant. It is experimentally found that $\mathcal{R} = i M_0 = 8.314$ J/mole·°K, where M_0 is the molecular weight of the gas and $\mathcal{R}$ is the universal gas constant. If $\mathcal{N}$ is the number of moles of the gas, then $\mathcal{N} = M/M_0$ and the equation of state becomes $p_r \mathcal{V} = \mathcal{N} \mathcal{R} T$. This equation represents the macroscopic state of an ideal gas. Microscopically an ideal gas is assumed to be in random motion, obeying Newton's laws of motion. Their total volume is a negligibly small fraction of the total volume of the gas. No appreciable forces act on the molecules except during collisions which are perfectly elastic and of negligible duration.

From these assumptions, the relationships between the microscopic characteristics and the macroscopic quantities p_r, $\mathcal{V}$, and T can be found. It is found that $p_r = \frac{1}{2} M n \langle v^2 \rangle$ where n is the number of particles per unit volume and $\langle v^2 \rangle$ is the particle mean square velocity. The density ρ of the gas is $\rho = Mn$ and the total mass M_c of the gas in the container is $M_c = NM$, where N is the total number of molecules in the container, $N = n \mathcal{V}$. By taking into consideration the equation of state, the total translational kinetic energy of the gas molecules is found to be directly proportional to the absolute temperature T:

$$ T = \frac{\frac{1}{2} M \langle v^2 \rangle}{\frac{3}{2} k}, $$

where k is Boltzmann's constant,

$$ k = \frac{\mathcal{R} \mathcal{N}}{N} = \frac{\mathcal{R}}{N_0} = 1.38 \times 10^{-23} \text{ J/°K}, $$

and N_0 is Avogadro's number,

$$ N_0 = \frac{N}{\mathcal{N}} = \frac{\mathcal{R}}{k} = 6.023 \times 10^{23} \text{ molecules/mole}. $$

The equation of state can now be written $p_r \mathcal{V} = NkT$. The internal energy of an ideal gas is found to depend only on temperature, $U = \frac{3}{2} NkT$. Introducing Eq. (2.3) into Eq. (2.4), the specific heats as defined by Eqs. (2.1) and (2.2), where the mass is expressed in number of moles ($M = \mathcal{N}$), become

$$ c_p - c_v = \mathcal{R}, \tag{2.5} $$

and the ratio of the specific heats is $\gamma = c_p/c_v$.

The theorem of equipartition of energy states that when the number of molecules is large and Newtonian mechanics holds, the available energy depends only on the temperature and distributes itself in equal parts to each degree of freedom of the molecule. If n_f is the number of degrees of freedom, then

$$ \gamma = \frac{2n_f + 3}{2n_f + 1}. $$

2.4 Thermodynamic Variables

It is apparent that there are two kinds of thermodynamic variables:

1. State variables, which are a function only of the system (the state).
2. Path variables, which are a function of both the state and the path.

There are two path variables: heat $\mathcal{Q}$ and work W. The state variables are the pressure p_r, the volume $\mathcal{V}$, the temperature T, and the internal energy U. It can be shown that there are only two *independent* state variables, from which all the others (e.g., pressure and volume) can be deduced. To the four state variables listed above, the *enthalpy* is added; by definition, the enthalpy $\mathcal{H}$ is

$$\mathcal{H} = U + p_r\mathcal{V}.$$

2.5 Reversible and Irreversible Processes

At a point of stable equilibrium, a system cannot undergo a change from its original state to another state without the aid of an external influence, such as a force or a reservoir at a different temperature. If a series of equilibrium points constitutes a process from an initial state to another state, all the properties between the system and its neighboring systems are exactly balanced.

The process which is conducted from one state to another under the condition of neutral equilibrium is termed a reversible process. Any process which is not reversible is called an irreversible process. The reversibility of a heat engine requires that

1. The engine should be able to operate as a heat engine as well as a refrigerator.
2. Its exchange ratio $(\mathcal{Q}_c/\mathcal{Q}_h)$ should be the same for both directions of operation.
3. The efficiency should be maximum.

2.6 The Carnot Cycle

In a simple heat engine, heat is extracted from a heat source at one temperature and rejected to another source at a lower temperature, with a useful work output. In Fig. 2.3, the heat $\mathcal{Q}_h$ from the boiler, which is at temperature T_h, is transferred to a heat engine, and heat $\mathcal{Q}_c$ is transferred from the engine to the radiator, which is at a lower temperature T_c. If the heat source (boiler) were in contact with the heat sink (radiator), heat would be transferred directly from the source to the sink and no work would be done. The use of the heat engine makes possible the production of useful work.

For an engine to operate indefinitely, it should return to its initial state periodically after following a certain cycle. Consider the system of Fig. 2.3; from the first law $(dU = 0)$.

$$W = \mathcal{Q}_h - \mathcal{Q}_c, \qquad (2.6)$$

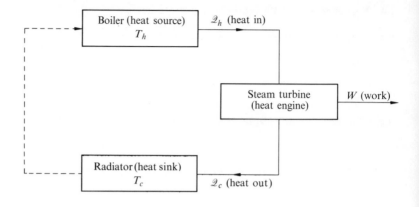

Figure 2.3. Example of a heat engine.

where W is the net work accomplished when the engine returns to its initial state after having completed its cycle.

The thermal efficiency η is the ratio of the output to the input. Thus $\eta = W/\mathcal{Q}_h$ and, introducing Eq. (2.6),

$$\eta = 1 - \frac{\mathcal{Q}_c}{\mathcal{Q}_h}. \tag{2.7}$$

Increasing $\mathcal{Q}_h$ and decreasing $\mathcal{Q}_c$ results in a more efficient engine. If the engine is now run in reverse so that $\mathcal{Q}'_c$ is extracted from the sink and forced into the source by the engine, then the engine is operating in a refrigeration mode by doing work W', where $W' = \mathcal{Q}'_h - \mathcal{Q}'_c$. The ratio of the useful heat extracted to the required work expenditure is called the performance coefficient $\mathscr{C}' = \mathcal{Q}'_c/W'$. It can be shown that, unlike η, $\mathscr{C}'$ may exceed unity in some instances.

There are an infinite variety of cycles possible and many different kinds are utilized for engines. A particularly important one, of profound theoretical significance, is the Carnot cycle. This is an idealized cycle for converting heat into mechanical energy, and vice versa, and is illustrated by the $p - \mathscr{V}$ diagram of Fig. 2.4 for the case of a gas. The cycle is composed of four paths delineated by the intersection of a pair of adiabatics ($\mathcal{Q} = $ constant) and a pair of isotherms ($T = $ constant):

1. Path a-b: *isothermal expansion.* Heat $\mathcal{Q}_h$ is transferred from the source at temperature T_h so that the working gas undergoes an isothermal expansion.
2. Path b-c: *adiabatic expansion.* The working gas is allowed to change adiabatically until its temperature decreases to T_c, doing useful work.
3. Path c-d: *isothermal compression.* Heat $\mathcal{Q}_c$ is transferred out of the working gas into the sink isothermally at temperature T_c.

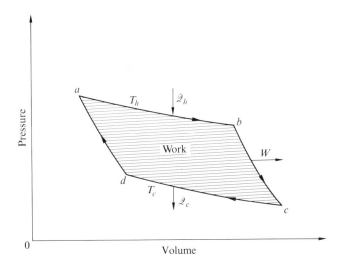

Figure 2.4. The p_r-$\mathscr{V}$ diagram of the Carnot cycle of a gas. The cycle is composed of four paths delineated by the intersection of a pair of adiabatics and a pair of isotherms.

4. Path *d-a*: *adiabatic expansion*. The gas is allowed to change adiabatically until its temperature increases to T_h.

For an ideal gas subjected to the Carnot cycle, it is easy to show that $\mathscr{Q}_h/T_h = \mathscr{Q}_c/T_c$, and that from Eq. (2.7) the Carnot efficiency is $\eta_c = 1 - T_c/T_h$, where the temperatures are in degrees Kelvin. It can be shown that for any real engine operating between T_h and T_c, the efficiency cannot exceed the Carnot efficiency, and always $\eta \leqslant \eta_c$.

This is a serious limitation on efficiency, since T_c can never reach $0°$K and T_h is limited by obvious technological problems.

2.7 The Second Law of Thermodynamics and Entropy

The first law is simply the conservation of energy. However, many processes cannot convert all energy from one form into another—particularly all heat into work. The second law deals with this limitation.

There are several alternative forms for the second law and each form is derivable from any other. The forms stated by Clausius and by Kelvin are widely used. Clausius stated that "it is impossible for any self-acting machine to convey heat continuously from one body to another at a higher temperature." Lord Kelvin stated that "a transformation whose only final result is to transform into work the heat extracted from a source which is at the same temperature throughout, is impossible."

To express the second law in a quantitative form, a quantity which can measure the ability of a system to do work is needed. This physical quantity is called *entropy* and is a measurable property. For an infinitesimal segment of a reversible path, the infinitesimal change in entropy $d\mathscr{S}$, is defined by $d\mathscr{S} = \delta\mathscr{Q}/T$ and the change in entropy from state $\mathscr{S}_1$ to state $\mathscr{S}_2$ is

$$\Delta\mathscr{S} = \mathscr{S}_2 - \mathscr{S}_1 = \int_1^2 \frac{\delta\mathscr{Q}}{T}. \tag{2.8}$$

This quantity is independent of the path chosen between 1 and 2 provided it is reversible, and for an entire reversible *cycle*, the change in the entropy is zero.

It can be shown that a natural process always takes place in such a direction as to cause an increase in the entropy of the system plus the environment. This is because of the irreversibility of natural processes. In the case of an isolated system, it is the entropy of the system which increases, and the second law can be stated in terms of the entropy as $\Delta\mathscr{S} \geqslant 0$. The entropy of a closed physical system never decreases.

Finally, the specific Gibbs free energy $\mathscr{G}$ is defined in terms of the entropy and the enthalpy by the relation

$$\mathscr{G} = \mathscr{H} - T\mathscr{S}, \tag{2.9}$$

and, for constant T,

$$\Delta\mathscr{G} = \Delta\mathscr{H} - T\Delta\mathscr{S}.$$

Since $\Delta\mathscr{H}$ can be visualized as the heat of combustion, it is clear that $\Delta\mathscr{G}$ is the maximum useful work which can be obtained from a system, the ideal efficiency of which can be written as

$$\eta_{\text{ideal}} = 1 - T\frac{\Delta\mathscr{S}}{\Delta\mathscr{H}}.$$

2.8 The Third Law of Thermodynamics

The first and second laws of thermodynamics cannot determine the numerical values of entropy and Gibbs free energy for a given system. The latter are defined in terms of differential quantities. For instance, the change in entropy from a state $\mathscr{S}_0$ to any other state is given by Eq. (2.8), or similarly $\Delta\mathscr{S} = \mathscr{S} - \mathscr{S}_0$ and the value of $\mathscr{S}$ is determined only if $\mathscr{S}_0$ is known. In itself, this is not a serious handicap for entropy or even for free energy, since these quantities are always needed in their differential forms in most of the applications. The same cannot be said for Gibbs free energy, the value of which can be obtained only if the absolute value of $\mathscr{S}$ is determined. Indeed, from Eq. (2.9) a change in Gibbs free energy is deduced; it is equal to

$$\Delta\mathscr{G} = \Delta\mathscr{H} - T\Delta\mathscr{S} - \mathscr{S}\Delta T \tag{2.10}$$

when the temperature T is not a constant quantity.

Planck [11] studied this problem and reached the following conclusion: "As the temperature of a system tends to absolute zero, its entropy tends to a constant $\mathscr{S}_0$ which is independent of pressure, state of aggregation, etc." Nernst [9] continued the work and introduced a new variable equal to the differential *free energy* $\mathscr{F}$. The free energy was defined as $\mathscr{F} = U - T\mathscr{S}$. It can be shown that at constant volume

$$\mathscr{S} = -\frac{\partial \mathscr{F}}{\partial T}\bigg)_V.$$

Thus

$$\Delta\mathscr{F} - \Delta U = T\,\Delta\left(\frac{\partial \mathscr{F}}{\partial T}\right)_V \tag{2.11}$$

at constant temperature. Nernst proposed what amounts to the third law of thermodynamics, i.e.,

$$\lim_{T\to 0}\frac{d(\Delta\mathscr{F})}{dT} = \lim_{T\to 0}\frac{d(\Delta U)}{dT} = 0.$$

Introducing this into Eq. (2.11), one can make the following deduction readily: "As the temperature of a system tends toward zero, its entropy tends toward zero."

This is the general form under which the third law of thermodynamics is usually stated. From this law, the following conclusions can be drawn:

1. The coefficient of thermal expansion as well as the specific heat coefficients c_p and c_v tend toward zero as the temperature tends toward zero.

2. The point of absolute zero temperature can be attained only asymptotically.

B. QUANTUM MECHANICS

2.9 Introduction

In the nineteenth century, all dynamic systems were explained by Newton's principles. Those principles were defied in the twentieth century by the discovery of new, baffling phenomena which led to the birth of two new major fields in physics: the theory of relativity to explain the behavior of particles approaching the speed of light and quantum mechanics to explain the behavior of particles of molecular dimensions.

2.10 Duality of Light and Matter

From the classical wave theory of light, $v = c/\lambda$, where λ is the wavelength, v is the frequency, and c is the speed of light in a vacuum ($c = 3 \times 10^8$ m/sec). It has also been found that in photoelectric experiments, $\mathscr{E}_{\text{kin}} = hv - W$, where hv is the energy of the incident photon, $\mathscr{E}_{\text{kin}}$ is the kinetic energy of the emitted electron, W is the work function characteristic of a given metal (the work needed to remove an electron from the metal's surface), and h is a universal constant known as Planck's constant ($h = 6.6254 \times 10^{-34}$ J · sec).

From the special theory of relativity, the energy of the photon is given by Einstein's relation $\mathscr{E} = Mc^2$. Therefore, photons can be endowed with a mass M and a momentum $\dot{p}$. Taking into consideration Eq. (2.10), it is found that $M = h\nu/c$, and

$$\dot{p} = Mc = \frac{h\nu}{c} = \frac{h}{\lambda}.$$

Collisions of photons with other particles, such as Compton photon-electron collisions, become possible and theoretically understandable. The laws of conservation of energy and of momentum can be applied with

$$\mathscr{E} = h\nu, \tag{2.12}$$

and

$$\dot{p} = h/\lambda. \tag{2.13}$$

The unit of energy used is the electron-volt, which is the amount of energy acquired by one electron when it is accelerated through a potential difference of 1 V: $1 \text{ eV} = 1.6 \times 10^{-19}$ J.

In 1924, Louis de Broglie pointed out that material particles, like photons, might have a dual nature. A material particle is usually defined by its mass M, momentum $\dot{p}$, and energy $\mathscr{E}$ but the argument can be reversed in such a way that a wavelength λ and a frequency ν can be associated with the particle: $\lambda = h/\dot{p}$ and $\nu = \mathscr{E}/\lambda$. If v is the velocity of the particle, then $\lambda = h/Mv$, and $\nu = \frac{1}{2}Mv^2/h$.

All radiations and particles (photons, electrons, neutrons, etc.) move as if they were guided by waves. Classical mechanics becomes inadequate on the atomic scale, and it becomes more useful to look upon these radiations and particles as wave packets. These packets are emitted and absorbed as units, and their wavelengths determine their behavior in diffraction experiments.

2.11 The Uncertainty Principle

The many individual inadequacies and false predictions of classical mechanics led to the concept of Heisenberg's uncertainty principle. This principle follows from experimental evidence through the wave packet concept. It is concerned with the simultaneous measurement of certain *pairs* of variables, and consequently has two aspects:

1. The simultaneous measurement of momentum $\dot{p}_x$ and position x of a particle. The uncertainty principle states that a given experiment cannot determine $\dot{p}_x$ and x to an unlimited precision; position is determined within a range Δx and momentum within a range $\Delta \dot{p}_x$ such that $\Delta x \cdot \Delta \dot{p}_x \geqslant \frac{1}{4}h$. The restriction is not on Δx and $\Delta \dot{p}_x$ separately, but on their product.

2. Similarly, on the problem of the simultaneous measurement of energy $\mathscr{E}$ and time t $\Delta\mathscr{E} \cdot \Delta t \geqslant \frac{1}{4}h$.

The uncertainty principle concept leads to the probabilistic description of the position and motion of a particle. A probability density function $\rho_p(\mathbf{r}, t)$ is used.

The probability of finding a given particle in a volume element $d\mathcal{V}$ will be $\rho_p(\mathbf{r}, t) \, d\mathcal{V}$. The quantity $\rho_p(\mathbf{r}, t)$ can be readily modified to describe the wave nature of a particle. Since it is always positive, it can be related to a wave function ψ such as

$$|\psi(\mathbf{r}, t)|^2 = \rho_p(\mathbf{r}, t).$$

2.12 Schrödinger's Wave Equation

In classical mechanics, to describe the motion of a particle means to study the trajectory $\mathbf{r}(t)$. Similarly, in quantum mechanics, the problem is to find the differential equation for $\psi(\mathbf{r}, t)$ and its solution. Consider a plane wave in the x-direction with frequency ν and wavelength λ; its wave function is known to be

$$\psi = \psi_0 \exp\left[j(k_p x - \omega t)\right], \tag{2.14}$$

where

$$k_p = \frac{2\pi}{\lambda}, \tag{2.15}$$

and

$$\omega = 2\pi\nu. \tag{2.16}$$

The total energy $\mathscr{E}$ is equal to the sum of the potential energy $\mathscr{E}_{pot}$ and the kinetic energy. For a one-dimensional case,

$$\mathscr{E} = \mathscr{E}_{pot}(x, t) + \frac{1}{2}\frac{p^2}{M}. \tag{2.17}$$

From Eqs. (2.13) and (2.14), it is found that

$$\frac{\partial \psi}{\partial x} = jk_p \psi = \frac{2\pi j}{h} p\psi.$$

Therefore

$$\frac{h}{2\pi j}\frac{\partial}{\partial x}\psi = p\psi,$$

and taking as the p-operator,

$$p = -j\hbar\frac{\partial}{\partial x},$$

where

$$\hbar = h/2\pi. \tag{2.18}$$

Similarly, using Eqs. (2.12) and (2.14), it is found that

$$\frac{\partial \psi}{\partial t} = -j\omega\psi = -\frac{2\pi j}{h}\mathscr{E}\psi.$$

Therefore

$$-\frac{h}{2\pi j}\frac{\partial}{\partial t}\psi = \mathscr{E}\psi,$$

and the $\mathscr{E}$-operator's form is

$$\mathscr{E} = jh\frac{\partial}{\partial t}.$$

Replacing $\dot{p}$ and $\mathscr{E}$ by their operators in Eq. (2.17) and applying it to ψ leads to Schrödinger's wave equation:

$$jh\frac{\partial\psi}{\partial t} + \frac{1}{2}\frac{\hbar^2}{M}\frac{\partial^2\psi}{\partial x^2} = \mathscr{E}_{pot}(x,\,t)\psi. \tag{2.19}$$

With Eq. (2.17) there are three conditions which determine ψ:

1. $\int_{-\infty}^{+\infty} |\psi|^2\,dx$ is finite,
2. ψ must be continuous and single valued,
3. $\partial\psi/\partial x$ must be continuous.

2.13 Discrete Energy States

One of the most important results of Schrödinger's equation is the discrete properties of matter. From the wave function $\psi(x,\,t)$ given by Eq. (2.14), the variables x and t can be separated by writing

$$\psi = \psi(\mathbf{r}) \cdot \exp\left(-j\frac{\mathscr{E}}{\hbar}t\right) \tag{2.20}$$

where ω is replaced by its value as deduced from Eqs. (2.20), (2.12), and (2.18). Introducing Eq. (2.20) into Eq. (2.19) leads to Schrödinger's spatial equation:

$$\mathscr{E}\psi(\mathbf{r}) = \mathscr{E}_{pot}(\mathbf{r})\psi(\mathbf{r}) - \frac{1}{2}\frac{\hbar^2}{M}\nabla^2\psi(\mathbf{r}), \tag{2.21}$$

where generalization to the three-dimensional form has been made. Solutions of $\psi(\mathbf{r})$ that satisfy the three conditions listed above exist only for discrete values of $\mathscr{E}$. These values are called *eigenvalues*, and the corresponding $\psi(\mathbf{r})$ solutions are called *eigenfunctions*.

2.14 Electron in Field-Free Space

Consider the simple, special case of an electron traveling in a region of constant potential energy $\mathscr{E}_{pot}$ and with a constant momentum $\dot{p}$. Introducing this in Eq. (2.21), for the case of one dimension, leads to

$$\frac{d^2\psi}{dx^2} + \frac{2M}{h^2}(\mathscr{E} - \mathscr{E}_{pot})\psi = 0. \tag{2.22}$$

The coefficient of ψ is a constant. Equation (2.22) is a wave equation, the solution of which is

$$\psi = \mathscr{A}_1\exp\left(\frac{j}{h}\sqrt{2M(\mathscr{E} - \mathscr{E}_{pot})}\,x\right) + \mathscr{A}_2\exp\left(-\frac{j}{h}\sqrt{2M(\mathscr{E} - \mathscr{E}_{pot})}\,x\right),$$

where $\mathscr{A}_1$ and $\mathscr{A}_2$ are constants to be evaluated by the specific boundary conditions. Introducing this equation into Eq. (2.20) yields

$$\psi(x, t) = \mathscr{A}_1 \exp\left[-j\left(\frac{\mathscr{E}}{\hbar} t - \frac{\sqrt{2M(\mathscr{E} - \mathscr{E}_{pot})}}{\hbar} x\right)\right]$$

$$+ \mathscr{A}_2 \exp\left[-j\left(\frac{\mathscr{E}}{\hbar} t + \frac{\sqrt{2M(\mathscr{E} - \mathscr{E}_{pot})}}{\hbar} x\right)\right]. \qquad (2.23)$$

Comparison of Eq. (2.23) with Eqs. (2.14), (2.15), and (2.16) leads to

$$\lambda = \frac{h}{\sqrt{2M(\mathscr{E} - \mathscr{E}_{pot})}},$$

and

$$\mathscr{E} = h\nu.$$

Now suppose that the total energy of the electron drops from $\mathscr{E}_1$ to $\mathscr{E}_2$ by emitting radiation. The emitted photon will then have a frequency equal to the difference in frequencies before and after the change in $\mathscr{E}$; i.e.,

$$(\mathscr{E}_1 - \mathscr{E}_2) = h(\nu_1 - \nu_2) = h\nu_{21},$$

which is independent of the choice of any reference potential energy. To describe the motion of a single electron, a wave packet of finite length and spread in λ and $\dot{p}$ would have to be used. If this electron is subjected to electromagnetic fields in a laboratory vacuum tube, then quantum mechanics will predict its diffraction in the same way as would classical mechanics. The results do not differ because electromagnetic fields do not vary in any appreciable way over distances of the order of magnitude of a wave packet's size.

2.15 Electron in a Square Well Potential

An atom may be schematically viewed as a number of orbital electrons moving in a potential well resulting from the atomic nucleus and from appropriate groups of coresident electrons. This potential is represented as a function of the distance in Fig. 2.5. Consider the bound states for which $\mathscr{E} < \mathscr{E}_0$. This problem can be solved by applying Schrödinger's equation to each of the three regions shown in Fig. 2.5: $\mathscr{E} = \mathscr{E}_0$ for regions I and III, and $\mathscr{E} < \mathscr{E}_0$ for region II. The constants of integration will be calculated by using the following boundary conditions:

$$\psi_{II}(+x_0) = \psi_{II}(+x_1),$$
$$\psi_I(-x_0) = \psi_{II}(-x_1),$$
$$\frac{d\psi_{II}}{dx}(+x_0) = \frac{d\psi_{III}}{dx}(+x_0), \qquad (2.24)$$
$$\frac{d\psi_I}{dx}(-x_0) = \frac{d\psi_{II}}{dx}(-x_0).$$

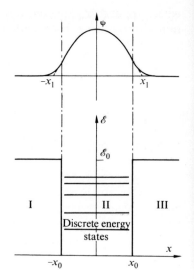

Figure 2.5. Potential energy as a function of distance for the square well atomic model.

For region II, Eq. (2.21) becomes

$$\frac{d^2\psi}{dx^2} + \frac{2M}{\hbar^2}\, \mathscr{E}\psi = 0,$$

the solution of which is

$$\psi_{\mathrm{II}} = \mathscr{A}_{\mathrm{II}} \sin\left(\frac{\sqrt{2M\mathscr{E}}}{\hbar}\, x\right) + \mathscr{B}_{\mathrm{II}} \cos\left(\frac{\sqrt{2M\mathscr{E}}}{\hbar}\, x\right).$$

For region III, Eq. (2.21) becomes

$$\frac{d^2\psi}{dx^2} - \frac{2M}{\hbar^2}(\mathscr{E}_0 - \mathscr{E})\psi = 0,$$

the solution of which is

$$\psi_{\mathrm{III}} = \mathscr{A}_{\mathrm{III}} \exp\left(-\frac{\sqrt{2M(\mathscr{E}_0 - \mathscr{E})}}{\hbar}\, x\right) + \mathscr{B}_{\mathrm{III}} \exp\left(\frac{\sqrt{2M(\mathscr{E}_0 - \mathscr{E})}}{\hbar}\, x\right).$$

The coefficients of x in the exponents are real and since ψ_{III} cannot go to infinity because of condition (1), $\mathscr{B}_{\mathrm{III}}$ must equal zero.

For region I, the solution is similar to that of region III and

$$\psi_{\mathrm{I}} = \mathscr{A}_{\mathrm{I}} \exp\left(-\frac{\sqrt{2M(\mathscr{E}_0 - \mathscr{E})}}{\hbar}\, x\right) + \mathscr{B}_{\mathrm{I}} \exp\left(\frac{\sqrt{2M(\mathscr{E}_0 - \mathscr{E})}}{\hbar}\, x\right).$$

Since ψ has to remain finite, $\mathscr{A}_\mathrm{I}$ should equal zero. Thus

$$\psi_\mathrm{I} = \mathscr{B}_\mathrm{I} \exp\left(\frac{\sqrt{2M(\mathscr{E}_0 - \mathscr{E})}}{\hbar} x\right),$$

$$\psi_\mathrm{II} = \mathscr{A}_\mathrm{II} \sin\left(\frac{\sqrt{2M\mathscr{E}}}{\hbar} x\right) + \mathscr{B}_\mathrm{II} \cos\left(\frac{\sqrt{2M\mathscr{E}}}{\hbar} x\right),$$

$$\psi_\mathrm{III} = \mathscr{A}_\mathrm{III} \exp\left(-\frac{\sqrt{2M(\mathscr{E}_0 - \mathscr{E})}}{\hbar} x\right).$$

These three equations must be joined by the boundary conditions of Eqs. (2.24). The general shape of ψ is represented by a cosine function in region II with exponential tails in regions I and III (see Fig. 2.5). For smaller $\mathscr{E}_0 - \mathscr{E}$, these tails have larger amplitudes and fall less rapidly with distance away from the well. The fact that ψ is not zero in regions I and II is a prediction that the electron will penetrate some distance at the edge of the well. This well will be similar to a well having a width equal to $2x_1$ and infinitely high sides. In that case the solution of the Schrödinger equation will be

$$\psi = \mathscr{A} \sin\left(\frac{\sqrt{2M\mathscr{E}}}{\hbar} x\right) + \mathscr{B} \cos\left(\frac{\sqrt{2M\mathscr{E}}}{\hbar} x\right).$$

The values of $\mathscr{A}$ and $\mathscr{B}$ are given by the boundary condition $\psi(\pm x_1) = 0$. The solution is, for n an odd integer,

$$\psi = \frac{1}{\sqrt{x_1}} \cos\left(n \frac{\pi}{2} \cdot \frac{x}{x_1}\right),$$

and for n an even integer,

$$\psi = \frac{1}{\sqrt{x_1}} \sin\left(n \frac{\pi}{2} \cdot \frac{x}{x_1}\right),$$

where n is called a quantum number and is given by

$$n = \frac{4x_1}{\hbar} \sqrt{2M\mathscr{E}_n},$$

or

$$\mathscr{E}_n = n^2 \frac{\hbar^2}{32Mx_1^2}. \qquad (2.25)$$

An important feature of this result is that solutions are possible only if the energy levels are discrete. An electron with energy $\mathscr{E}_n$ is said to be in the nth quantum state and the quantum state with the lowest energy $\mathscr{E}_1$ is called the *ground state*.

C. SOLID STATE

2.16 The Electronic Structure of Matter

The electronic structure of matter forms the basic source of electrons which do the work necessary in the conversion of energy. All matter can be described by systems of atoms consisting of positively charged nuclei and a certain number of electrons orbiting around each nucleus. The number of electrons is such that the total atomic structure is electrically neutral. Each electron is in a certain state defined by the energy, momentum, and spin of the electron, and, as expressed by Pauli's exclusion principle, no two electrons in the atom may simultaneously exist in the same state. Electron energies vary in discrete steps, and because of this the states associated with a certain energy level are grouped into *shells*.

The most energetic electrons are those existing in the outer shell. They are called *valence* electrons and are the easiest to be detached to become mobile particles. Elements having the same number of valence electrons have similar electronic characteristics, and they are arranged in the periodic table in groups labeled with numbers equal to the number of valence electrons.

The material structure is made up of a system of atoms which are bound together and interact with each other, in the case of a solid. Because of Heisenberg's uncertainty principle, the electrons must be associated with groups of atoms. Because of Pauli's exclusion principle and atomic interactions, the electronic energies are altered and grouped in energy *bands* corresponding to the energy levels of a single atom.

2.17 The Fermi Factor

The famous Maxwell-Boltzmann distribution law states that the probability of a subsystem being in the state n having an energy $\mathscr{E}_n$ is

$$p_b(n) = \mathscr{A} \exp\left(-\mathscr{E}_n/kT\right), \tag{2.26}$$

where $\mathscr{A}$ is a proportionality constant and k is Boltzmann's constant. Since the subsystem must be in one state or another, the total probability $\Sigma_n\, p_b(n)$ must be unity; this determines $\mathscr{A}$ so that

$$p_b(n) = \frac{\exp\left(-\mathscr{E}_n/kT\right)}{\Sigma_n \exp\left(-\mathscr{E}_n/kT\right)}. \tag{2.27}$$

To calculate the average number of particles in each state, consider the ith state as a subsystem having an energy $\mathscr{E}_i$. The energy of one particle will be as referred to some reference energy level $\mathscr{E}_0$:

$$\mathscr{E} = \mathscr{E}_i - \mathscr{E}_0.$$

The energy of the subsystem with n particles is, therefore,

$$\mathscr{E}_n = n(\mathscr{E}_i - \mathscr{E}_0), \tag{2.28}$$

and introducing Eq. (2.28) into Eq. (2.27) yields

$$p_b(n) = \frac{\exp\left[-n(\mathscr{E}_i - \mathscr{E}_0)/kT\right]}{\Sigma_n \exp\left[-n(\mathscr{E}_i - \mathscr{E}_0)/kT\right]}.$$ (2.29)

The average number of particles in the state S_i is

$$\bar{n}_i = \sum_n np(n),$$

or, taking into consideration Eq. (2.29),

$$\bar{n}_i = \frac{\Sigma_n \, n \exp\left[-n(\mathscr{E}_i - \mathscr{E}_0)/kT\right]}{\Sigma_n \exp\left[-n(\mathscr{E}_i - \mathscr{E}_0)/kT\right]}.$$ (2.30)

There are two kinds of particles—*fermions* and *bosons*—distinguished by their state occupation rules. There can be only one fermion (e.g., electrons, protons, neutrons) in each separate state. There can be any number of bosons in the same state.

Thus, for fermions, n is equal to 0 or 1, leading to the famous Fermi factor $f_F(\mathscr{E}_i) = \bar{n}_{iF}$, as deduced from Eq. (2.30)

$$f_F(\mathscr{E}_i) = \frac{1}{\exp\left[(\mathscr{E}_i - \mathscr{E}_F)/kT\right] + 1},$$ (2.31)

where the reference energy level is called the Fermi level $\mathscr{E}_F$. The behavior of $f_F(\mathscr{E}_i)$ is shown in Fig. 2.6, where the temperature is a parameter. For absolute

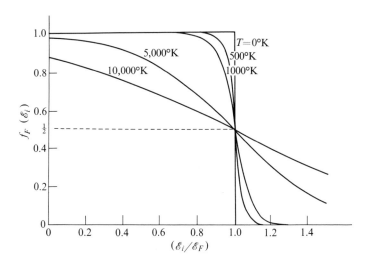

Figure 2.6. The Fermi factor as a function of the energy, with temperature as parameter. (For this figure, $\mathscr{E}_F = 1.725$ eV.)

zero temperature, the curve is unity below the Fermi energy level, and zero above. For higher temperatures, the curve is rounded in the vicinity of the drop.

For bosons, n may be any nonnegative integer, and Eq. (2.30) becomes

$$f_B(\mathscr{E}_i) = \frac{1}{\exp\left[(\mathscr{E}_i - \mathscr{E}_0)/kT\right] - 1}. \tag{2.32}$$

Due to degeneracy, more than one state, i.e., n_{gi} states, may have the same energy level $\mathscr{E}_i$. The average number of particles with energy $\mathscr{E}_i$ is then $n_{gi} f(\mathscr{E}_i)$.

2.18 Energy Bands

As was mentioned earlier, an atom can be viewed as a number of electrons moving in the potential well resulting from the atomic nucleus and coresident electrons. When two atoms are far apart, the valence electron of one atom does not feel the force of the other atom, so that each atom remains unaffected by the other, each atom having its own energy levels and states, except in the case where the atoms are of the same element. In this case, the state and energy levels are the same for both atoms.

If the two atoms approach each other, the forces of one affect the other and the probability of interchange and sharing of electrons, especially the valence electrons,

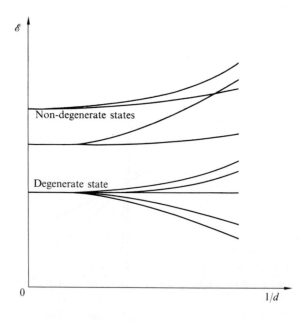

Figure 2.7. The splitting of energy levels as two atoms are brought closer to each other (d is the distance between the two atomic centers).

between them increases. Thus, it is necessary to refer to energy levels of a combination of two atoms rather than those of one atom alone. As the interatomic distance decreases, each original level splits into two levels. In Fig. 2.7, the variation of the energy levels is shown as a function of $1/d$, where d is the distance between the two atomic centers. The closer the atomic centers are to each other, the more the states differ from the states of an individual atom. Note that there can be considerable overlapping in which the original discreteness becomes lost.

If there are more than two atoms, say N, each original level splits into N levels. In the case of a crystal, the atoms are packed close to each other with a regular pattern. The interatomic distances are of the order of magnitude of the atom itself. In such a situation a valence atom is so close to neighboring atoms that it becomes strongly tied to it. For a large assembly of atoms packed so closely together, the wave functions and energy levels for each atom lose their meaning and one must consider a more accurate description in which the wave functions and energy levels are referred to the crystal as a whole.

The interatomic distances in each type of crystal have definite values which are independent of the size of the material. In turn, the pattern of states and energy level distributions for a crystal is also largely dependent on the material size, for all but the smallest samples. In many cases, the allowed energy levels exhibit a grouping tendency leading to the mentioned concept of energy bands.

2.19 Intrinsic Semiconductors

Semiconductors are materials having conductivities between those of conductors and those of insulators. Most of these are elements in group IV of the periodic table, germanium (Ge) and silicon (Si) being the most used. When these materials are chemically pure and have no imperfections in their crystal structure, they are called *intrinsic semiconductors.*

The atomic structure of an intrinsic semiconductor can thus be represented by a pure, perfect crystal in which the positions of the ions (nuclei plus all the inner electronic shells) form the lattice sites. Valence electrons from two adjacent atoms are shared mutually by the atoms in a covalent bonding. The crystal structure is thus formed and no mobile charges can be generated without the supply of a certain amount of energy.

The energy band diagram of an intrinsic semiconductor is shown in Fig. 2.8. In this diagram, the valence band represents the energy band of the valence electrons, and the conduction band is the energy band in which the electrons are practically free of atomic bonding and are able to move freely. These two bands are separated by a forbidden band in which no electron states are possible. For a valence electron to be able to jump to the conduction band, it is necessary to supply it with an amount of energy $\mathscr{E}_c - \mathscr{E}_v$, where $\mathscr{E}_c$ is the lowest energy level in the conduction band and $\mathscr{E}_v$ is the largest in the valence band.

At absolute zero, all valence band states are filled and no electron exists in the conduction band. If the temperature rises, the thermal energy given to the

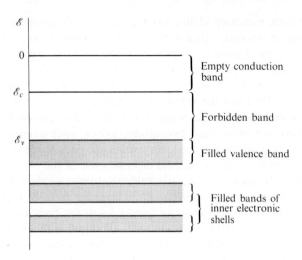

Figure 2.8. Energy band diagram for an intrinsic semiconductor at absolute zero temperature.

system causes the electrons and ions to vibrate and some covalent bonds are broken, generating electron-hole pairs. A hole—actually an absence of a valence electron—behaves as a positive charge. Some electrons will now have enough energy to occupy states in the conduction band. Electrons and holes are generated in equal numbers; they may be influenced by external forces to create electrical currents.

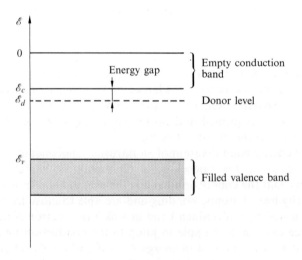

Figure 2.9. Energy band diagram of an n-type extrinsic semiconductor at absolute zero temperature.

2.20 Extrinsic Semiconductors

To produce more mobile charge carriers, it is often practical to "dope" the intrinsic semiconductor by introducing an impurity element from group III or group V of the periodic table. These doped semiconductors are called *extrinsic semiconductors*. The doping levels are generally extremely low (one impurity atom per 10^6 to 10^8 host atoms), and the physicochemical behavior of the extrinsic material is essentially the same as that of the intrinsic material.

When the dopant is a group V element, such as arsenic (As) and antimony (Sb), which replaces some silicon atoms, four of the dopant's valence electrons will fill the covalent bond. The fifth becomes loosely bound and moves with the addition of only a small amount of energy. These group V impurities are called *donors*, because they donate additional electrons for the electronic current. The extrinsic semiconductors doped with group V elements are *n-type* in which the negatively charged particles (electrons) are the majority carriers and the holes are the minority carriers.

Similarly, when the dopant is from group III, such as gallium (Ga) or boron (B) only three electrons fill the covalent bond of the replaced host atom. The vacancy acts as a hole and moves by ionization transfer. These impurities are called *acceptors*. Semiconductors doped with group III elements are *p-type* in which the positive charges (holes) are the majority carriers.

The energy band diagram of an n-type semiconductor at absolute zero is shown in Fig. 2.9 in which $\mathscr{E}_d$ is the energy level of the donor. $\mathscr{E}_c - \mathscr{E}_d$ is the ionization energy necessary to free the loosely bound fifth valence electron. This gap energy is much smaller than $\mathscr{E}_c - \mathscr{E}_v$. For instance, for silicon, $\mathscr{E}_c - \mathscr{E}_v = 1.0$ eV, and when the impurity is arsenic $\mathscr{E}_c - \mathscr{E}_d = 0.04$ eV, allowing the generation of a sufficient amount of charge carriers at room temperatures. In the case of a p-type semiconductor at $0°K$, the energy band diagram is as shown in Fig. 2.10. $\mathscr{E}_a$ is the energy level associated with the vacant valence band and $\mathscr{E}_a - \mathscr{E}_v$ is the ionization energy.

2.21 pn Junctions

The pn junction is the basic component of all semiconductor devices. In direct energy conversion, it is the seat of the photovoltaic effect. By definition, a pn junction is the transition region separating the p-region from the n-region in a semiconductor. Such a junction can be obtained physically by doping one end of an intrinsic semiconductor crystal with a donor and the other end with an acceptor. The host material remains the same, e.g., germanium.

The energy band diagram of a pn junction is shown in Fig. 2.11. In this figure it is seen that the Fermi level remains the same for both the p-region and the n-region. Some electrons in the p-region will diffuse through the junction to the n-region, giving rise to an electric current I_{no} in the opposite direction of the electronic flow. This current is proportional to the number of electrons

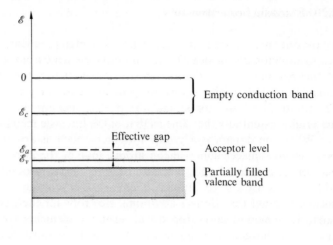

Figure 2.10. Energy band diagram of a p-type extrinsic semiconductor at absolute zero temperature.

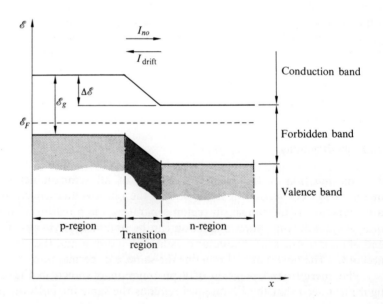

Figure 2.11. Energy diagram of a pn junction in equilibrium, i.e., no externally applied voltage.

in the conduction band of the p-region, which in turn is proportional to $\exp[-(\mathscr{E}_g - \mathscr{E}_F)/kT]$,* as can be deduced from Eq. (2.27). This is a thermally generated current which is often called the *diffusion current*.

The difference in overall polarity between the p- and n-regions gives rise to an electric field and consequently to a potential barrier. This produces a drift current of the electrons in the direction opposite the diffusion current. In equilibrium and in the absence of an external field, the sum of the diffusion and drift currents is equal to zero. The same reasoning can be made for the holes.

When an external voltage is applied, additional energy is provided. The applied voltage is called a *bias*. For forward biasing (i.e., $V > 0$) the potential barrier is lowered, leading to an increase in the number of drifting electrons. For reverse biasing (i.e., $V < 0$) it is heightened, resulting in a reduction in the number of drifting electrons. In both instances, a net current is then produced, since the diffusion current is not affected by this potential. The overall current is equal to the sum of the currents due to the holes and to the electrons. From the above it can be shown easily that the overall output current is

$$ I = I_0\left[\exp\left(\frac{eV}{kT} - 1\right)\right], \tag{2.33} $$

where I_0 is the sum of the diffusion currents due to both holes and electrons. It is called the *saturation* or *dark current*. From Eq. (2.33) it is seen that for a negative voltage, the current tends toward the saturation value. At a certain value of this voltage, however, collisions between the charge carriers result in a breakdown, called Zener breakdown, characterized by a very large increase in the negative current at the peak inverse voltage. The use of the pn junction for power generation will be seen in detail in the chapter on photovoltaic energy conversion.

2.22 Metals

The metallic structure consists of a number of ions surrounded by loosely bound valence electrons. The electrons move easily inside the crystal, explaining the high electrical conductivity of metals.

The energy band diagram of a metal is shown in Fig. 2.12. It is characterized by an overlap of the valence and conduction bands. When an electron reaches the zero energy level, it is emitted outside the crystal and becomes completely free from the metallic structure. The type of emission depends on the kind of energy used to bring about energy emission. The most important types are:

1. *Thermionic emission*, where the energy used is heat
2. *Photoelectric emission*, where the electromagnetic energy of light is used
3. *Field emission*, which is a result of an electric field of high intensity at the surface of the metal

---

* $\mathscr{E}_g$ is the forbidden band gap characteristic of the host material.

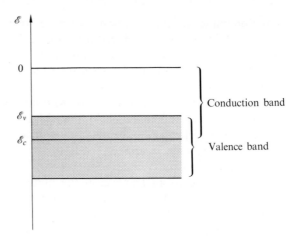

$\mathcal{E}$

0

$\mathcal{E}_v$

$\mathcal{E}_c$

Conduction band

Valence band

Figure 2.12. Energy band diagram of a metal at absolute zero temperature.

4. *Secondary emission,* caused by the bombardment of the metallic surface by other electrons or by positive ions.

D. PLASMA PHYSICS

2.23 Characteristics of a Plasma

A plasma is a conglomeration of positively and negatively charged particles. It is on the average neutral, so that the number of positive charges equals the number of negative charges. The plasma may contain neutral particles or may be fully ionized (when all the particles are charged).

The plasma state is one of the four different states of matter, the other three being the solid, liquid, and gaseous states. It is well known that when sufficient energy is added to a solid, the solid changes state and becomes a liquid, which in turn, with enough energy added to it, will become a gas. The molecules of such a gas assume a variety of degrees of freedom. If energy is added, the kinetic energy of the particles will rise and the molecular impacts will become so intense that dissociation between some of the electrons and the rest of each molecule will result. The gas is said to be *ionized*; it has become a plasma.

The plasma state is the most common form of matter (up to 99 % of the known universe). A material will require on the average 0.01 eV per particle to change from a solid to a liquid or from liquid to gas, whereas a change of state from gas to plasma requires an energy between 1 and 30 eV per particle, depending on the type of material.

The two independent characteristics of a plasma are the charged particle density n and the temperature T. Since the plasma is on the average neutral, the positively charged particle density n_p equals the negatively charged particle density n_e. When the plasma departs from neutrality, $\Delta n = n_p - n_e$ is defined.

Two other characteristics are of the utmost importance for the description of a plasma. These are the Debye radius r_D and the characteristic plasma frequency ω_p. The Debye radius is defined as the radius of a sphere cocentric to the charged particle such that at its surface, the kinetic energy $2\gamma kT$ of the particles equals eV, where V is the potential. By solving Poisson's law in spherical coordinates and considering only the effect of electrons ($\Delta n = -n_e$), the Debye radius is found to be equal to

$$r_D = \sqrt{\frac{\varepsilon_0 k T_e}{e^2 n_e}}.$$

(2.34)

To calculate the characteristic plasma frequency, consider a plasma region of thickness ℓ and displace all the electrons by a distance x. The heavy ions can be disregarded, since the light electrons will react much more quickly. The balance of forces requires that at equilibrium the force due to the electric field E be equal to Newton's force due to the acceleration. After using Poisson's law to replace E, a wave equation in x is obtained:

$$x + \frac{M\varepsilon_0}{n_e e^2} \frac{d^2 x}{dt^2} = 0.$$

(2.35)

The solution is $x = x_0 \exp(j\omega_p t)$, where ω_p is the characteristic plasma frequency for the electrons; it is equal to

$$\omega_p = \sqrt{\frac{en}{M\varepsilon}}.$$

The plasma will then oscillate with a frequency $f_p = \omega_p/2\pi$ around its steady state position. If the collisions are negligible, this oscillation will continue indefinitely.

A plasma cannot exist without charge separation, which takes place only if the collective effects are dominant compared to single particle effects; thus $r_0 > n^{-1/3}$. The second condition for existence of a plasma is that on the average the ionized gas should remain neutral; for this to happen, $r_0 < \ell$, where ℓ is the macroscopic length of the plasma region. In order that oscillations may develop in a plasma, the collisional damping frequency v_c should satisfy $\omega_p \gg v_c$.

The fields of application of plasma physics can now be represented in a diagram (Fig. 2.13) in which the logarithm of plasma temperature is plotted along the abscissa, the ordinate being the logarithm of the electron density. The values of the characteristic plasma frequency and the Debye radius are also shown.

2.24 Cross Section and Ionization

In 1903, Lenard [53] experimentally determined the attenuation of a beam of monoenergetic electrons by a gas. When the beam was passed into a field-free

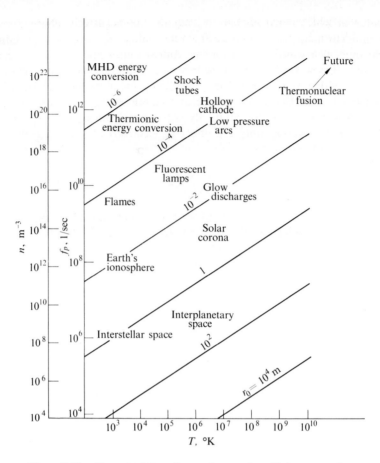

Figure 2.13. Characteristics of several natural and laboratory plasmas.

T = electron temperature

n = electron density

f_p = electron characteristic plasma frequency

r_0 = Debye radius

space, electrons were scattered out of the beam by collisions with the gas atoms. Lenard's simple apparatus collected all the deflected electrons. Electrons which suffered energy changes without being deflected by the beam would have been ignored, as if they did not suffer any collision with the gas atoms. The Lenard experiment led to a constant cross section of collision corresponding to the hard sphere model and equal to $Q_T = \pi r_0^2$, where r_0 is the radius of the atom.

However, the atom is actually a *system* of charges with their resultant electric fields. It does not agree with the hard sphere model, since an electron passing by the atom will be deflected by these fields. Ramsauer solved the problem by using an apparatus in which electrons are emitted by a photoelectric source and are bent in a uniform magnetic field in such a way that a collimated monoenergetic beam is formed at a source slit. Residual electron currents are measured in a collector and collisions occur in a field-free box. If an electron suffers any angular deflection greater than the angular aperture of the detector, it will be lost from the beam. Ramsauer found that the total scattering cross section for elastic collisions varies with the energy of the electrons. The extremely low values of Q for low energies is known as the *Ramsauer effect*.

The chemical properties of a neutral atom depend on the number n_z of the orbiting electrons and on their arrangement. If A is the atomic weight of an atom, there will be $A - n_z$ neutrons in the nucleus. The particles of the atom are bound together in a very small space by short range attractive forces which are much stronger than the electrostatic repulsions between the protons.

Excitation and ionization exist only when the speed of a projectile particle colliding with the atom is at least equal to the orbital speed of an atomic electron. Thus, heavy particles will not cause much ionization by impact unless their energy is very large.

An atom is said to be *excited* when an electron is lifted from a lower to a higher energy level. This may be due to an exchange of energy resulting from a mechanical collision with an electron, an ion, or a fast neutral particle, or by the absorption of a radiation. Ionization can be regarded as an extreme case of excitation, when the electron is given energy which is larger than the input excited level of the atom. The electron then escapes from the attraction of its parent nucleus and the atom becomes a positively charged ion.

After excitation, the electron will remain about 0.01 μsec in the higher energy level and will eventually fall back to a lower energy level, emitting radiation. The states into which the electrons are excited by the absorption of radiation are called *resonance states* and the radiation is called *resonance radiation*. If the electron is excited into a state from which it cannot fall spontaneously, it is said to be in a *metastable* state.

A plasma is deionized by electron-ion recombination processes which are of three types: radiative recombination, in which a positively charged ion becomes excited by absorbing an electron and emitting radiative energy; three-body recombination, in which a positive ion becomes excited by absorbing one electron and releasing another electron; and dissociative recombination, in which a positive molecular ion absorbs an electron and becomes an excited molecule. This excited molecule is very unstable and it dissociates almost instantaneously into an excited atom and a neutral one.

2.25 Forces Between Charge Carriers

There are two important classes of intermolecular forces: short-range forces and long-range forces. Short-range forces are repulsive because the atoms are so close to each other that their electron clouds overlap. Long-range forces such as those due to electrostatic, inductive, and dispersive effects are attractive. Many mathematical models have been proposed for the description of the long-range potential. A simple model is deduced from the hard sphere suggestion in which the potential is assumed equal to zero outside the hard molecule and becomes infinitely repulsive at its surface. A more realistic model has been proposed by Lennard-Jones [54] and it is widely used. It states that

$$eV(r) = 4\mathscr{E}_m[(r_m/r)^{12} - (r_m/r)^6].$$

This is shown in Fig. 2.14. $\mathscr{E}_m$ is the maximum energy of attraction which occurs at $r_0 = 2^{1/6}r_m$. This equation is valid for most spherical nonpolar molecules.

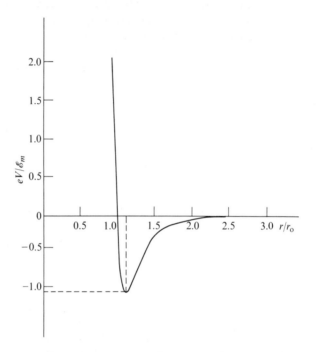

Figure 2.14. The Lennard-Jones potential, valid for most spherical nonpolar molecules.

In the presence of an electric field **E**, the force **F** acting on a medium of charge density ρ_e is $\mathbf{F} = \rho_e\mathbf{E}$, and in the presence of an electromagnetic field, the Lorentz force is

$$\mathbf{F} = \mathbf{J} \times \mathbf{B} + \rho_e\mathbf{E}. \tag{2.36}$$

2.26 Boltzmann Equation

In describing the behavior of a plasma which necessarily consists of a large number of particles, a statistical approach is used. This approach, known as classical statistics, is based on the laws of classical mechanics. It is valid only for systems in which the total volume is much larger than the individual volume of a particle. A Hamiltonian H is defined as the sum of the kinetic and the potential energies of a single charged particle. The equations of motion are then written as a function of the Hamiltonian and the generalized moments and coordinates. It is thus found, by differentiating the volume occupied by the particles in the momentum-space coordinates over time, that the phase volume occupied by a system of particles does not change as the particles move (Liouville's theorem). The number of such particles is equal to the product of a distribution function $f_e(v, \mathbf{r}, t)$ and the volume occupied by them in the momentum-space system of coordinates. If the effect of collisions is considered, the well-known Boltzmann's equation is found readily:

$$\frac{\partial f_e}{\partial t} + v \frac{\partial f_e}{\partial \mathbf{r}} + \frac{\mathbf{F}}{M} \frac{\partial f_e}{\partial v} = \left. \frac{\partial f_e}{\partial t} \right)_{\text{coll}} . \qquad (2.37)$$

The velocity distribution function f_e gives the probability that a particle of velocity v and the coordinate $\mathbf{r}$ be in the phase space. The left-hand side of the equation is the drift term and the right-hand side is the collision term. The force acting on the particle is $\mathbf{F}(\mathbf{r}, t)$. Boltzmann's equation is valid for each type of plasma particle separately. It is very important, since all the transport coefficients can be deduced directly from it. When the force is due to an electromagnetic field, its value becomes

$$\mathbf{F} = e(\mathbf{E} + v \times \mathbf{B}).$$

2.27 Transport Phenomena

Because of a combination of the Brownian motion and the effects of collision, particles are randomized in a plasma. In the absence of any external force, the random motion of the plasma particles will result in a particle current from one region to another only if there exist gradients of particle density, pressure, or temperature in the plasma. The net effect of an applied external force is to cause the particles to move in a direction determined by the force. The existence of such motions results in mass and charge transports which are the most elementary of the transport phenomena. These phenomena are simply described by five transport coefficients: diffusion $\mathcal{D}$, mobility ℓ, electrical conductivity σ, thermal conductivity κ, and viscosity v.

The process by which a net particle current results from the random motion of the individual particles is called *diffusion*. Assuming an isotropic distribution of the particles and neglecting collisions, the random current density can be defined by $\Gamma_r = \frac{1}{4}n\langle v \rangle$, where $\langle v \rangle$ is the velocity of the particles averaged over its distribution function, and n is the number density of the particles. For a more general case where the distribution function is the solution of Eq. (2.37), the particle current density is

$$\Gamma_r = \iiint_{\text{all } v} f_e v \, d^3 v.$$

In the case of free diffusion, this net current density becomes necessarily proportional to the concentration gradient, the constant of proportionality being by definition the diffusion coefficient $\mathscr{D}$. Thus

$$\Gamma_{\text{dif}} = -\mathscr{D} \, \nabla n. \qquad (2.38)$$

Under the effect of an externally applied force, the particles acquire a drift velocity v_d and are said to have a mobility ℓ. If the external force is due to an applied electric field $\mathbf{E}$, the resulting drift velocity of the particles can readily be obtained from the equation of balance of forces:

$$v_d = \frac{e}{M}\left\langle \frac{1}{v_c} \right\rangle \mathbf{E}, \qquad (2.39)$$

where $\langle 1/v_c \rangle$ is the inverse of the collision frequency averaged over the distribution function. The mobility is defined as the ratio of the drift velocity over the electric field, $\ell = v_d/E$, and, comparing this equation to Eq. (2.38), it is found that

$$\ell = \frac{e}{M}\left\langle \frac{1}{v_c} \right\rangle .$$

If the electric field is varying sinusoidally with a frequency ω, the mobility becomes

$$\ell = \frac{e}{M}\left\langle \frac{1}{v_c + \omega} \right\rangle .$$

One of the most important transport coefficients is the electrical conductivity defined as the net flow of charges in response to an applied electric field. It is found experimentally that the current density produced by an electric field is proportional to it, the constant of proportionality being by definition the electrical conductivity. Thus

$$\mathbf{J} = \sigma \mathbf{E}, \qquad (2.40)$$

which is known as Ohm's law. Since the electric field is nothing more than a potential gradient, Eq. (2.40) can be written in a form similar to Eq. (2.38), or

$$\mathbf{J} = -\sigma \, \nabla V, \qquad (2.41)$$

where V is the electrical potential. The current density of the electrons is expressed in terms of the distribution f_e by

$$\mathbf{J} = -e \iiint_{\text{all } v} f_e v \, d^3 v.$$

The result of this integration for different types of plasmas determines the value of the conductivity in terms of such quantities as the electronic temperature and density. This subject will be treated in more detail in the chapter on magnetohydrodynamics.

The mass transport results in a flow of kinetic energy or heat. This heat flow is given in terms of the distribution function f_e by

$$\frac{d\mathcal{Q}}{dS} = \tfrac{1}{2}M \iiint_{\text{all } v} f_e v^2 v \, d^3 v.$$

In the case of free diffusion, the heat flow is necessarily proportional to the temperature gradient in the plasma, the constant of proportionality being, in this case, defined as the thermal conductivity κ. Thus

$$d\mathcal{Q}/dS = -\kappa \, \nabla T. \tag{2.42}$$

Similarly, in some media, the current density Γ_{xy} in the plane xy is proportional to the gradient of the velocity v in the y-direction. The constant of proportionality is defined as the viscosity v. Thus

$$\Gamma_{xy} = -v \frac{dv}{dx}. \tag{2.43}$$

Other plasma physics phenomena will be considered in more detail in the following chapters, especially those dealing with fusion, magnetohydrodynamics, and thermionic power generation.

PROBLEMS

2.1 Prove Eq. (2.5).

2.2 Show that for an adiabatic process $p_r \mathscr{V}^\gamma = \text{constant}$.

2.3 Generator G is a perfect Carnot engine and operates between a source temperature of 350°C and a sink temperature of 20°C. (a) Assuming that in each cycle the generator receives 2000 calories of heat, how many calories are rejected to the sink? (b) If the engine G is made to work as a refrigerator receiving 2000 calories at the sink, how many calories will it deliver to the source? (c) Assuming now that the refrigerator is converting directly the mechanical work required for its operation into heat, what will be the number of calories produced?

2.4 In a 300 m^3 room, 10 persons work, producing an average total of 1000 calories per hour. It is desired to utilize this heat energy at about 20°C to power a perfect Carnot transducer working between 20°C and a sink temperature of -30°C. Calculate the electric power produced by this transducer.

2.5 What is the effect on the values of energy if the square well potential of Section 2.15 is made wider? or higher?

2.6 Assuming that the potential well of Section 2.15 has infinitely high walls, show that Eq. 2.25 becomes

$$\mathscr{E} = \tfrac{1}{8}n^2 h^2 / M x_1^2.$$

2.7 Solve Schrödinger's equation for a hydrogen atom. (Use spherical coordinates.)

2.8 Solve Schrödinger's equation for a potential ring of radius r having a potential $V = 0$ on the ring and $V = \infty$ outside it. Show that in this case

$$\mathscr{E}_n = \tfrac{1}{2}h^2 n^2 (n + 1)^2 / M r^2.$$

2.9 Using the balance of forces on an orbiting electron around the nucleus show that

$$r_n = \frac{n^2 h^2 \mathscr{E}_0}{\pi M e^2},$$

where n is the quantum number and r_n is the radius of the n-orbital. Use the fact that the electron can possess only those orbits having a radius

$$r = \tfrac{1}{2} n \lambda / \pi.$$

2.10 Consider Fig. 2.9, corresponding to an extrinsic semiconductor. By using the principle of conservation of charges, show that

$$n = N_c \exp{(\mathscr{E}_F - \mathscr{E}_c)}/kT,$$
$$p = N_v \exp{(\mathscr{E}_v - \mathscr{E}_F)}/kT,$$

where n is the electron density, p is the hole density, N_c is the effective density of states in the conduction band, and N_v is the effective density of states in the valence band.

2.11 Using Eq. (2.31) and the results of Problem 2.10, show that for an extrinsic semiconductor

$$N_c \{\exp{[(\mathscr{E}_F - \mathscr{E}_c)/kT]} - \exp{[(\mathscr{E}_v - \mathscr{E}_F)/kT]}\} = N_d,$$

when $M_e = M_h = M$, $N_c = N_v$, $\exp{[(\mathscr{E}_F - \mathscr{E}_d)/kT]} \ll 1$, and where N_d is the number of donors per unit volume.

2.12 Solve Problem 2.11 for the case of intrinsic germanium having $N_d = 10^{21}$ donors/m³, $\mathscr{E}_c - \mathscr{E}_v = 0.75$ eV, and $\mathscr{E}_c - \mathscr{E}_d = 0.01$ eV.

2.13 In a certain plasma, the Debye radius is equal to 1 m and the plasma frequency to 1 MHz. (a) Find the electronic temperature and the particle density. (b) What type of plasma is this?

2.14 From the definition of the Debye radius, deduce Eq. (2.34).

2.15 (a) Elaborate on the calculations leading to Eq. (2.35). (b) What is the plasma frequency in a plasma where the effect of ions is not negligible?

REFERENCES AND BIBLIOGRAPHY

A. Thermodynamics

1. Callen, E., *Thermodynamics*, Wiley, New York, 1960.
2. DeGroot, S. R., and P. Mazur, *Non-Equilibrium Thermodynamics*, Interscience, New York, 1962.
3. Denbigh, K. B., *The Thermodynamics of the Steady State*, Wiley, New York, 1951.
4. Eckert, F. R., and R. M. Drake, *Heat and Mass Transfer*, McGraw-Hill, New York, 1959.
5. Fermi, E., *Thermodynamics*, Dover, New York, 1937.
6. Jeans, J., *Kinetic Theory of Gases*, Cambridge University Press, Cambridge, 1946.
7. Miller, D. G., "Thermodynamics of Irreversible Processes. The Experimental Verification of the Onsager Reciprocal Relations," *Chemical Reviews*, Vol. 60, No. 1, p. 15, 1960.

8. Moore, W. J., *Physical Chemistry*, Allyn and Bacon, Boston, 1962.

9. Nernst, W., *Theoretical Chemistry*, Macmillan, New York, 1927 (English translation from German).

10. Osterle, J. F., "A Unified Treatment of the Thermodynamics of Steady-State Energy Conversion," *Appl. Scient. Res.*, Section A, Vol. 12, p. 425, 1964.

11. Planck, M., *Treatise on Thermodynamics, 1888*, Dover, New York, 1945.

12. Prigogine, I., *Introduction of Thermodynamics of Irreversible Processes*, Interscience, New York, 1961.

13. Rossini, R. D., *Chemical Thermodynamics*, Wiley, New York, 1950.

14. Soo, S. L., *Analytical Thermodynamics*, Prentice-Hall, Englewood Cliffs, N.J., 1962.

15. Wrangham, D. A., *The Theory and Practice of Heat Engines*, Cambridge University Press, Cambridge, 1960.

16. Zemansky, M. W., *Heat and Thermodynamics*, McGraw-Hill, New York, 1957.

B. and C. Quantum Mechanics and Solid State

17. Azaroff, L. V., *Introduction to Solids*, McGraw-Hill, New York, 1960.

18. Bitter, F., *Nuclear Physics*, Addison-Wesley, Reading, Mass., 1950.

19. Blanchard, C. H., et al., *Introduction to Modern Physics*, Prentice-Hall, Englewood Cliffs, N.J., 1958.

20. Bragg, W. H., and W. L. Bragg, *The Crystalline State*, Macmillan, New York, 1948.

21. Bube, R. H., *Photoconductivity of Solids*, Wiley, New York, 1960.

22. Dekker, A. J., *Solid State Physics*, Prentice-Hall, Englewood Cliffs, N.J., 1957.

23. Drabble, J. R., and H. J. Goldsmid, *Thermal Conduction in Semiconductors*, Pergamon Press, New York, 1961.

24. Eisberg, R. M., *Fundamentals of Modern Physics*, Wiley, New York, 1961.

25. Evans, R. D., *The Atomic Nucleus*, McGraw-Hill, New York, 1955.

26. Flügge, S. (ed.), *Handbuch der Physik*, Springer-Verlag, Berlin, 1957.

27. Goldstein, H., *Classical Mechanics*, Addison-Wesley, Reading, Mass., 1950.

28. Green, A. E., *Nuclear Physics*, McGraw-Hill, New York, 1955.

29. Heisenberg, W., *The Physical Principles of the Quantum Theory*, University of Chicago Press, Chicago, 1930.

30. Kemble, E. C., *The Fundamental Principles of Quantum Mechanics with Elementary Applications*, Dover, New York, 1937.

31. Kittel, C., *Introduction to Solid State Physics*, Wiley, New York, 1956.

32. Landau, L. D., and E. M. Lifshitz, *Statistical Physics*, Addison-Wesley, Reading, Mass., 1958.

33. Leighton, R. B., *Principles of Modern Physics*, McGraw-Hill, New York, 1959.

34. Lindsay, R. B., *Physical Mechanics*, Van Nostrand, Princeton, N.J., 1950.

35. Schiff, L. I., *Quantum Mechanics*, McGraw-Hill, New York, 1949.

36. Segre, E. (ed.), *Experimental Nuclear Physics*, Wiley, New York, 1953.

37. Semat, H., *Introduction to Atomic and Nuclear Physics*, Rinehart, New York, 1954.

38. Slater, J. C., *Modern Physics*, McGraw-Hill, New York, 1955.

39. Smith, L., *Semiconductors*, Cambridge University Press, Cambridge, 1959.

40. Sproull, P. L., *Modern Physics*, Wiley, New York, 1956.

D. Plasma Physics

41. Arzimovich, L. A., *Elementary Plasma Physics*, Blaisdell, New York, 1965.

42. Brown, S. L., *Basic Data of Plasma Physics*, MIT Press, Cambridge, Mass. and Wiley, New York, 1959.

43. Cambel, A. B., *Plasma Physics and Magnetofluidmechanics*, McGraw-Hill, New York, 1963.

44. Chapman, S., and T. C. Cowling, *The Mathematical Theory of Non-uniform Gases*, Cambridge University Press, Cambridge, 1939.

45. Cobine, J. D., *Gaseous Conductors*, Dover, New York, 1941.

46. Cowling, T. C., *Magnetohydrodynamics*, Interscience, New York, 1957.

47. Delcroix, J. L., *Introduction to the Theory of Ionized Gases*, Interscience, New York, 1960.

48. Denisse, J., and J. L. Delcroix, *Theory of Waves in Plasmas*, Interscience, New York, 1960.

49. Drummond, J. E., *Plasma Physics*, McGraw-Hill, New York, 1961.

50. Francis, G., *Ionization Phenomena in Gases*, Butterworths, London, 1960.

51. Hughes, W. F., and F. J. Young, *Electromagnetodynamics of Fluids*, Wiley, New York, 1967.

52. Kunkel, W. B. (ed.), *Plasma Physics in Theory and Application*, McGraw-Hill, New York, 1966.

53. Lenard, P., "Production of Cathode Rays by Ultra-Violet Light," *Annalen der Physik*, Vol. 2, No. 2, p. 359, 1900.

54. Lennard-Jones,* J. E., "On the Determination of Molecular Fields II. From the Equation of State of a Gas," *Proc. Roy. Soc.* (*London*), Vol. A106, p. 463, 1924.

55. Loeb, L. B., *Basic Processes of Gaseous Electronics*, Cambridge University Press, Cambridge, 1960.

56. Spitzer, L., *Physics of Fully Ionized Gases*, Interscience, New York, 1956.

57. Stix, T. H., *The Theory of Plasma Waves*, McGraw-Hill, New York, 1962.

58. Sutton, C. W., and A. Sherman, *Engineering Magnetohydrodynamics*, McGraw-Hill, New York, 1965.

59. Tanenbaum, B. S., *Plasma Physics*, McGraw-Hill, New York, 1967.

60. Thompson, W. B., *An Introduction to Plasma Physics*, Pergamon Press, New York, 1967.

61. Uman, A. M., *Introduction to Plasma Physics*, McGraw-Hill, New York, 1964.

---

*   This author used the name J. E. Jones in his early papers.

3 FUSION POWER

3.1 Introduction

Energy can be released by the nuclear fusion of light elements. When two light nuclei combine to form a heavier one, their total mass is greater than the mass of the resulting heavy particle. The balance has been transformed into heat. The reaction is therefore exothermic or exoenergetic; it releases energy in the form of heat. The relation between the missing mass and the amount of energy released is given by the famous Einstein relation, $\mathscr{E} = Mc^2$, where c is the speed of light in vacuum.

By the end of the first third of this century [23], it had been determined that the conversion of hydrogen into helium by the fusion of two hydrogen nuclei is the main reaction occurring in almost all the stars, including the sun. Indeed, helium, of mass four, is an extremely stable element, and reactions having it as an end product are extremely energetic. The task of scientists and engineers is to imitate in the laboratory and eventually on an industrial scale such phenomena which are and will remain the main source of energy in the universe.

Fusion reactions on a very small scale were performed in the laboratory in the 1930's [90]. Particle accelerators used by physicists can accelerate protons to high energies. If these energies are high enough, the proton will be able to break the nuclear electrical repulsion and fuse with the light nucleus of a target.

The direct conversion of fusion energy to electricity is a remote goal, but is already being considered [7, 25, 26, 29]. Although the research is in its infancy, the prospects are very bright.

3.2 Conditions for Nuclear Fusion

The most practical fusion fuels are the hydrogen isotopes, deuterium (D_1^2) and tritium (T_1^3). Deuterium is a relatively cheap material and, because of this, it is often used alone in fusion experiments. However, because of the high critical ignition temperature of the deuterium-deuterium reaction (see Table 3.1), tritium is also used. The most important reactions are listed in Table 3.1, where n indicates a neutron.

The fusion cross section in Table 3.1 can be defined as a probability of interaction. Indeed, the probability per unit time that a projectile nucleus undergoes a fusion is $p_{\text{fusion}} = nQv$, where n is the density of the target particles in the plasma, v is the velocity of the projectile nucleus, and Q is the area presented by a single target nucleus to the incoming projectile. This area is by definition the fusion cross section.

Table 3.1

Thermonuclear reactions

Reaction number	Reaction	Fusion cross section, cm^2
1	$D_1^2 + D_1^2 \rightarrow He_2^3$ (0.82 MeV) + n (2.45 MeV)	10^{-26}
2	$D_1^2 + D_1^2 \rightarrow T_1^3$ (1.01 MeV) + H_1^1 (3.02 MeV)	10^{-26}
3	$D_1^2 + T_1^3 \rightarrow He_2^4$ (3.5 MeV) + n (14.10 MeV)	2×10^{-24}
4	$D_1^2 + He_2^3 \rightarrow He_2^4$ (3.6 MeV) + H_1^1 (14.7 MeV)	2×10^{-27}
5	$T_1^3 + T_1^3 \rightarrow He_2^4$ (3.8 MeV) + 2n (7.6 MeV)	
6	$He_2^3 + He_2^3 \rightarrow He_2^4$ (4.3 MeV) + $2H_1^1$ (8.5 MeV)	
7	$D_1^2 + Li_3^6 \rightarrow 2He_2^4$ + 22.4 MeV	
8	$H_1^1 + Li_3^7 \rightarrow 2He_2^4$ + 17.4 MeV	
9	$Li_3^6 + n \rightarrow He_2^4 + T_1^3$ + 4.6 MeV	
10	$Li_2^7 + D_1^2 \rightarrow Be_4^8 + n$ + 15.0 MeV	

Reactions 1 and 2 are almost equally probable. The energy released from either reaction is expressed in terms of megaelectron-volts (MeV). The total energy released by reaction 1 is therefore equal to (2.45 + 0.82) = 3.27 MeV, whereas reaction 2 releases (3.02 + 1.01) = 4.03 MeV. In the first reaction, the products of fusion are a neutron and a helium-3 nucleus, and in the second reaction, they are a proton (H_1^1) and a triton (T_1^3). The latter will fuse almost immediately with deuterium, leading to the formation of a neutron and a helium-4 nucleus, as in reaction 3. The amount of energy released in this reaction is equal to 17.6 MeV. Another reaction is possible between the helium-3 and the deuterium nuclei, as stated by reaction 4. However, the probability of occurrence of this reaction is fairly low, due to its small cross section. In the deuterium-tritium reaction (3), it is necessary to obtain the tritium independently from the deuterium-deuterium reaction (2). This can also be achieved by using lithium-6, as shown by reaction 9. There is a need for a primary source of neutrons in this type of reaction. This need may be fulfilled by the use of a nuclear fission reactor. It is important here to note that once the deuterium-tritium reaction is started, a new source of neutrons will be available. In this reaction, the role of impurities becomes very important; a small percentage of oxygen, argon, or xenon would quench the field of neutrons.

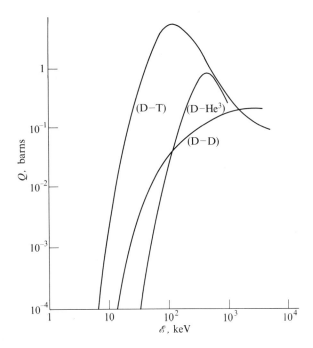

Figure 3.1. Cross sections for deuterium-tritium, deuterium-deuterium, and deuterium-helium-3 reactions. Each reaction has a maximum cross section at a certain energy. The energy $\mathscr{E}$ corresponds to the temperature T following the relation $\mathscr{E} = \frac{3}{2}kT$. (1 barn $= 10^{-24}$ cm².)

Figure 3.1 shows the variations of the different cross sections with temperature for reactions 1, 3, and 4. Note that the cross sections are extremely small at low energies. They are given by a Gamow factor,

$$Q = \frac{\mathscr{A}_1}{\mathscr{E}} \exp\left(-\frac{\mathscr{B}_1}{\sqrt{\mathscr{E}}}\right),$$

where $\mathscr{A}_1$ and $\mathscr{B}_1$ are constants. This cross section rises rapidly until a maximum value is reached, and then decreases for very high energies.

The rate of thermonuclear energy release increases rapidly with the ionic temperature. Allis [2, p. 2] gives the following practical formula for its calculation:

$$\dot{\mathscr{V}} = \mathscr{A} n_1 n_2 (\mathscr{B}/T)^{2/3} \exp\left[-(\mathscr{B}/T)^{1/3}\right], \tag{3.1}$$

where $\dot{\mathscr{V}}$ is the reaction rate given in cubic centimeters per second, $\mathscr{A}$ and $\mathscr{B}$ are constants particular to a given reaction, n_1 and n_2 are the concentrations per cubic centimeter of relevant species of ions, and T is the ionic temperature in electron-volts. The values of the constants $\mathscr{A}$ and $\mathscr{B}$ are, for deuterium-deuterium

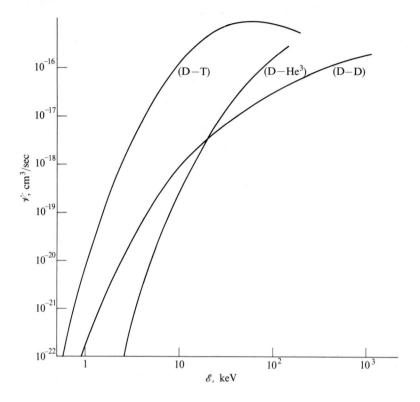

Figure 3.2. Reaction rates for deuterium-tritium, deuterium-deuterium, and deuterium-helium-3 reactions ($\mathscr{E} = \frac{3}{2}kT$).

reactions, $\mathscr{A} = 75 \times 10^{-22}$ cm³/sec and $\mathscr{B} = 6.602 \times 10^{6}$ eV, and for deuterium-tritium reactions, $\mathscr{A} = 11 \times 10^{-16}$ cm³/sec and $\mathscr{B} = 15.763 \times 10^{6}$ eV. Figure 3.2 shows the temperature variations of $\mathscr{V}$ for deuterium-deuterium, deuterium-tritium, and deuterium-helium-3 reactions.

A serious problem is obtaining environmental conditions which lead to a self-sustaining reaction. These conditions may be summarized as follows:

1. Attainment of a minimum ignition temperature. This is the temperature at which the energy production equals the energy loss. If only the loss by brehms-strahlung* is considered, this temperature will be 46×10^{6} °K for the deuterium-tritium reaction and 410×10^{6} °K for the deuterium-deuterium reaction. This is illustrated in Fig. 3.3. Actually, the temperature of ignition is somewhat higher

---

* Radiation resulting from interaction between fast charge carriers in the plasma, especially electrons colliding with ions.

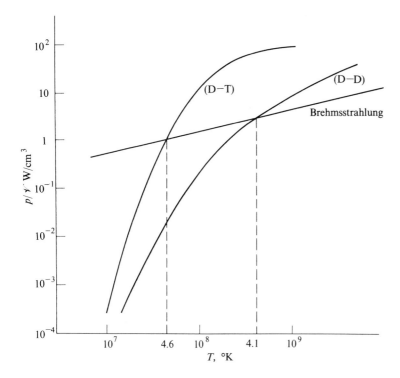

Figure 3.3. Evaluation of the critical conversion temperature for deuterium-tritium and deuterium-deuterium (total) reactions. The deuterium-tritium reaction is more advantageous because of its lower ignition temperature.

than these values because of the existence of other losses. Therefore, the environmental energy loss should be reduced at least to the rate of energy release through fusion.

2. Attainment of a critical fuel mass. The temperatures attained in these reactions are considerably higher than those required to ionize most of the elements involved. Since the elements are created above their ionization potentials, the plasma will become fully ionized. The pressure should be at least as high as 10^{-3} mm Hg at the start of the experiment. Because of the high temperature, this leads to a particle density of the order of 10^{14} particles/cm^3.

3.3 Plasma Confinement by Magnetic Fields

It would seem that because of the extremely high temperature involved, any known substance would be evaporated if it were to remain in such an environment. At the present state of the art, because of the relatively low densities which are attained, the total energy of the plasma is insufficient to cause such damage. Even so, the loss of energy by the fuel nuclei will be extremely large if they are permitted to strike the containing walls. For this reason, there is a great necessity to seek other forms of confinement. The magnetic field confinement is the most promising method.

It is known that a charged particle moving in a magnetic field is deflected in a direction perpendicular to both the velocity of the particle and the direction of the magnetic field. To confine a plasma by magnetic fields, it is necessary to solve three problems:

1. An equilibrium state between the gas pressure and the magnetic pressure should be reached. The pressure of the plasma is proportional to its particle density and its temperature. This pressure is outward and it should be balanced by an equal magnetic pressure if the particles are to stay confined.

2. The particles should not escape from the volume enclosed by the magnetic field. For this reason, the magnetic field must maintain a suitable configuration. Different configuration possibilities will be discussed later.

3. The confinement should be stable. Due to the fact that a collection of charged particles acts collectively, the plasma will have a tendency to become unstable. In such a case, it will escape from the confining magnetic "bottle," if allowances are not made to prevent this escape.

3.4 Plasma Instabilities [1; 50; 59; 77, Chap. XII]

Consider a system with K degrees of freedom and a number of Lagrangian coordinates $q_1, q_2, \ldots, q_K$. Assuming that all the forces derive from a potential energy $\mathcal{E}_{pot}$, the condition of equilibrium is $\partial \mathcal{E}_{pot}/\partial q_j = 0$ with $j = 1, 2, \ldots, K$. The equilibrium is said to be stable if any increment or decrement of a general coordinate corresponds to an increase in potential energy, i.e., if

$$\frac{\partial \mathcal{E}_{pot}}{\partial q_j^2} > 0. \tag{3.2}$$

Then, since the sum of the kinetic energy and the potential energy is constant, any increase in potential energy will be counterbalanced by a decrease in kinetic energy, and vice versa. If the kinetic energy remains below a certain limiting value, the system will return to its equilibrium position. If Eq. (3.2) is not satisfied, for any degree of freedom, the system will depart further and further from its equilibrium position and will never return to it.

Figure 3.4 shows examples of stable and unstable equilibria.

Taking into account the electromagnetic and the kinetic energies, similar considerations apply to equilibrium in plasmas. In this case, the degrees of freedom become extremely numerous and it becomes impossible to consider the problem. No complete treatment of plasma instabilities so far exists, and the existing

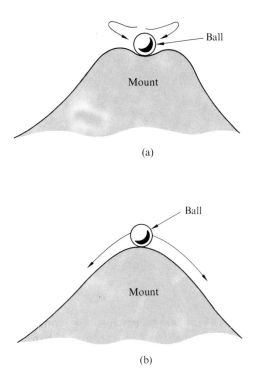

Figure 3.4. Examples of stable and unstable equilibria. (a) Stable equilibrium: under the effect of an energy lower than a certain critical energy, the ball will always return to its initial position. (b) Unstable equilibrium: the slightest force on the ball makes it depart without return from its initial position.

partial theories are often nonrigorous or exceedingly complicated. There are, however, three levels of approximation in the general theory of plasma instabilities:

1. *The MHD approximation,* where the plasma is considered as a continuum in the manner of a liquid metal, can be applied to hydromagnetic instabilities.

2. *The wave approach,* where instabilities are considered as self-growing waves, can be applied to microinstabilities.

3. *The Boltzmann equation approach,* which is more rigorous than the other two, will not be examined, because it is outside the scope of this chapter.

The instabilities which can exist in a confined plasma can be divided into two categories: the electrostatic instabilities and the hydromagnetic instabilities. The *electrostatic instabilities* are due to the rise of electric fields inside the plasma which tend to drive the charged particles outside the magnetic confinement. The character of these instabilities has yet to be clarified. The *hydromagnetic*

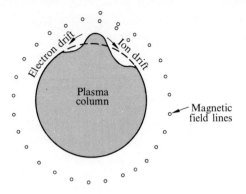

Figure 3.5. The flute instability in a plasma column is due to an increase of the magnetic pressure over the plasma pressure (cross section of column).

instabilities are due to the mutual motion of the confined plasma and the magnetic field. These instabilities, which are fairly well understood, fall into several categories, among which are the flute instability, the kink instability, and the sausage instability:

a) *The flute instability* (Fig. 3.5) occurs in many confined plasma systems. It arises whenever the plasma pressure becomes lower than the magnetic pressure.

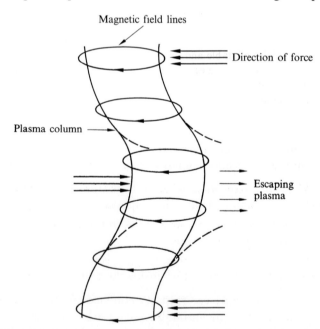

Figure 3.6. The kink instability in a plasma column. The magnetic pressure becomes greater inside the kink and smaller outside it, thus increasing the instability until final break of the column (side view of column).

If a steady-state plasma confined by a magnetic field is considered, a displacement of the plasma will become possible when the interchange of the magnetic lines of force is such that the magnetic field strength remains unchanged. If the displacement takes the form of a decrease of plasma energy, the system becomes unstable. This instability may be eliminated by an increase in plasma energy.

b) *The kink instability* (Fig. 3.6) originates as a kind of kink in a plasma column. The column retains its circular cross section, but the magnetic lines of force become more concentrated on the concave portion of the column and less on the convex portion. Therefore, the magnetic pressure becomes greater inside and smaller outside. This tends to amplify the kink and eventually break the column.

c) *The sausage instability* (Fig. 3.7) results from a localized azimuthally symmetrical constriction or expansion of the plasma column. The law of Biot-Savart states that a current of I amperes in the axial direction of a plasma column produces an azimuthal field B_θ given by $B_\theta = \mu_0 I/2\pi r$, where μ_0 is the permeability of free space and r is the radius of the column. Therefore, the magnetic pressure becomes greater where the discharge is more constricted and smaller where it is less constricted. As a result, any irregularity in the column diameter tends to be amplified.

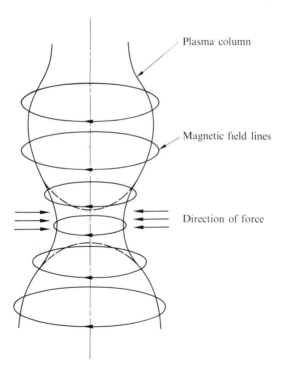

Figure 3.7. The sausage instability in a plasma column. The magnetic pressure becomes stronger all around the constriction until final break of the column (side view of column).

The stabilization of the plasma column is at least partially possible. This has been demonstrated by an extensive theoretical and experimental investigation. For instance, the sausage instability can be eliminated by "hardening" the column. This is done by creating an internal magnetic field with its lines of force parallel to the plasma column. Such a magnetic field is obtained by passing an electrical current through a solenoid around the plasma column. This method will also inhibit the development of kink instabilities.

3.5 Different Approaches to Nuclear Fusion

The first proposals for a type of fusion reactor led to the *linear pinch*, a linear device in which the plasma is confined by magnetostriction. As a result of the problems of stability which arose, other configurations were (and are) being investigated. A list of the important fusion reactors is presented in Table 3.2. Each of these configurations will be studied separately.

Table 3.2

Thermonuclear reactors

Reactors	Countries
Closed magnetic line fusion reactors	
Linear and toroidal pinches	U.S., U.K., U.S.S.R., France, Sweden, Germany
Stellarators	U.S.
Mirror machines	U.S., U.S.S.R.
Astron	U.S.
Open magnetic line fusion reactors	
Theta pinches	U.S.
Homopolar machines	U.S.

The pinch [15]. An electric current in a conductor creates an azimuthal magnetic field. When the conductor is an electrical discharge (plasma), the magnetic field produced by the current will compress and thus confine the plasma. The plasma is also heated by this compression effect. This phenomenon is known as the *pinch effect*. The pinch can be either linear or toroidal; in the latter case it is an electrodeless discharge.

Pinch experiments were started simultaneously in several countries during the early 1950's [61]. It does not seem probable at this time that the pinch geometry, because of its instabilities, will be the final solution to the plasma confinement problem.

It is possible to study the pinch effect by assuming that its unstable state is a succession of quasi-stable ones. In this case, one can assume that at any point on the surface of the discharge the electromagnetic force equals the plasma force.

It follows that for a plasma containing n_i ions/m^3 and n_e electrons/m^3,

$$\nabla p_i - n_i e(\mathbf{E} + v_i \times \mathscr{B}) = 0,$$
$$\nabla p_e + n_e e(\mathbf{E} + v_e \times \mathscr{B}) = 0,$$

(3.3)

where ∇p_i and v_i represent the pressure gradient and the average velocity of the ions, ∇p_e and v_e represent those of the electrons, and $\mathbf{E}$ and $\mathscr{B}$ are the electric and magnetic fields, respectively. By making the assumptions of cylindrical symmetry, the existence of an azimuthal component of the magnetic field, plasma neutrality, and a Maxwellian distribution of the velocities leading to $p_r = nkT$, Eqs. (3.3) become, when added to each other,

$$k(T_i + T_e)\frac{dn}{dr} - ne(v_i - v_e)B_\theta = 0,$$

(3.4)

where T_i and T_e are the temperatures of the ions and the electrons, respectively. However, the ionic velocity is negligible compared to the electronic velocity. Furthermore, the axial component of the current density is given by Maxwell's equation. The axial component is also proportional to the product of the plasma density and the electronic velocities; thus,

$$J_z = \mu_0 \frac{1}{r} \cdot \frac{\partial}{\partial r}(rB_\theta) = env_e.$$

Therefore,

$$\frac{\partial}{\partial r}(rB_\theta) = \frac{en}{\mu_0}rv_e.$$

Introducing all these considerations in Eq. (3.4) yields

$$\frac{d}{dr}\left(\frac{r}{n} \cdot \frac{dn}{dr}\right) + \frac{e^2 v_e^2}{\mu_0 k(T_e + T_i)}nr = 0.$$

(3.5)

Assuming that $dn/dr = 0$ and $n = n_0$ at the boundary $r = 0$, the solution of Eq. (3.5) becomes

$$\frac{n}{n_0} = \frac{1}{[1 + \mathscr{K}(r/r_0)^2]^2},$$

(3.6)

where r_0 is the radius of the container and $\mathscr{K}$ is a constant given by

$$\mathscr{K} = \frac{e^2 v_e^2 n_0 r_0^2}{8k\mu_0(T_e + T_i)}.$$

Equation (3.6), plotted in Fig. 3.8, is known as the Bennett [7] distribution function. The plasma is constricted toward the axis of the discharge but goes to zero only at an infinite radius.

The snow-plow model of the pinch. The above assumption of quasi-equilibrium applies only when the current through the discharge is practically constant. This is not the case in most of the experimental pinches. Most of the pinches are "fast,"

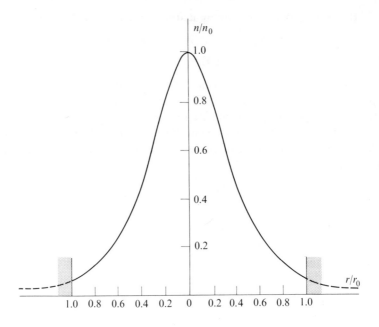

Figure 3.8. The pinch effect as demonstrated by the Bennett distribution function for $\mathcal{K} = 3$.

i.e., the pinch is established relatively quickly as compared to the field diffusion and the sound transit times.

Now, consider a pinched column of fully ionized plasma. The column is assumed to be an ideal conductor in such a way that the penetration depth can be neglected and the current density can be considered as confined to the surface. The purely external magnetic field exerts a pressure on the column, the value of which is

$$p_r = -\tfrac{1}{2}B_\theta^2/\mu_0, \tag{3.7}$$

where p_r is the inward pressure. The azimuthal magnetic field B_θ is given by

$$B_\theta = \frac{\mu_0 I}{2\pi r}, \tag{3.8}$$

where r is the radius of the external plasma sheath and I is the current carried by it. On the other hand, the momentum balance of the surface of the pinch requires that

$$2\pi r p_r = d(Mv)/dt, \tag{3.9}$$

where M is the inward accelerated mass of the particles per unit length and $v = -dr/dt$. Furthermore, if ρ is the initial mass density and r_0 is the total radius of the column, then

$$M = \pi(r_0^2 - r^2)\rho. \tag{3.10}$$

Introducing Eqs. (3.8), (3.9), and (3.10) into Eq. (3.7) yields

$$\frac{\mu_0 I^2}{4\pi r} = n \frac{d}{dt}\left[\rho(r_0^2 - r^2)\frac{dr}{dt}\right]. \tag{3.11}$$

Assume, now, that the discharge current $I_0 \sin \omega t$ increases linearly with time, for small time t. Then $I \cong I_0\omega t$. Defining $m_p = r/r_0$ as the pinch ratio and m_t as

$$m_t = t\left(\frac{\mu_0 I_0^2 \omega^2}{4\pi^2 \rho r_0^4}\right)^{1/4},$$

one has for Eq. (3.11),

$$\frac{d}{dm_t}\left[(1 - m_p^2)\frac{dm_p}{dm_t}\right] = -\frac{m_t^2}{m_p}. \tag{3.12}$$

The solution of this equation, $m_p = f(m_t)$, is plotted in Fig. 3.9.

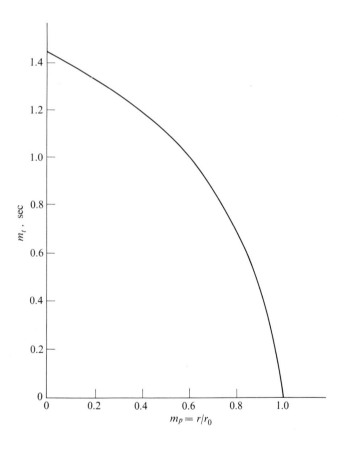

Figure 3.9. Time dependence of the pinch ratio in a stellarator.

Other pinch geometries. Many experiments were performed on the pinch effect in straight geometries. To avoid losses at the ends of the straight geometries, toroidal geometries were used. In a toroidal geometry, the magnetic field pressure compresses the plasma so that the current flows in a narrow ring near the center of the tube in such a way as to keep the plasma away from the walls. Due to instabilities, this pinched plasma persists only a few microseconds. No significant amount of energy could be produced in such a short time. To avoid the instabilities, it is necessary to stabilize the plasma by using magnetic fields parallel to the current in the ring.

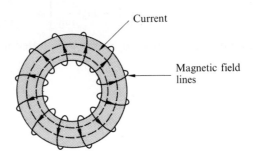

Figure 3.10. Inhomogeneous axial magnetic field in a toroidal plasma.

Other solutions to this problem were investigated. The hard core pinch is one of the most important of those solutions. It consists of an insulated metal ring located in the center of the toroidal tube. A current in the ring creates an azimuthal magnetic field. The plasma is surrounded and compressed by an axial field although this field is not trapped in the plasma. The plasma is consequently confined between and compressed by the two fields. The current heating the plasma flows around the axis of the tube. This current, it was found, became a source of instabilities, leading scientists to investigate other methods of heating, including the use of radio-frequency energy.

The stellarator. As in the study of the pinch, in order to avoid the end losses, a toroidal configuration was devised. This can be achieved by applying an axial magnetic field around the plasma by winding a coil around a toroidal tube. An important difficulty arises when this is done; the axial magnetic field strength increases from the outer to the inner wall of the torus. This leads to a charge separation, since charge carriers tend to drift perpendicularly to the nonuniform magnetic field lines: ions in one direction, electrons in the other. The electric field thus created will interact with the magnetic field and force the plasma particles to the wall of the torus. In such a torus, each magnetic line of force closes upon itself after one circuit around the torus, as shown in Fig. 3.10. These are called *degenerate* lines of force. Charge neutralization is therefore impossible in this configuration. But if the magnetic field is distorted in such a way that when the lines of force make a complete circuit they will not close upon themselves, the plasma drift to the walls will be reduced drastically. The fields are said to have a rotational transform and are no longer considered degenerate. This is the basic principle of the stellarator system, and there are several methods of achieving it.

Production of rotational transforms. One of the methods used to produce a rotational transform is to twist the torus out of a single plane. The first distortion proposed is a twist of the toroidal tube into a figure 8. A modification of this geometry was obtained in later models (such as the model B-64 stellarator) by making all the bends of the 8-shape equal to 90°. This model was designed primarily as a flexible experimental device.

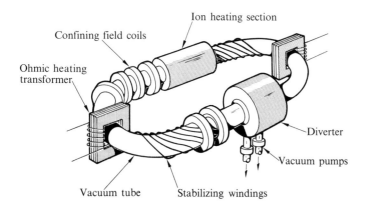

Figure 3.11. The stellarator (model C) after Glasstone [30].

More recently, the production of a rotational transform was obtained by the use of a closed plasma tube and two sets of magnetic field coils: confining field coils to produce an axial field, and helical stabilizing windings with the electric current flowing in the opposite direction to the former. This configuration is shown in Fig. 3.11.

In other words, it can be said that a rotational transform ϕ_r can be obtained in an undistorted toroidal discharge by passing a current having a component parallel to the magnetic axis in the discharge. Consider such a toroid with the radius of curvature of the magnetic lines of force equal to r_0 (see Fig. 3.12) and let ϕ be the angle existing between a reference radial axis going through the center of the toroid and an axis going through a point P on a line of force around the z-axis of rotational symmetry. The toroidal cylinder will have r, θ, and z as

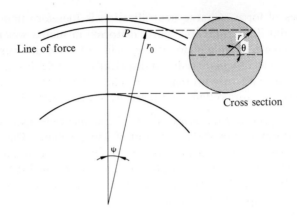

Figure 3.12. The rotational transform and the axial current in a toroidal configuration.

cylindrical coordinates. For small changes of θ and ϕ, it can be written

$$\frac{r\,d\theta}{B_\theta} = \frac{r_0\,d\phi}{B_\phi}, \tag{3.13}$$

where B_θ and B_ϕ are the radial and axial components of the magnetic field at P. From Eq. (3.13),

$$d\theta = \frac{r_0}{r}\frac{B_\theta}{B_\phi}\,d\phi.$$

The rotational transform angle ϕ_r is equal to the total rotation of θ for P. Thus, its value is

$$\phi_r = \int_{\phi=0}^{\phi=2\pi} d\theta(\phi) = 2\pi\frac{r_0}{r}\cdot\frac{B_\theta}{B_\phi}.$$

Since $\ell = 2\pi r_0$ is the total axial length around the center of the torus, this rotational transform becomes

$$\phi_r = \frac{\ell}{r}\cdot\frac{B_\theta}{B_\phi}. \tag{3.14}$$

The value of this angle determines how far a line of force in a fusion reactor may go before it cuts the reactor wall. It can be seen that the larger the angle ϕ_r, the longer the line of force and consequently the higher the probability for the plasma to be kept in confinement.

Other approaches. Another way to prevent the drift of particles to the wall is by using the so-called "corrugated" magnetic fields. The path of a charged particle subjected to such a field will be circular and will rotate about the axis of the field.

The use of the so-called "bumpy" torus was also proposed. It is made up of a number of circular current loops around the torus.

It has also been suggested that the particle drift could be reduced if each curved end section of the discharge were replaced by a series of short alternate curved pieces. These pieces are called "scallops," and the magnetic field lines inside them have opposite but equal curvatures.

The problem of impurities. The existence of energetic ions which are heated by the thermonuclear reaction will increase drastically the energy loss to the walls. It is also necessary to remove the products of the thermonuclear reactions and to reduce the influx of impurities into the discharge. A device known as a *divertor* was built for these purposes. A magnetic field diverts the impurity ions from the discharge into a side chamber from which they are pumped away.

The mirror machines. This configuration was proposed by Post [71] in 1951 and was a major preoccupation of the Sherwood research team at Livermore. The starting point of this machine is the injection of an already hot plasma which will be trapped in the magnetic mirror system. The system consists of an open-ended vessel used as a container and magnetic field coils which surround the vessel in such a way as to provide much stronger fields at the ends than in the middle. The strong axial magnetic fields at the ends of the tube constitute the magnetic mirrors, since they prevent the plasma from escaping. The charged particles will slow down as they move toward the ends because of the stronger magnetic field and will eventually turn back at the mirrors. The same field confines the plasma in the central region, as shown in Fig. 3.13.

The mirror ratio. It can be shown that a particle subjected to a quasi-static magnetic field moves in a circular path, the radius of which is

$$r = \frac{M v_\perp}{eB}, \tag{3.15}$$

where $v_\perp$ is the velocity of the particle in the plane perpendicular to the magnetic field. The angular (Larmor) frequency is

$$\omega = eB/M. \tag{3.16}$$

If the field frequency is small compared with the Larmor frequency given by Eq. (3.16), the particle's angular momentum will be an adiabatic invariant in time; thus, $M v_\perp r = $ constant. Replacing r by its value from Eq. (3.15) yields

$$m = M v_\perp^2 / B = \text{constant}$$

where m is the magnetic moment of the particle. Therefore m is an adiabatic invariant of the particle. Assume that at the center of the mirror machine the magnetic field is equal to B_0 and that it increases to a maximum value B_m at the mirror (see Fig. 3.13). A charged particle moving in the magnetic field will have at the origin a component of velocity in the plane parallel to the magnetic field equal to $v_{\|0} = v \cos \theta_0$. Also $v_{\perp 0} = v \sin \theta_0$.

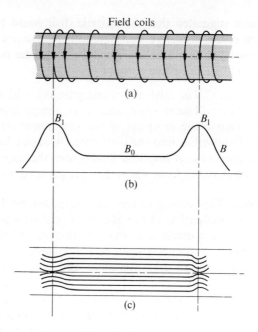

Figure 3.13. The magnetic mirror: (a) windings of the field coils, (b) magnetic field strength distribution, (c) magnetic lines of force.

Since the total kinetic energy is conserved,

$$\tfrac{1}{2}Mv_{\parallel}^2 + \tfrac{1}{2}Mv_{\perp}^2 = \tfrac{1}{2}Mv^2 = \text{constant.} \tag{3.17}$$

On the other hand, the magnetic moment is constant,

$$\frac{Mv_{\perp}^2}{B} = \frac{Mv_{\perp 0}^2}{B_0} = \frac{Mv^2}{B_0}\sin^2\theta_0.$$

Dividing Eq. (3.17) by B, and rearranging, yields

$$\left(\frac{v_{\parallel}}{v}\right)^2 = 1 - \frac{B}{B_0}\sin^2\theta_0. \tag{3.18}$$

Therefore, it can be seen that $v_{\parallel}$ will decrease with increasing B, i.e., as the charged particles approach the mirror region. If the angle θ_0 is large enough, $v_{\parallel}$ will go to zero and the particle will be reflected; in this case $\sin^2\theta_c = B_0/B_1$. The mirror ratio is defined as $m_m = B_1/B_0$. Therefore, the condition of reflection for the charged particles is

$$\sin\theta_c \geqslant (m_m)^{-1/2}. \tag{3.19}$$

Injection of plasma. Since the trapped plasma is obtained by injection of hot particles from the outside of the tube, these particles should have an appreciable velocity component perpendicular to the magnetic field in the center of the mirror machine. If they do not, they will just cross the configuration without being trapped.

Once the plasma has been introduced between the magnetic mirrors, it can be further heated in two sequences: first, through a radial compression, by increasing the magnetic field strength in the central as well as in the mirror regions; and second, through an axial compression, by bringing the magnetic mirrors closer together.

The plasma will then become randomized, leading to the occurrence of the nuclear reaction. If the plasma is allowed to expand back against the fields, currents will be generated in the field coils. This is a possible method of converting the thermal energy of the plasma directly into electrical energy.

Problems of stability The most important instability in a mirror machine is the flute instability. Other instabilities will occur as the plasma temperature and density increase. To stabilize the geometry, it was proposed to superimpose a multiple magnetic field on the original mirror field. This can be achieved by placing currents parallel to the central axis of the mirror machine. The current should then flow through alternate conductors, called Ioffe [41] bars.

Astron system. The astron is a promising method for confining a plasma capable of sustained thermonuclear reaction. It was first proposed by Christofolios [18] in 1953 and later investigated in the Livermore Laboratories [19]. A beam of high energy electrons is injected into one end of an evacuated cylinder in which an axial field is maintained by external solenoidal coils. The latter are wound so as to form a small bump in the magnetic field lines of force at each end of the cylinder. Due to the action of the axial magnetic field, the electrons form a circulating sheet of current around the central axis of the tube. The current of this layer, known as the E-layer, creates a magnetic field in such a way as to form closed magnetic field lines of force, thus confining a plasma of cold deuterium or deuterium-tritium gas. This gas will be heated as a result of collisions with the electronic layer until it reaches temperatures capable of sustaining thermonuclear reactions. This is shown in Fig. 3.14. At first sight, the astron may seem very similar to the mirror machine. However the mirrors at both ends of the tube here do not confine the plasma, but serve mainly to contain the electrons in the E-layer. After injection, the charged particles will gyrate around the central axis while traveling axially from one mirror to the other.

Electron injection. The E-layer electrons in the astron geometry lose energy to the plasma. To maintain the plasma, it is necessary to continually inject relativistic electrons. Thus, electron bunches should be introduced into the E-layer region in an irreversible way without disturbing the electrons already injected and without being disturbed by the presence of the layer. Electrons are introduced at the top of the well, and by decreasing their axial energy they proceed to the bottom of the magnetic well, their axial momentum becoming small enough to prevent them from escaping from the confining mirror field.

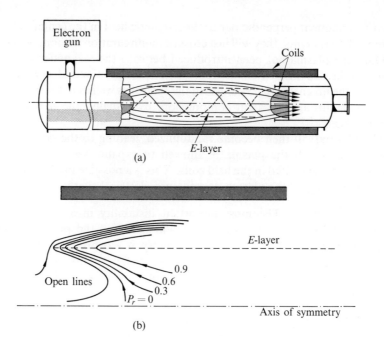

(a)

(b)

Figure 3.14. The astron geometry and its E-layer: (a) schematic field configuration with one relativistic electron spiraling, (b) the E-layer and the magnetic lines of force. From Christofolios [18].

For an injection system to be feasible, high electron energies (5 MeV) are required, because the axial defocusing force must be smaller than the decellerating force. The defocusing force to decellerating force ratio is independent of the current, and the electronic energy increases with decreasing defocusing force. Therefore, an energy increase will yield $F_{def} \ll F_{dec}$ as required. The magnetic field of the E-layer would, however, tend to attract the incoming electron bunches by creating a deeper potential well. This would affect the decellerating field during the buildup of the E-layer and is therefore not desirable. To avoid this, coils were provided in the injection region to cancel the effect of the E-layer field.

Confinement in the E-layer. Once the electron bunch reaches the E-layer, it will spread out within the mirror configuration in the axial direction. The energy stored in the field of the electron bunch is thus transformed into kinetic energy. Injection of subsequent bunches of electrons will increase the energy of the E-layer electrons up to a value limited by the rate of escape from the mirror fields. As the number of electrons approaches a critical value n_c, the magnetic field inside the volume enclosed by the E-layer will go to zero.

The value n_c can be easily calculated. Let H be the circulating sheet current per unit length and B the created induction, $B_{ind} = \mu_0 H$, and

$$H = \frac{nec}{2\pi r_E},$$

where r_E is the E-layer radius, c is the speed of light, and n is the electronic number density. For $n = n_c$, $B_{ind} = B_{ext}$. Thus

$$n_c = \frac{2\pi r_E B_{ext}}{\mu_0 ec},$$

and since, from Eq. (3.15),

$$r_E = \frac{M_e v_\perp}{eB},$$

then

$$n_c = \left(\frac{2\pi M_e}{\mu_0 e^2 c}\right) v_\perp. \tag{3.20}$$

Increasing the number of electrons beyond n_c causes a reversal of the magnetic field lines forming the closed pattern around the E-layer.

The astron concept seems to provide high hopes for a useful thermonuclear reactor, despite the many problems which are still awaiting solution.

Other configurations

The cusped geometry [6, 28]. In search of a more stable confinement, it is theorized that magnetic configurations in which the confining magnetic lines of force are convex should be inherently stable. Completely new configurations entered the competition as a result of this theory. One such configuration is the so-called *picket fence* shown in Fig. 3.15. It is also called the *cusped geometry*. It can be obtained by two parallel coils which carry currents in opposite directions. Their magnetic fields will be in opposition, and the magnetic field strength in the central region between the two coils would equal zero.

For a low-density plasma, a new problem arises: there will be an unacceptable loss of particles through the cusps. However, if the density is made higher, it was shown by Grad [32] that the rate of leakage is reduced drastically.

The problem of creating an initial sharp boundary for the cusp geometry was tackled by using shock techniques. Although the final success of this geometry seemed promising at the beginning of the work, the competition of other geometries hindered its continuation.

The theta pinch. The confinement and heating of a plasma is achieved here by an initial rapid compression phase (shock heating) followed by a slower adiabatic one. Such a confinement can be obtained by applying a rapidly increasing magnetic field in the axial direction. This will produce a current in the θ-direction (azimuthal) in the discharge. The plasma will be driven toward the axis of the discharge and will thus be heated and confined.

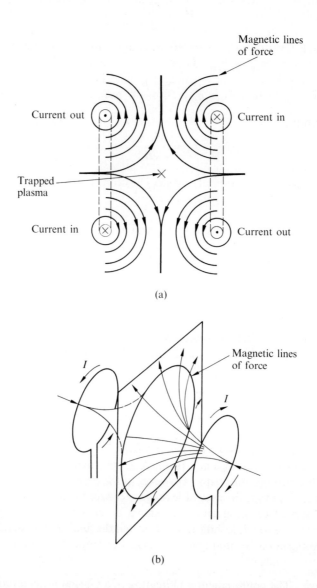

Figure 3.15. Representation of the magnetic lines of force in the cusped geometry: (a) cross section in the case of a high density plasma, (b) spatial representation.

The homopolar system. In their search for a solution to the instability problem, scientists thought that the gyroscopic effect of a rotating plasma might be useful. The first apparatus based on this concept is the homopolar device [20].* In this device, an electric field is applied between the inner and the outer radii of an annular discharge, in addition to an axial magnetic field. This causes the plasma particles to rotate around the central axis of the chamber. This rotation creates a centrifugal force which bends the magnetic field lines outward. The plasma is thus confined and heated. It is interesting to note that the homopolar device is able to act as a capacitor by storing energy which can be recovered under some favorable conditions.

The ixion device. This is the result of another experiment using crossed electric and magnetic fields. It combines the features of both the mirror machine and the homopolar device. First a cloud of ionized deuterium gas is injected in a discharge chamber along its axis, creating a central plasma electrode. Then a transverse electric field is applied between this plasma and the outer conducting wall. Due to this transverse field and to the axial magnetic field, an energetic plasma rotating about the axis is created [13].

The magnetron device. The Berkeley ion magnetron device [20] is an example of this method of plasma confinement. The device consists of a long solenoidal magnetic field between two mirrors and a central metallic anode. A radial electric field is applied between this anode and the outer wall. Because of the low pressure of the discharge, an anode plasma sheath is formed. The magnetron device operates under continuous conditions because of this sheath. Energetic electrons rotate around the anode in the sheath, leading to ionization. The ions formed are then accelerated out of the sheath following magnetron-like orbits. Due to the axial magnetic field, these ions will not reach the outer wall, and thus will become effectively confined.

Radio-frequency confinement. All the methods treated so far have used magnetic fields as the means of confinement. It has been suggested that the application of radio-frequency fields could confine a plasma. If the electromagnetic radiation frequency is smaller than the characteristic plasma frequency, the radiation is indeed reflected by the plasma, as is shown by the study of electromagnetic waves. This can be visualized as the effect of a radiation pressure which will be opposed to the plasma pressure. The plasma would thus be confined when these two pressures become equal.

Internal electrostatic plasma confinement. Electrical fields have been used for confinement purposes in conjunction with magnetic fields in devices using rotating plasma. The use of electric fields alone seems unrealistic, since this would seemingly lead to a charge separation of the plasma. This separation can be avoided by the use of a nonequilibrium system for the electrons and a potential well for the ions.

---

* Developed at the University of California, Radiation Laboratory, Livermore, California.

3.6 Formation of Plasmas

Each atom of an ordinary gas is normally electrically neutral. It consists of a positively charged nucleus surrounded by a negatively charged cloud of electrons. If one of these electrons is given enough energy to escape from the gas atom, the remainder of the atom becomes positively charged, whereas the free electron is negatively charged. This is the process whereby a gas is said to be *ionized*. A fully ionized gas is called a *plasma*.

In the case of thermonuclear fusion, the gas is usually a deuterium-tritium mixture at low pressure. An initial ionization of this gas is necessary. This ionization can be achieved by a radio-frequency electrical discharge of moderate voltage. Because of their low mass, the electrons move more rapidly under the effect of an electric field than do the ions. They will therefore gain more energy and, by their collisions with neutral atoms, they create new charge carriers. To fully ionize the gas, it is necessary to supply a great amount of energy to the particles. Therefore, large electrical currents have to go through the plasma. Due to the *skin effect*, radio-frequency methods cannot be utilized for full ionization. It is instead necessary to use one of three methods: (1) magnetic compression heating, (2) cyclotron resonance heating, or (3) ohmic heating.

Magnetic compression heating. The gas can be heated by applying an increasing magnetic field, as explained in the pinch and mirror systems. If the increase is relatively slow, there will result an adiabatic heating of the gas.

The compression may be radial, i.e., the radius of the system decreases under the effect of the confining field. This method is more useful than the axial compression which can be achieved by bringing the magnetic mirrors of a mirror machine closer together.

In the case of magnetic compression, the neutral atoms are not affected by the compressing field. They will be left behind and will become an important source of loss for the confined plasma.

When the compression becomes rapid, a shock heating may follow. This may be useful in a mirror machine. The temperature of the plasma will also increase if it is subjected to a magnetic field which is alternately compressed and expanded at an appropriate frequency. This method is called heating by *magnetic pumping*.

Cyclotron resonance heating. A principle similar to that used in a cyclotron to accelerate charge carriers can be used for increasing the energy of the ions. This is known as ion cyclotron resonance heating. Ions will gyrate around the lines of force of an applied magnetic field B with a frequency equal to

$$f_i = \frac{eB}{2\pi M_i} \text{ Hz,}$$

where M_i is the mass of the ion. If A is the ionic mass number, $f_i = 1.54B/A$ kHz. For a deuterium plasma, $A = 2$.

Now, if an alternating electric field of frequency f_i is superimposed upon the static magnetic field, the ions would acquire energy from the oscillating field.

It is obvious that for the ion to acquire enough energy, the gyration frequency of the ions should be much larger than both their collision frequency and their transit frequency through the applied static magnetic field.

Ohmic heating. This is an important method of supplying energy to the charged particles of a plasma. After the initial ionization (breakdown), the plasma acquires a conductivity σ. If a current density J passes through this plasma, heat will be dissipated at a rate J^2/σ per unit volume.

If the plasma discharge is toroidal, a current will pass through it, the toroid serving as the secondary of an iron-core transformer.

Because of their smaller mass, electrons absorb almost all the energy taken from the electric field. They will be continuously accelerated, and will give off energy resulting from their collisions with the ions. The distribution of the electronic velocities will become Maxwellian, leading to a favorable condition of Ohmic heating. This may, however, be hindered by the "run away" of electrons if the frequency of collisions is not high enough.

3.7 Losses and Efficiency in Fusion Devices

Brief comments on the problem of direct conversion of thermonuclear energy to electricity have appeared in the European and American literatures [11, 44, 49, 64]. The fusion converter may operate in the following manner: a plasma is confined and compressed by an increasing magnetic field. Due to the compression, the density of the particles increases. The power released increases as the square of the density. The pressure will also increase, causing the plasma to expand against the magnetic lines of force. This will lead to a variation in the magnetic field, which in turn could induce voltages in neighboring circuits, from which power could be extracted. Therefore a direct thermonuclear converter is theoretically feasible.

However, despite the tremendous progress in the field of fusion, the practical form of the future thermonuclear converter is not yet evident. Nevertheless, the basic considerations related to the problem of extracting electrical energy from a thermal fusion reactor are very clear.

Thermonuclear power production. The energy released by a fusion reaction is generally the kinetic energy of the reaction products. The products of a reaction are of two different types: the charged particles and the neutral particles. Direct energy conversion is possible for the charged particles only. The charged particles may be deenergized by interaction with an electromagnetic field. As for the energy of the neutral particles, it can be recovered by conventional heat engines.

The most practical fuel for the fusion reactions is a deuterium gas. A pair of deuterons can fuse according to two reactions having almost the same probability of occurrence:

$$1. \quad D + D \rightarrow (He^3 + 0.82 \text{ MeV}) + (n + 2.45 \text{ MeV}),$$
$$2. \quad D + D \rightarrow (T + 1.01 \text{ MeV}) + (H + 3.02 \text{ MeV}).$$

The tritium fuses readily with the deuterons giving:

$$3. \quad D + T \rightarrow (He^4 + 3.5 \text{ MeV}) + (n + 14.1 \text{ MeV}).$$

This secondary reaction has a much higher probability of occurrence. The energy released per unit time per unit volume, i.e., the fusion power density, is

$$P_f/\mathscr{V} = \tfrac{1}{2} n_D n_D \dot{\mathscr{V}}_1 \mathscr{E}_1 + \tfrac{1}{2} n_D n_D \dot{\mathscr{V}}_2 \mathscr{E}_2 + n_D n_T \dot{\mathscr{V}}_3 \mathscr{E}_3, \qquad (3.21)$$

where $\mathscr{E}_1$, $\mathscr{E}_2$, and $\mathscr{E}_3$ are the energies released by reactions 1, 2, and 3, respectively, n_D and n_T are the number densities of fuel deuterium and fuel tritium, respectively, and $\dot{\mathscr{V}}_1$, $\dot{\mathscr{V}}_2$, and $\dot{\mathscr{V}}_3$ are the average products over the velocity space of the product of the cross section Q and the relative velocity v between reacting nuclei for the reactions 1, 2, and 3, respectively. Figure 3.2 shows plots of $\dot{\mathscr{V}}_{12}$ $(= \dot{\mathscr{V}}_1 + \dot{\mathscr{V}}_2)$ and $\dot{\mathscr{V}}_3$.

The average lifetimes of deuterons and tritons are inversely proportional to their reaction rate parameters. The lives of tritons are consequently very short compared to the lives of deuterons. An equilibrium condition is thus quickly reached, in which the rate of consumption of tritons in reaction 3 equals the rate of their formation in reaction 2. Thus

$$\tfrac{1}{2} n_D n_D \dot{\mathscr{V}}_2 = n_D n_T \dot{\mathscr{V}}_3,$$

and

$$n_T = \tfrac{1}{2} n_D \dot{\mathscr{V}}_2 / \dot{\mathscr{V}}_3.$$

Replacing n_T in Eq. (3.21) by this value and assuming that $\dot{\mathscr{V}}_{12} = 2\dot{\mathscr{V}}_1 = 2\dot{\mathscr{V}}_2$, yields

$$P_f/\mathscr{V} = \tfrac{1}{2} n_D^2 \dot{\mathscr{V}}_{12} \bar{\mathscr{E}}, \qquad (3.22)$$

where

$$\bar{\mathscr{E}} = \tfrac{1}{2}(\mathscr{E}_1 + \mathscr{E}_2 + \mathscr{E}_3).$$

Losses in the thermonuclear reactor. There are several sources of loss in a thermonuclear reactor, which can be listed as follows:

1. Loss of the *neutrons* and their kinetic energy, since they cannot be confined by electromagnetic fields.

2. The existence of instabilities, since magnetic confinement may be destroyed locally, allowing a bunch of charge carriers to escape.

3. Different local deficiency *holes* due to the configuration of the confinement allow the escape of the charged particles along the magnetic lines of force.

4. The existence of *porosities*. No matter how high the magnetic field may be, the transverse ambipolar diffusion* is never zero. Thus local porosities may exist.

5. Losses due to thermal conductivity. Note also that the presence of impurities significantly increases the loss and should be avoided.

6. Radiation, by far the most important source of loss in a thermonuclear reactor at the present state of the art. Both *brake radiation* (Brehmsstrahlung) and *cyclotron radiation* are manifestations of the same effect: radiation of electrostatic charges submitted to electrostatic (brake) and magnetic (cyclotron) acceleration.

a) *Brehmsstrahlung*. The power radiated by an electric charge undergoing an acceleration is

$$P = \frac{e^2 a^2}{6\pi\varepsilon_0 c^3},\tag{3.23}$$

where e is the charge of the particle, c is the speed of light in vacuum, ε_0 is the permittivity of free space, and a is the acceleration of the particle. If n_z is defined to be the atomic number of the completely stripped ion and r to be the distance between the ion and the colliding electron, the electronic acceleration becomes

$$a = \frac{n_z e^2}{4\pi\varepsilon_0 M r^2}.$$

Replacing a by its value in Eq. (3.23) yields

$$P = \frac{n_z^2 e^6}{96\pi^3 \varepsilon_0^3 M^2 c^3} \cdot \frac{1}{r^4}.\tag{3.24}$$

The orbit of the electron can be regarded as a straight line, since the dominant type of interaction has here a small scattering angle. Thus if x_b is the impact parameter† and s the abscissa along the path, $r^4 = (x_b^2 + s^2)^2$, and at the point of minimal approach $s = 0$ and $r = x_b$. Thus, as s increases, P decreases very rapidly. Therefore P can be visualized as a square pulse having the same height and width equivalent to the true pulse. The width can be evaluated by considering the points at which P is equal to half the maximum value:

$$\ell = (\tfrac{3}{2})^{1/2} x_b,$$
$$\tau = (\tfrac{3}{2})^{1/2} x_b / v,$$

where ℓ is the pulse width, τ is the average time between two collisions, and v is the electronic velocity. Therefore, the energy radiated by one electron in one

---

* When, under the effect of an electric field, the diffusion rates of particles of different signs become equal, the corresponding total diffusion of the plasma is called *ambipolar* diffusion.

† The impact parameter is the distance between the initial velocity vector of the projectile and a line parallel to it and passing through the center of the target.

collision is

$$\mathscr{E}_b = P\tau = \left(\frac{3}{2}\right)^{1/2} \frac{n_z^2 e^6}{96\pi^3 \varepsilon_0^3 c^3 x_b^2 M^2 v}. \tag{3.25}$$

The number of interactions per unit volume and per unit time between N^- missile electrons of speed v and N^+ target ions of cross section $2\pi x_b$ is given by

$$dv = N^- N^+ v 2\pi x_b \, dx_b$$
$$= n_z N^2 v 2\pi x_b \, dx_b.$$

The power density of the interactions corresponding to all the possible values of x_b is $d(P_b/\mathscr{V}) = \mathscr{E}_b \, dv$. Consequently

$$\frac{P_b}{\mathscr{V}} = \int_{x_b, \min}^{\infty} \left(\frac{3}{2}\right)^{1/2} \frac{n_z^3 N^2 e^6}{48\pi^2 \varepsilon_0^3 M^2} \cdot \frac{dx_b}{x_b}. \tag{3.26}$$

The upper limit is the Debye radius, but since the integral converges rapidly, infinity can be used. The lower limit of the integral is the electron Compton radius given by

$$x_{b,\min} = \frac{h}{2\pi M v},$$

where h is Planck's constant.

The average over the velocities, assuming a Maxwellian distribution, has been taken. Then

$$\frac{P_b}{\mathscr{V}} = \left(\frac{e^6}{12\pi \varepsilon_0^3 c^3 M h}\right)\left(\frac{3k}{\pi M}\right)^{1/2} n_z^3 N^2 T^{1/2},$$

or

$$P_b/\mathscr{V} = \tfrac{1}{2}\mathscr{C}_\beta n_z^3 N^2 T^{1/2}, \tag{3.27}$$

where $\mathscr{C}_\beta$ is a constant equal to 0.96×10^{-36} in the MKSC system. As is clear from this formula, since $P_b/\mathscr{V}$ is directly proportional to n_z^3, heavy impurities have a catastrophic effect on the losses.

b) *Cyclotron radiation.* Rose and Clark [77]* gave the following approximate expression for the power radiated by an electric charge submitted to a magnetic acceleration:

$$\frac{P_c}{\mathscr{V}} = \left(\frac{e^2 \omega_c^2}{3\pi \varepsilon_0 c}\right)\left(\frac{NkT}{Mc^2}\right)\left(1 + \frac{5kT}{2Mc^2} + \cdots\right), \tag{3.28}$$

where ω_c is the cyclotron frequency given by $\omega_c = eB/M$. Limiting the expression to its first factor only, by assuming that the higher order terms are negligible, one has

$$\frac{P_c}{\mathscr{V}} = \left(\frac{e^4 k}{3\pi \varepsilon_0 c^3 M^3}\right) B^2 N T,$$

---

* Equation (11.82).

or

$$P_c/\mathscr{V} = \tfrac{1}{2}\mathscr{C}_\gamma B^2 N T, \tag{3.29}$$

where $\mathscr{C}_\gamma$ is a constant equal to 0.668×10^{-4} in the MKSC system.

In the case of an ideal confinement, with no magnetic field inside the confined plasma, where the center fringe of the plasma would be in a region of a strong heterogeneous magnetic field, it is not yet clear whether or not this would give less cyclotron radiation. Eq. (3.29) is valid only for a reasonably homogeneous plasma.

Efficiency of a fusion converter. If it is assumed that the fuel gas is deuterium, $N = n_D$ and $n_z = 1$. In this case, Eqs. (3.27) and (3.29) become

$$P_b/\mathscr{V} = \tfrac{1}{2}\mathscr{C}_\beta n_D^2 T^{1/2}$$

and

$$P_c/\mathscr{V} = \tfrac{1}{2}\mathscr{C}_\gamma n_D^2 B^2 T.$$

If only the losses due to diffusion, cyclotron, and braking radiations are considered, it is clear that the efficiency will be

$$\eta = \frac{P_f - \mathscr{C}_\alpha P_f - P_b - P_c}{P_f},$$

where $\mathscr{C}_\alpha$ is the fraction of total fusion appearing as kinetic energy of the neutrons. Thus

$$\eta = 1 - \mathscr{C}_\alpha - \frac{\mathscr{C}_\beta n_D T^{1/2} + \mathscr{C}_\gamma B^2 T}{n_D \mathscr{V}_{12}\mathscr{C}}. \tag{3.30}$$

Note that if the effect of the cyclotron radiation is disregarded, the efficiency η will be a function of temperature only. The energy production of the three important reactions, as well as the Brehmsstrahlung are shown in Fig. 3.3.

3.8 Economic Aspects and Future Trends [59, 73]

Since the early 1950's, when the research in fusion started with a few scientists and limited funds, the fusion research program has grown tremendously. This fact is made easily apparent by an account of the annual rate of expenditures. In the U.S. alone, expenses which began at several hundred thousand dollars in the early 1950's were counted in tens of millions of dollars in the early 1960's.

Research in this field is still in its infancy, the end is not yet clear and its form cannot yet be known. The success of a particular fusion device will depend greatly on whether or not the causes of loss can be reduced. Simultaneously, before a final form of a fusion reactor can be sketched, a satisfactory solution to the problem of instabilities should be sought. A glance at the work which has been done in the field shows that tremendous progress has been achieved in these problems and in others. Important progress has been made both theoretically in the understanding of plasmas and experimentally through the perfecting of diagnostic techniques and in all other fields of material science and technology.

Whether the fusion method of converting energy will become competitive with the conventional techniques cannot yet be guessed. However, for the first time in history, man is faced with the prospect of opening for himself an unlimited source of energy. The gain is indeed as exciting a prospect for mankind as it is a stimulating challenge to the scientists.

PROBLEMS

3.1 At the steady state and in a neutral plasma, show that

$$\nabla \cdot (v \times \mathbf{B}) = 0.$$

Does this relation stand when the plasma departs from neutrality?

3.2 The long rectangular channel shown in Fig. 3.16 is ended by perfectly conducting electrodes. Inside the channel, there is a conducting flow through which a constant current I flows, from an external source, perpendicularly to an externally applied magnetic field. Is the static solution the only possible one or can the fluid become unstable and break up into vortices? (Natural example: Gulf Stream in the Atlantic.)

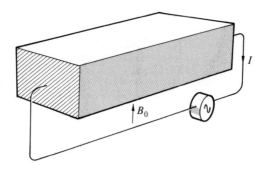

Figure 3.16. Confined conducting flow under the effect of crossed electric and magnetic fields.

3.3 One of the configurations proposed for the confinement of a plasma is the octahedron geometry. (a) show that a single wire can be wound around the edges, making the curvature of the magnetic lines of force favorable for stability. (b) Is it really necessary to use this configuration, or would a simpler geometry be as stable?

3.4 Taking a device somewhat similar to the Zeta machine, calculate the equilibrium pinch radius by using the following assumptions: (a) Current flows in a thin layer at the pinch surface. (b) No toroidal curvature. Numerical example: $r_0 = 0.5$ m, $B = 0.02$ Wb/m^2, $I = 10^5$ A, $p_{gas} = 10^{-4}$ mm D$_2$.

3.5 Study the path of the electrons in the coaxial triax pinch shown in Fig. 3.17 [77, p. 368].

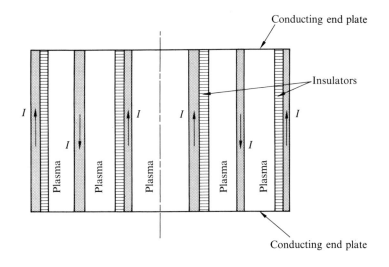

Figure 3.17. The coaxial triax pinch. From Rose and Clark [77, p. 368].

3.6 Show by using Maxwell's equations (Chapter 4) that the plasma cannot be confined in the configuration shown in Fig. 3.18. [*Hint:* Study the forces on the particles.]

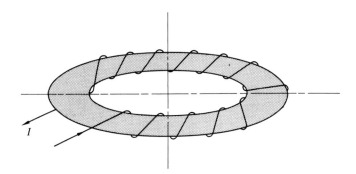

Figure 3.18. Toroidal plasma in an azimuthal magnetic field.

3.7 In a helically wound stellarator, the induction must be a cyclic function of $(n\theta - z/d_0)$ where d_0 is related to the pitch d of the helical winding by the relation

$$d = 2\pi n d_0,$$

the pitch being the axial distance between similar positions of one wire. If the induction is expanded in an orthogonal Fourier series, it is obtained:

$$B_r = \sum_v f_r(n, r) \exp \left[jv(n\theta - z/d_0) \right],$$

$$B_\theta = \sum_v f_\theta(n, r) \exp \left[jv(n\theta - z/d_0) \right],$$

$$B_z = \sum_v f_z(n, r) \exp \left[jv(n\theta - z/d_0) \right],$$

where n is an integer number. Use the appropriate boundary conditions and the relations $\nabla \cdot \mathbf{B} = \nabla \times \mathbf{B} = 0$ to solve B_θ.

3.8 In the mirror machine of Fig. 3.13 assume that the mirror ratio is given by

$$m_m = 1 + (z/z_0)^2,$$

where z_0 is a constant value and z is the coordinate in the axial direction. Find the relationship between the radial and the axial compression ratios.

3.9 (a) From Maxwell's equations, find the solution for the vector potential $\mathbf{A}$ in the case of a cylindrical astron device. (b) For an astron power reactor having the characteristics $B_{ext} = 4$ Wb/m^2 and $r_E = 0.5$ m, find the electronic number density.

3.10 Determine the magnetic field configuration in the case of the cusp geometry shown in Fig. 3.15. Use Maxwell's equations of the magnetic field in the case of a steady state under the following conditions: (a) $\mathbf{n} \cdot \mathbf{B} = 0$ at the surface of the plasma, where $\mathbf{n}$ is a unit vector normal to the surface, (b) $B = B_0$ near both exterior conductors, (c) constant magnetic pressure. (Characteristic values of the cusp are $r_0 = 10$ cm = radius, $T = 10$ eV = temperature of the electrons, $B = 1$ Wb/m^2, $\tau = 0.01$ sec = confinement time.)

3.11 (a) Study the orbits of a charged particle in the presence of a magnetic field. (b) What will be the effect of this field on a bipolar plasma?

3.12 Determine the electronic temperature and density of a fusion device having a 1% efficiency and producing 500 MW of energy. Use the best reactions technically possible as explained in Section 3.2.

3.13 Prove Eq. (3.23).

3.14 Study the feasibility of a direct fusion energy converter working on the principle mentioned at the beginning of Section 3.8.

3.15 Consider the deuterium plasma in which there is as much cyclotron radiation as Brehmsstrahlung radiation. This plasma is also characterized by a cyclotron frequency $\omega_c = 10^{22}$ Hz and an electronic number density $n_D = 10^{-3}$ particles/m^3. Find (a) the average energy of the electrons, (b) the cyclotron radiation.

REFERENCES AND BIBLIOGRAPHY

1. Abraham, E. E., L. W. Crawford, and D. M. Mills, "Observations of a Hose, Instability of an Electron Beam in a Plasma," *J. Appl. Phys.*, Vol. 38, No. 2, p. 911, 1967.

2. Allis, W. P., *Nuclear Fusion*, Van Nostrand, Princeton, N.J., 1960.

3. Artsimovich, L. A., "Research on Controlled Nuclear Fusion and the Physics of High-Temperature Plasma in the U.S.S.R.," *J. Nuclear Eng.*, Part C, Vol. 7, p. 477, 1955.

4. Artsimovich, L. A., et al., "Insulation of Plasma in 'Tokamak' Installation," *Atomic Energy (USSR)*, Vol. 22, p. 259, 1967 (Russian).

5. Balebamov, V. M., and N. N. Semashko, "Azimuthal Drift of Charged Particles in Axially-Symmetric Magnetic Field with Mirrors," *Atomic Energy (USSR)*, Vol. 21, p. 500, 1966 (Russian).

6. Belitz, H. J., and E. Kugler, "Investigation of the Compression and the Containment of a Plasma in Cusp Geometry," *Kernforschungsanlag*, Juelich, West Germany. Institut fuer Plasmaphysik, CONF-661041-3, Jan. 1967.

7. Bennett, W. H., *Device for the Thermonuclear Generation of Power*, U.S. Patent 3,120,475, Oct. 10, 1957.

8. Becker, E. W., R. Klingehofer, and P. Lohse, "On the Possibility of the Directed Introduction of Deuterium and Tritium in Nuclear Fusion Experiments," *Z. Naturforsch.*, Vol. 15a, p. 64, 1960.

9. Bickerton, R. J., "Status of Work Fusion," *Nucleonics*, Vol. 20, No. 2, p. 55, 1962.

10. Bickerton, R. J., and A. Gibson, "Shear Attainable in Stellarator Systems," *Phys. Fluids*, Vol. 10, No. 3, p. 682, 1967.

11. Bishop, A. S., *Project Sherwood—The U.S. Program in Controlled Fusion*, Addison-Wesley, Reading, Mass., 1958.

12. Bockesten, et al., *Controlled Thermonuclear Fusion Research*, Review Series No. 17, International Atomic Energy Agency, Vienna, 1961.

13. Boyer, K., et al., *Theoretical and Experimental Discussion of Ixion, a Possible Thermonuclear Device*, Paper P/1418, Second United Nations Conference on Peaceful Uses of Atomic Energy, Geneva, 1958.

14. Brown, D. E., and H. G. Loos, *Toroidal Plasma Containment with Rotating Magnetic Field*, Contract AF49(638)-1539, AD632379, April 1966.

15. Caruso, A., and P. Giupponi, "The Pinch Effect in a Low Pressure Discharge," *Nuovo Cim.*, Vol. 38, p. 255, 1965.

16. Chapman, J. L., "Uncanny World of Plasma Physics," *Harpers*, Vol. 223, No. 10, p. 64, 1961.

17. Chen, F. F., "Leakage Problem in Fusion Reactors," *Scientific American*, Vol. 217, No. 7, p. 76, 1967.

18. Christofolios, N. C., *Astron Thermonuclear Reactor*, Paper P/2446, Second United Nations Conference on Peaceful Uses of Atomic Energy, Geneva, 1958.

19. Christofolios, N. C., *Note on the Stability of the E-Layer*, University of California, Livermore, Lawrence Radiation Laboratory, Contract W-7405-eng-48, UCID-15067, Jan. 1967.

20. Colgate, S. A., *A Summary of the Berkeley and Livermore Pinch Programs*, Paper P/1064, Second United Nations Conference on Peaceful Uses of Atomic Energy, Geneva, 1958.

21. Colgate, S. A., and H. P. Furth, "Thermonuclear Plasma," *International Science and Technology*, Vol. 2, No. 2, p. 34, 1962.

22. D'Atkinson, R. E., and F. G. Houtermans, "Building Up of the Elements in Stars," *Zs. Phys.*, Vol. 54, No. 9, p. 10, 1929.

23. Dougal, A. A., *Problems and Progress In Control of Thermonuclear Fusion for Electrical Power Production*, Proceedings of the 2nd Annual Energy Conversion and Storage Conference, Oklahoma State University, Oct. 12–13, 1964.

24. Fowler, T. K., and R. F. Post, "Progress Toward Fusion Power," *Scientific American*, Vol. 215, No. 6, p. 21, 1966.

25. Fuechsel, K. M., "Direct Conversion of Nuclear Energy into Electrical or Thrust Energy," *Trans. American Nuclear Soc.*, Vol. 4, No. 2, p. 335, 1961.

26. Fuechsel, K. M., "Direct Conversion of Nuclear Energy into Electrical or Thrust Energy," *Nuclear Engineering*, Vol. 7, No. 8, p. 306, 1962.

27. Galeev, A. A., "Concerning Anomalies in the Escape of Dense Plasma from a Magnetic Mirror System Owing to the Presence of the Loss Core," *Zh. Tekh. Fiz.*, Vol. 36, No. 11, p. 1959, 1966 (Russian).

28. Gallagher, C. C., et al., *Design and Construction of a Cusp Magnetic Field Device of Plasma Containment*, Air Force Cambridge Research Laboratories, L. G. Hanscom Field, Bedford, Mass., Conf. 661016, 1966.

29. Gardner, J. W., "Direct Conversion and Fusion Reactors," *Brit. Power Eng.*, Vol. 5, No. 7, p. 24, 1962.

30. Gibson, A., et al., "Plasma Confinement During a Period of Reduced Fluctuations in Zeta," *Plasma Phys.*, Vol. 9, p. 1, 1967.

31. Glasstone, S., "Controlled Nuclear Fusion," U.S. Atomic Energy Commission Series entitled *Understanding of the Atom*, June 1964.

32. Glasstone, S., and R. H. Lovberg, *Controlled Thermonuclear Reactors*, Van Nostrand, Princeton, N.J., 1960.

33. Golgov-Savel'ev, G. G., et al., *Investigation of a Toroidal Discharge in a Magnetic Field*, AEC-TR-4189, NSA 15-15143, 1960.

34. Grad, H., "Plasma Trapping in Cusped Geometries," *Phys. Rev. Letters*, Vol. 5, No. 5, p. 222, 1960.

35. Green, T. S., *Thermonuclear Power*, Newnes, London, 1963.

36. Hass, G. M., and R. A. Dandl, "Observation of Ion Heating Using a Modulated Electron Beam in a Magnetic Mirror Field," *Phys. Fluids*, Vol. 10, No. 3, p. 678, 1967.

37. Hellund, E. J., *The Plasma State*, Reinhold, New York, 1962.

38. Hinnov, E., et al., *Interpretation of Atomic Hydrogen Light in the C-Stellarator*, MATT-270, 1964.

39. Hoyaux, M. F., "Recent Progress in Nuclear Fusion and Future Prospects," *Assoc. Ing. Elec. Sortis de l'Inst. Electrotech. Monteflore Bull.*, Vol. 73, p. 449, 1960.

40. Hurwitz, H. J., *Electrical Power Problems in Fusion Research*, General Electric Co. NP-12054, GP-96, Sept. 1958.

41. Iiyoshi, A., H. Yamato, and S. Yoshikawa, "Limitation of Ion Cyclotron Heating in the Local Mirrors of the C-Stellarator," *Phys. Fluids*, Vol. 10, No. 4, p. 49, 1967.

42. Ioffe, M. S., et al., *Investigation of the Containment of a Plasma in a Trap with Magnetic Mirrors, Parts I and II*, AEC-TR-4217, NSA 15-16563, 1960.

43. Ioffe, M. S., et al., "On Escape of a Plasma From a Magnetic Trap. II," *Zhur. Eksptl. i Teoret. Fiz.*, Vol. 40, No. 1, p. 40, 1961 (Russian).

44. Ionescu, G., "On the Possibility of Utilizing Controlled Thermonuclear Reactions to Obtain Electric Energy," *Trans. Electrotechnica (Rumania)*, Vol. 8, No. 8, p. 75, 1960.

45. Ito, H., "On the Theory of Cusp Losses," *Kakuyugo Kenkyu (Japan)*, Vol. 14, p. 364, 1965.

46. Jephcott, D. F., "The Salzburg Conference on Plasma Physics and Controlled Nuclear Fusion Research (4th to 9th September, 1961)," *Contemp. Phys.*, Vol. 4, No. 10, p. 49, 1962.

47. Johns, T. F., *Prospects for Thermonuclear Power*, Harrap, London, 1962.

48. Jukes, J. D., "Possibilities of Direct Energy Conversion from Fusion Reactors," *Proc. IEE*, Vol. 106A, No. 4, p. 173, 1959.

49. Jukes, J. D., *Man Made Sun: The Story of Zeta*, Viking Press, New York, 1961.

50. Kadomtsev, B. B., and P. O. Rogutse, "Flute Instability of a Plasma in Toroidal Geometry," *Dokl, Acad. Nauk, SSSR*, Vol. 170, No. 10, p. 811, 1966.

51. Karchevskii, A. I., *Injection of Plasmoids in a Mirror Trap with Adiabatic Plasma Compression 'ASPA' Apparatus*, Paper CN-21/162, Conference on Plasma Physics and Controlled Nuclear Fusion Research, Culham, England, Sept. 6–10, 1965.

52. Kettani, M. A., and M. F. Hoyaux, "Fast Randomization of an Electron Gas by Trapped Electroacoustic Waves," *Phys. Fluids*, Vol. 11, No. 1, p. 143, 1968.

53. Kever, H., and K. Schindler, "Experimente Zur Kontrollierten Kernfusion," *Kerntechnik*, Vol. 4, No. 12, p. 562, 1962.

54. Khapaev, M. M., "On the Focusing of Fast Charged-Particle Beams by Magnetic Fields of Stellarator-Type," Vestnik, Moskow Univ. Sec. 3, *Fis. Astronact*, No. 3, p. 57, 1965.

55. Kimmitt, M. F., "Far Infrared Measurements on Thermonuclear Plasmas," *J. R. Radar Estb.*, Vol. 52, p. 113, 1965.

56. Kurchatov, I. V., "Kurchatov on Thermonuclear Reactors," *Engineering*, Vol. 185, No. 4804, p. 431, 1958.

57. Kurchatov, I. V., *Research on Controlled Thermonuclear Reactions at the Atomic Energy Institute of the USSR Academy of Sciences*, Second United Nations Conference on Peaceful Uses of Atomic Energy, Geneva, 1958.

58. Kurchatov, I. V., "Certain Results of Investigations on Controlled Thermonuclear Reactions in the Soviet Union," *Uspekhi Fiz. Nauk*, Vol. 73, No. 4, p. 605, 1961.

59. Kvartskava, I. F., et al., "Instability of the Induction (Theta) Pinch," *Zhur. Eksptl. Fiz.*, Vol. 38, p. 1641, 1960.

60. Little, E. M., et al., "Plasma End Losses and Heating in the Low-Pressure Regime of a Theta Pinch," *Phys. Fluids*, Vol. 8, No. 6, p. 1168, 1965.

61. Longmire, C., J. L. Tuck, and W. B. Thompson (eds.), *Plasma Physics and Thermonuclear Research*, Pergamon Press, New York, 1959 (Papers of 2nd Geneva Conference).

62. Macklin, R. J., Jr., and P. R. Bell, "High-Energy Injection for Controlled Fusion," *Nucleonics*, Vol. 20, No. 11, p. 58, 1962.

63. Maslen, S. H., "Fusion for Space Propulsion," *Trans. IRE*, Vol. MIL-3, No. 4, p. 52, 1959.

64. Meyer, K., "Methods for Direct Conversion of Nuclear Fusion Energy into Electrical Energy," *Energietechnik*, Vol. 12, No. 11, p. 496, 1962.

65. Mills, R. G., *Thermonuclear Power and Super-Conductivity*, 2nd Symposium on the Engineering Aspects of MHD, Philadelphia, March 1961.

66. Mills, R. G., *Four Lectures on Fusion Power*, MATT-145, Princeton University, 1962.

67. Nowak, K., "A Project for the Obtention of Controlled Nuclear Fusion," *Neue Physik*, Vol. 2, p. 90, 1960 (German).

68. Ornstein, L. T., "Controlled Thermonuclear Reactions. Parts I and II," *Atomenergie Haar Toepassingen (Holland)*, Vol. 3, No. 4, p. 57, and No. 5, p. 77, 1961 (Dutch).

69. Osovets, S. M., "Dynamic Stabilization of a Plasma Ring," *Zhur. Eskptl. i Teoret. Fiz.*, Vol. 39, No. 8, p. 311, 1960 (Russian).

70. Parvan, R., "High Velocity Electrons of the Positive Column of Continuous Current Discharge in Gases," *Compt. Rend. Ser. A and B*, Vol. 264, p. 1231, 1967 (French).

71. Post, R. F., *Sixteen Lectures on Controlled Thermonuclear Reactions*, University of California Radiation Laboratory, Livermore, California, UCRL-4231, Feb. 1954.

72. Post, R. F., "Fusion Power," *Scientific American*, Vol. 197, No. 12, p. 73, 1957.

73. Post, R. F., *Some Aspects of the Economics of Fusion Reactors*, 2nd Symposium on the Engineering Aspects of MHD, Philadelphia, March 9–10, 1961.

74. Pyle, R. V., "Controlled Thermonuclear Fusion," in *Plasma Physics in Theory and Application*, W. B. Kunkel (ed.), McGraw-Hill, New York, 1966.

75. Quinn, W. E., et al., *Scylla Theta-Pinch Experiments: Status, Plans, and Proposal for a Closed Toroidal Theta Pinch (Scyllac)*, LA-3289-MS, 1965.

76. Ribe, F. L., et al., *Feasibility Study of a Pulsed Thermonuclear Reactor (Thetatrons)*, LA-3294-MS, 1965.

77. Rose, D. J., and M. Clark, Jr., *Plasmas and Controlled Fusion*, MIT Press, Cambridge, Mass., 1961.

78. Rye, B. J., et al., "Schlieren Photography of the First Half-Cycle of a Theta Pinch Discharge," *Brit. J. Appl. Phys.*, Vol. 16, p. 1404, 1965.

79. Sato, N., et al., *Electron Heating Effects and Universal Stability in Cesium Plasma*, Vol. 24A, p. 293, 1967.

80. Saxe, R. F., *Approaches to Thermonuclear Power*, Temple Press, London, 1962.

81. Seren, L., "Fusion Reactors, Promising Nuclear Power Plants," *Aviation Age*, Vol. 28, No. 7, p. 42, 1967.

82. Siambis, J. G., and T. G. Northrop, "Magnetic Field Geometry and the Adiabatic Invariants of Particle Motion," *Phys. Fluids*, Vol. 9, No. 10, p. 2001, 1966.

83. Simon, A., *An Introduction to Thermonuclear Research*, Pergamon Press, New York, 1959.

84. Sinel'nikov, K. D., *Plasma Physics and Controlled Thermonuclear Fusion*, Proceedings of the 4th Conference on Plasma Physics and Controlled Nuclear Fusion, Kharkov, 1963, Naukova Dumka (USSR), 1965.

85. Sleator, D. E., and R. A. Rudd, *Survey Spectra of a Pinched Argon Plasma*, Ballistic Research Laboratories, Aberdeen Proving Ground, BRL-MR-1847, June 1967.

86. Slepian, J., "The Ionic Centrifuge and Fusion Nuclear Power," *National Acad. Sci. Proc.*, Vol. 47, No. 3, p. 313, 1961.

87. Spitzer, L., Jr., "The Stellarator," *Scientific American*, Vol. 199, No. 10, p. 28, 1958.

88. Spitzer, L., Jr., *Physics of Fully Ionized Gases*, Interscience, New York, 1962.

89. Takeda, S., and S. Shina, "The Initial-Pressure Dependence of the Instabilities in a Low B Theta Pinch," *J. Phys. Soc. Japan*, Vol. 20, p. 1275, 1965.

90. Thonemann, P. C., *Controlled Thermonuclear Research in the United Kingdom*, Paper P/78, Second United Nations Conference on Peaceful Uses of Atomic Energy, Geneva, 1958.

91. Toschi, R., "Production of High Pulsed Currents Through Experiments on Controlled Thermonuclear Fusion," *Electrotechnica (Rome)*, Vol. 53, p. 647, 1966 (Italian).

92. Tuma, D. T., and A. J. Lichtenberg, "Electron-Cyclotron Heating in a Mirror Machine," *Plasma Phys.*, Vol. 9, p. 87, 1967.

93. Van Atta, C. M., et al., *Recent Research in Controlled Nuclear Fusion*, Review Series No. 4, International Atomic Energy Agency, Vienna, 1960.

94. Van Goeler, S., and R. W. Motley, "Direct Measurement of the Confinement Time of Cesium Plasma in a Q-Device," *Phys. Fluids*, Vol. 10, No. 6, p. 1360, 1967.

95. Voronov, A. Y., "Confined Plasma Ellipsoid," Vestnik Moskov. Univ. Ser. III, *Fiz. Astronact.*, Vol. 15, No. 5, p. 74, 1960 (Russian).

96. Ware, A. A., "The Quest for Controlled Thermonuclear Power," *Electron and Power*, Vol. 11, No. 1, p. 12, 1965.

97. Wolczek, O., "Cold Fusion Reactors and Their Application for the Propulsion of Space Vehicles," *Raketentech. U. Raumfahrtforsch.*, Vol. 2, p. 93, 1958 (German).

98. Yamato, H., et al., "Confinement of Energetic Plasma in Magnetic Mirrors in the Model C Stellarator," *Phys. Fluids*, Vol. 10, No. 9, p. 756, 1967.

99. Yatsue, K., and Y. Inuishi, "Nonlinear Effects of Helical Instability in Short Magnetized Positive Columns," *J. Phys. Soc. Japan*, Vol. 22, No. 2, p. 626, 1967.

100. Zykov, V. G., et al., "Volume Polarization Interaction of Plasmas in a Multipole Magnetic Field," *Zh. Tekh. Fiz.*, Vol. 36, No. 11, p. 1971, 1966 (Russian).

4 MAGNETOHYDRODYNAMIC POWER GENERATION

4.1 Introduction

A magnetohydrodynamic (MHD) power generator is a device which converts the kinetic energy of a conducting material flowing in the presence of a magnetic field directly into electricity. The converted energy may be random, drift, or both. This type of power generation is based on the Faraday effect in which a voltage is induced in a circuit when the flux of induction through the circuit is changed.

This effect was first noticed by Michael Faraday [31] in 1831 when he made mercury flow through the field of a magnet. An electric field was produced in a direction perpendicular to both the magnetic field and the direction of the mercury flow. Five years later, Faraday [31] demonstrated in a more striking manner that the principle of magnetohydrodynamics can be used for power generation. He discovered that a current was produced in a galvanometer connected to two electrodes placed on the banks of the Thames River. Because the Thames River is mixed with salt water, it has a certain conductivity. This conducting fluid interacted with the earth's magnetic field to produce a small current.

Conventional electromechanical power converters use the same principle. However, instead of using a solid metal, a flowing ionized gas (plasma) is utilized in MHD power generation. This method has the advantage of avoiding any moving solid parts, leading to better reliability. With the advance in understanding of plasma discharges, by early scientists such as William Crookes [21] (1879) and Irving Langmuir [18] (1920–30), the possibility of using an ionized gas as the working fluid became extremely attractive. Most of the energy converted in this case is thermal. Recently, some consideration has been given to the use of liquid metal flows.

Industrial research was carried out for the first time by Karlovitz [46] in the 1930's. Karlovitz did extensive work at Westinghouse on an MHD generator that, unfortunately, failed to operate, mainly because of the extremely low conductivity of the working gas.

Due to the attractive possibility of using nuclear energy in a more direct way, extensive work started again in the later 1950's in the United States [32] and abroad [20]. Many problems have to be solved before MHD power stations become feasible, namely higher conductivities for the working fluids, stronger magnetic fields, and better refractive materials. These and other problems will be discussed in this chapter.

4.2 Gas Conductivity [8, 32]

To be utilized as a working fluid, a gas should be sufficiently conducting, therefore, ionized. Ionization is a process in which electrons are removed from an atom. Conduction is due to the free electrons and positive ions which move under the effect of a magnetic field. Their motion is continuously accelerated and is hampered only by interactions with neutral particles and positive ions.

The electrical conductivity of an ionized gas may be calculated approximately by considering the current density $\mathbf{J}$ resulting from an applied electric field $\mathbf{E}$. The conductivity σ is defined by Ohm's law:

$$\mathbf{J} = \sigma\mathbf{E}. \tag{4.1}$$

If these vectors are not in the same direction, then σ is a tensor. The current density can also be defined as a function of the velocity of the particles which move under the effect of the electric field. It is known that

$$\mathbf{J} = ne\upsilon, \tag{4.2}$$

where n is the electron density, e is the electronic charge, and υ is the velocity of the electrons. The mobility ℓ is defined by

$$\upsilon = \ell\mathbf{E}. \tag{4.3}$$

Introducing Eq. (4.3) into Eq. (4.2) and comparing Eqs. (4.1) and (4.2) yields

$$\sigma = ne\ell. \tag{4.4}$$

In the absence of an electric field, the paths of the particles are straight lines disturbed only after collisions, in a random manner. The effect of the electric field is to create a drift velocity of the particles in the direction of the field after each collision. If τ is the average collision time of a particular electron, the increase of its mean relative velocity will be from zero to $(eE\tau/M)$. Therefore, the average velocity will be $\langle\upsilon\rangle = eE\tau/2M$ and the mobility will be $\ell = e\tau/2M$. Replacing ℓ by its value in Eq. (4.4), we have

$$\sigma = \tfrac{1}{2}ne^2\tau/M. \tag{4.5}$$

Taking into account the fact that there is a random distribution about the mean collision time τ, the coefficient $\tfrac{1}{2}$ should be replaced by a more general factor $\mathscr{C}_\lambda$. It is also often practical to use the collision frequency v_c instead of τ, when v_c is defined as $v_c = 1/\tau$, and Eq. (4.5) becomes

$$\sigma = \frac{\mathscr{C}_\lambda n e^2}{M v_c}.$$ (4.6)

In their motion under the effect of the electric field, the electrons may collide with any type of particle existing in the plasma, such as electrons, ions, and neutral particles. The effect of electron-electron collisions can be neglected, since they will not change the average energy of the electrons in an appreciable manner. If v_{en} and v_{ei} are the collision frequencies of the electron with the neutral particles and the ions, respectively, the total collision frequency can be written $v_c = v_{en} + v_{ei}$. Replacing v_c by its value in Eq. (4.6), it is found that $1/\sigma = 1/\sigma_{en} + 1/\sigma_{ei}$, where σ_{en} and σ_{ei} are the electron-neutral particle and electron-ion conductivities, respectively:

$$\sigma_{en} = \frac{\mathscr{C}_\lambda n_e e^2}{M v_{en}},$$ (4.7)

and

$$\sigma_{ei} = \frac{\mathscr{C}_\lambda n_e e^2}{M v_{ei}}.$$ (4.8)

The collision frequencies are related to the collision cross sections Q_{en} of electrons with neutral particles, and Q_{ei} of electrons with ions, according to the relations

$$v_{en} = n_n \langle v \rangle Q_{en},$$

and

$$v_{ei} = n_i \langle v \rangle Q_{ei}.$$

If a magnetic field is applied, the electrons will drift in a direction influenced by the magnetic field and not necessarily parallel to the electric field, as shown by Eq. (2.36). Under these conditions the electrical conductivity becomes a tensor. On the other hand, since the Hall parameter for ions is smaller than that for electrons, due to their higher ionic mass, the ion current is not as greatly affected by the presence of the magnetic field. As the magnetic field is increased, the electron current decreases to the point where it becomes of the same order of magnitude as the ionic current. This phenomenon is called *ion slip*, and arrangements should be taken to avoid it. There are several types of ionization:

1. Thermal ionization
2. Magnetically induced ionization
3. Radio-frequency wave induced ionization
4. Radioactivity
5. Photoionization
6. Electron-beam ionization
7. Flames.

Thermal ionization. This is the most important method of ionizing a plasma. In this case, ionization is obtained by giving enough thermal energy to the gas. By assuming that the gas is in thermodynamic equilibrium at a temperature T, the equilibrium of the ionization reaction, $A \rightleftharpoons A^+ + e$ can be calculated by using statistical mechanical principles. This has been done by Saha [77], leading to his famous equation

$$\frac{n_e n_i}{n_n} = 2 \frac{P_i}{P_n} \left(\frac{2\pi MkT}{h^2} \right)^{3/2} \exp\left(-\frac{eV_i}{kT} \right), \tag{4.9}$$

where n_e, n_i, and n_n are the electron, ion, and neutral particle densities, P_i and P_n are the internal partition functions for the ion and the neutral particles, respectively, and V_i is the ionization potential of the gas. If it is assumed that the plasma is, on the average, neutral and only partially ionized, then $n_e = n_i \ll n_n$. Introducing this in Eq. (4.9), we get

$$n_e = \left(\frac{2 n_n P_i}{P_n} \right)^{1/2} \left(\frac{2\pi MkT}{h^2} \right)^{3/4} \exp\left(-\frac{\frac{1}{2}eV_i}{kT} \right), \tag{4.10}$$

and by replacing n_e by its value in Eqs. (4.7) and (4.8), one can obtain the expression for the conductivities. Note here that the conductivity is inversely proportional to the square root of the gas density n_n. The ratio (P_i/P_n) depends on the nature of the material and on the temperature. This ratio is either equal to $\frac{1}{2}$ for some alkalies or 2 for some other materials at $0°K$. To have a large value of σ, n_e should be large. From Eq. (4.10), it follows that the degree of ionization will be much larger for small values of the ionization potential. This effect of V_i on the electron density is shown in Fig. 4.1.

The ionization potential of the air, combustion gases, and noble gases used for MHD power generation is usually very high. This fact led to the necessity of seeding the gas with material with a low V_i, such as cesium or potassium which have ionization potentials of 4.34 and 3.89 V, respectively. For the case of a seeded gas, it can be found for the two conductivities σ_{ei} and σ_{en} (in the MKSC system) that [21]

$$\sigma_{ei} = \frac{1.5 \times 10^{-2} T^{3/2}}{\ln (8.7 \times 10^6 T^{3/2}/n_e^{1/2})},$$

and [21]

$$\sigma_{ei} = 7.05 \times 10^{-13} \frac{T^{3/4}}{p_r^{1/2} \exp (eV_i/kT)} \cdot \frac{i^{1/2}}{\bar{Q}},$$

where p_r is the pressure, i is the ratio of the partial pressure of the seed vapor over the partial pressure of the carrier gas, and $\bar{Q}$ is the average cross section, $\bar{Q} = iQ_{seed} + Q_{gas}$.

Magnetically induced ionization. The application of a dc voltage across a gas at reduced pressure creates an electric field which will supply energy to electrons. Because of the electron-ion mass ratio, only a small fraction of the electron energy will be transferred during an elastic collision with heavy particles.

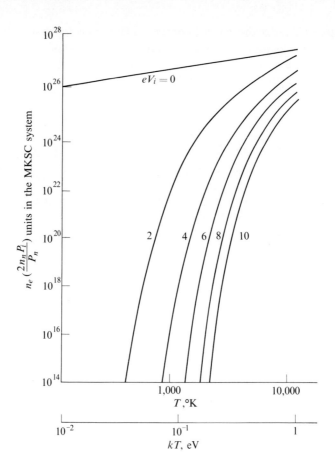

Figure 4.1. Saha's equation for a neutral, partially ionized gas.

It was demonstrated [51, 62] that at high temperatures the same effect can be achieved in seeded noble gases. However, as of 1967, experiments aimed at achieving this have been only partially successful. The lack of complete success is due mainly to nonuniformities in the gas, to fluctuations, and to local Joulean heating.

Radio-frequency waves. Radio-frequency waves can be used to produce a low amount of ionization in a gas. However, the radio-frequency field cannot produce the required high degree of ionization for MHD generators, because this ionization will lead to an extremely small skin depth.

Radioactivity. Radioactivity can also be utilized, although a practical method has not yet been found.

Photoionization. This requires a very long light path in seeded gases in order that the ionizing light be absorbed. This is due to the small ionization cross sections of most of the atoms and molecules.

Electron-beam ionization. This process may become efficient if the recombination rate is sufficiently low in the gas. This was shown experimentally by Shair and Sherman [78], who have preionized a 1500°K stream of argon and cesium at atmospheric pressure. They obtained a resulting conductivity of 6 mhos/m about 20 cm downstream of the electron beam.

Flames. Flames [1] have shown the existence of a certain amount of chemi-ionization which is, however, still insufficient.

4.3 Magnetohydrodynamic Equations

The general MHD equations can be derived by postulating a continuum plasma with given characteristics. The plasma is in the presence of a magnetic field and composed of several types of particles. Since the plasma is a conductor, the magnetic field will create body forces acting on the fluids leading to an energy exchange with the fluid.

Equation of continuity. The equation of continuity is based on the principle of conservation of mass. If $\rho(\mathbf{r}, t)$ is the mass density of the fluid at the point r at time t, and $v(\mathbf{r}, t)$ is its velocity, then the rate of increase of mass within a fixed volume $\mathscr{V}$ will be

$$\frac{d}{dt} \int_{\mathscr{V}} \rho \, d\mathscr{V} = \int_{\mathscr{V}} \frac{\partial \rho}{\partial t} \, d\mathscr{V}. \tag{4.11}$$

But mass will be leaving the volume at the rate $\rho v \, d\mathbf{S}$ through the boundary surface S. The mass conservation principle states that

$$\int_{\mathscr{V}} \frac{\partial \rho}{\partial t} \, d\mathscr{V} = - \int_{S} \rho v \, d\mathbf{S}.$$

Applying Gauss's theorem to the second term and simplifying yields

$$\frac{\partial \rho}{\partial t} + \nabla \cdot (\rho v) = 0, \tag{4.12}$$

which is the continuity equation.

Equation of motion. The equation of motion is based on the principle of conservation of momentum. If a moving volume whose surface always encloses the same amount of fluid is considered, then the forces acting on the fluid can be divided into surface forces and body forces. Newton's second law yields

$$\int_{\mathscr{V}} \rho \frac{Dv}{Dt} \, d\mathscr{V} = \int_{S} \mathbf{f} \, dS + \int_{\mathscr{V}} \mathbf{F} \, d\mathscr{V}, \tag{4.13}$$

where the operator D/Dt is

$$D/Dt = \partial/\partial t + v \cdot \nabla.$$

The surface force $\mathbf{f}$ can be described in terms of a stress tensor, with a scalar viscosity v,

$$p_{ij} = [p_r + \tfrac{2}{3}v(\nabla \cdot v)]\,\delta_{ij} - v\left(\frac{\partial v_i}{\partial x_j} + \frac{\partial v_j}{\partial x_i}\right),$$

where p_r is the pressure. The force f_i acting on dS can then be written

$$f_i\,dS = -\sum_j p_{ij}\,dS_j,$$

where dS_j are components of dS. The total force acting on the surface S is, after applying Gauss's theorem,

$$\int_S \mathbf{f}_i\,dS = -\int_{\mathscr{V}} \left(\sum_j \partial p_{ij}/\partial x_j\right) d\mathscr{V}. \tag{4.14}$$

Introducing Eq. (4.14) in Eq. (4.13), and simplifying yields

$$\rho\,\frac{Dv_i}{Dt} = -\sum_j \frac{\partial p_{ij}}{\partial x_j} + F_i. \tag{4.15}$$

Writing Eq. (4.15) in vectorial notation gives

$$\rho\,Dv/Dt = -\nabla p_r + \boldsymbol{\psi} + \rho_e \mathbf{E} + \mathbf{J} \times \mathbf{B}, \tag{4.16}$$

which is the equation of motion. The first term on the right-hand side of Eq. (4.15) has been expressed in terms of a scalar pressure p_r and a vectorial pressure $\boldsymbol{\psi}$ such as

$$-\sum_j \frac{\partial p_{ij}}{\partial x_j} = -\nabla p_r + \boldsymbol{\psi},$$

and

$$\boldsymbol{\psi} = -\tfrac{2}{3}\nabla(v\,\nabla \cdot v) + v[\nabla v + \nabla \cdot (\nabla \cdot v)]$$
$$+ 2[(\nabla v) \cdot \nabla]v + (\nabla v) \times (\nabla \times v).$$

After neglecting the body forces due to the magnetization of the plasma, it is found for the total force that $\mathbf{F} = \rho_e \mathbf{E} + \mathbf{J} \times \mathbf{B}$, where $\mathbf{E}$ is the electric field, $\mathbf{B}$ is the magnetic field, $\mathbf{J}$ is the current density, and ρ_e is the charge density.

Equation of energy. The rate of energy increase of the fixed amount of fluid moving with the flow can be written as

$$\text{energy increase} = \int_{\mathscr{V}} \tfrac{1}{2}\rho\,\frac{Dv^2}{Dt}\,d\mathscr{V} + \int_{\mathscr{V}} \rho\,\frac{DU}{Dt}\,d\mathscr{V},$$

where U is the internal energy of the plasma. Following the principle of conservation of energy, this energy increase should be equal to the energy entering the volume per unit time from different sources. These sources are electromagnetic, heat conduction, species diffusion, and pressure. Equating the energy rate entering the volume to the energy increase gives

$$\int_{\mathscr{V}} \tfrac{1}{2}\rho \frac{Dv^2}{Dt}\, d\mathscr{V} + \int_{\mathscr{V}} \rho \frac{DU}{Dt}\, d\mathscr{V} = \int_{\mathscr{V}} \mathbf{E} \cdot \mathbf{J}\, d\mathscr{V}$$

$$+ \int_{\mathscr{V}} \nabla \cdot (\kappa\, \nabla T)\, d\mathscr{V} - \sum_{s} \int_{\mathscr{V}} \nabla \cdot (v_s \rho_s \mathscr{H}_s)\, d\mathscr{V}$$

$$- \int_{\mathscr{V}} \sum_{i,j} \frac{\partial}{\partial x_j}(v_i p_{ij})\, d\mathscr{V}, \tag{4.17}$$

where κ is the thermal conductivity, v_s is the species diffusion velocity, and $\mathscr{H}_s$ is the enthalpy. Using the scalar pressure, and after simplification, Eq. (4.17) becomes

$$\tfrac{1}{2}\rho \frac{Dv^2}{Dt} + \rho \frac{DU}{Dt} = \mathbf{E} \cdot \mathbf{J} + \nabla \cdot (\kappa\, \nabla T) - \sum_{s} \nabla \cdot (v_s \rho_s \mathscr{H}_s) - \nabla \cdot (pv) + \mathscr{E}_v, \tag{4.18}$$

where

$$\mathscr{E}_v = \tfrac{2}{3}v(\nabla \cdot v) - v[\nabla(v) - (\nabla \times v) - 2v \cdot \nabla v].$$

Equation (4.18) is the energy conservation equation.

Maxwell's equations. Maxwell's equations are

$$\nabla \cdot \mathbf{D} = \rho_e,$$
$$\nabla \times \mathbf{E} = -\partial \mathbf{B}/\partial t,$$
$$\nabla \cdot \mathbf{B} = 0,$$
$$\nabla \times \mathbf{H} = \mathbf{J} + \partial \mathbf{D}/\partial t.$$

Ohm's law with ion slip and relativistic effects neglected can be written as

$$\mathbf{J} = \sigma(\mathbf{E} + v \times \mathbf{B}). \tag{4.19}$$

MHD equations and MHD approximations. The equations listed in this section are known as the MHD equations. In MHD engineering, three approximations, known as the MHD approximations, are made. Two of the approximations are due to the fact that the displacement and the charge transport currents are neglected in Maxwell's equations. The third approximation is made by neglecting the term $\rho_e \mathbf{E}$ in the equation of motion. By putting Maxwell's equations together in one induction equation, eliminating $\mathbf{J}$ from the above relations by using Eq. (4.19), and assuming the MHD approximation, one obtains the equations for the conservation of mass,

$$\frac{\partial \rho}{\partial t} + \nabla \cdot (\rho v) = 0, \tag{4.20}$$

conservation of momentum,

$$\rho \frac{Dv}{Dt} = -\nabla p + \psi + (\nabla \times \mathbf{B}) \times \frac{\mathbf{B}}{\mu_0}, \tag{4.21}$$

conservation of energy,

$$\rho \frac{DU}{Dt} = -p \, \nabla \cdot v - \mathscr{E}_v + \nabla \cdot (\kappa \, \nabla T) + \frac{(\nabla \times \mathbf{B}) \cdot (\nabla \times \mathbf{B})}{\mu_0^2 \sigma}, \qquad (4.22)$$

and induction equation,

$$\frac{\partial \mathbf{B}}{\partial t} = \nabla \times (v \times \mathbf{B}) - \frac{1}{\mu_0} \nabla \times \left(\frac{\nabla \times \mathbf{B}}{\sigma} \right).$$

To these equations, the following relations are added:

equation of state,

$$p_r = p_r(\rho, T), \qquad (4.23)$$

calorific equation,

$$U = U(\rho, T), \qquad (4.24)$$

and relations for the transport properties,

$$\kappa = \kappa(\rho, T),$$
$$\eta = \eta(\rho, T), \qquad (4.25)$$
$$\sigma = \sigma(\rho, T).$$

4.4 The Operating Range of an MHD Duct [13]

The electrical power density can be derived from Poynting's law:

$$P' = -\mathbf{J} \cdot \mathbf{E}. \qquad (4.26)$$

A loaded generator does have a braking effect on the working gas. The braking power density is given by

$$P'_r = -v \cdot (\mathbf{J} \times \mathbf{B}). \qquad (4.27)$$

Considering the homogeneous MHD duct shown in Fig. 4.2, and taking account of the Hall effect, one has

$$
\begin{aligned}
\mathbf{E} &= \mathbf{i}E_x + \mathbf{j}E_y, \\
\mathbf{J} &= \mathbf{i}J_x + \mathbf{j}J_y, \\
v &= \mathbf{i}v, \\
\mathbf{B} &= \mathbf{k}B,
\end{aligned}
\qquad (4.28)
$$

where $\mathbf{i}$, $\mathbf{j}$, and $\mathbf{k}$ are the unit vectors in the x-, y-, and z-directions, respectively; or in complex notations

$$\bar{i} = \frac{J_x}{\sigma v B} + j \frac{J_y}{\sigma v B} = x + jy, \qquad (4.29)$$

and

$$\bar{\varepsilon} = \frac{E_x}{v B} + j \frac{E_y}{v B} = \alpha + j\beta. \qquad (4.30)$$

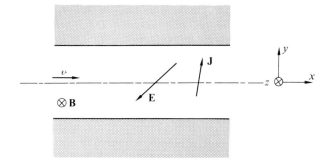

Figure 4.2. Homogeneous MHD duct with the Hall effect taken into consideration.

Thus, by using Eqs. (4.28), (4.29), and (4.30) in Eqs. (4.26) and (4.27),

$$P' = -\sigma v^2 B^2 (\alpha x + \beta y), \qquad (4.31)$$

and

$$P'_r = -\sigma v^2 B^2 y. \qquad (4.32)$$

From the generalized Ohm's law given by Eq. (4.19),

$$x = \alpha,$$
$$y = \beta - 1.$$

Eliminating α and β from Eqs. (4.31) and (4.32) and letting

$$P'_{max} = \tfrac{1}{4}\sigma(vB)^2, \qquad (4.33)$$

one obtains

$$x^2 + (y + \tfrac{1}{2})^2 = \tfrac{1}{4}(1 - P'/P'_{max}), \qquad (4.34)$$

and

$$P'_r/P'_{max} = -4y. \qquad (4.35)$$

An MHD duct can function as a generator, a brake, or a pump, depending on the sign of the power generated and the braking power. Indeed, it is,

1. A generator for $P' > 0$ and $P'_r > 0$,
2. A brake for $P' < 0$ and $P'_r > 0$,
3. A pump for $P' < 0$ and $P'_r < 0$.

The case of an MHD pump is widely used in MHD propulsion experiments. This field is closely related to MHD generation and is subjected to a great amount of research. In the $\bar{i}$ plane represented by Eq. (4.29) these ranges are limited by the x-axis (i.e., for no current in the y-direction) and by the limiting circle $P = 0$ as given by Eqs. (4.35) and (4.34), respectively. Fig. 4.3 illustrates the three different ranges named above.

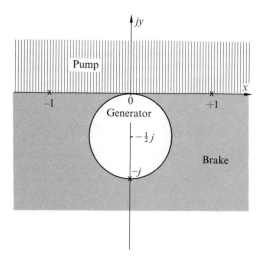

Figure 4.3. Operating ranges of an MHD duct.

4.5 Different Types of MHD Generators

There are two important types of MHD generators: dc generators and ac generators.

DC power generation. The dc generators can be divided into four categories: linear generators, vortex generators, radial outflow generators, and annular Hall generators.

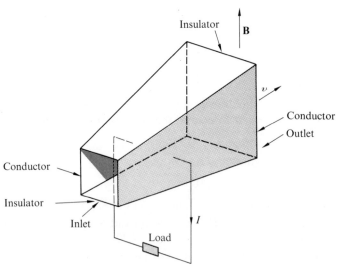

Figure 4.4. Linear MHD generator with variable cross section.

Linear generators. In this simple case, a conducting gas flows in a linear channel perpendicular to an applied magnetic field. Electrodes are placed on both sides of the channel and connected through a load in such a way that an electric current will flow through the load. This geometry is shown in Fig. 4.4. There are several types of linear MHD generators, differing in how the electrodes are connected. Considering the current density and the electric field to be in the x- and y-directions, as shown in Fig. 4.2, there are four important types of linear generators:

1. $E_x = 0$ for continuous electrode generators,
2. $J_x = 0$ for parallel connected, segmented electrode generators,
3. $E_y = 0$ for Hall generators,
4. $E_y = -\alpha E_x$ for series connected, segmented electrode generators.

These are shown in Fig. 4.5.

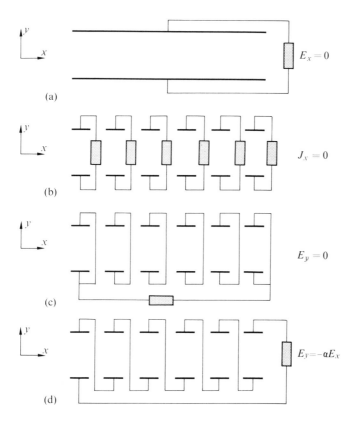

Figure 4.5. Electrode configuration for various linear generator types: (a) continuous electrode generator, (b) parallel connected, segmented electrode generator, (c) Hall generator, (d) series connected, segmented electrode generator.

The first generator is practical only if the Hall current is negligible. The Hall current is perpendicular to the normal Faraday current. The occurrence of this current can be explained as follows: an electron subjected to the effect of a magnetic field will follow a circular path about the lines of force in the plane perpendicular to the magnetic field and parallel to the velocity of the gas. The angular frequency of this circular path is called the *cyclotron frequency* and, as seen in Chapter 3, it is given by $\omega_c = eB/M$. But, as the electron moves in its circular path, it will collide with other particles of the gas with a frequency v_c, causing a motion in the direction of the fluid. This is the Hall effect. The angle through which the electrons rotate in the magnetic field between two collisions will be, on the average, equal to ω_c/v_c. If ω_c/v_c is much smaller than 1, the Hall current becomes negligible and continuous generators become practical. When the Hall effect becomes important, the power output of the generator is reduced by the factor $1/[1 + (\omega_c/v_c)^2]$. This will create an unacceptable amount of loss.

To avoid these losses, segmented electrodes are used, leading to the second type of generator. This generator has, however, the disadvantage of multiplicity of loads, each at a different potential. This can be corrected if the separate loads are made into one complete load, leading to the series connected electrode generators. When the Hall effect becomes very important, the transverse current components may be short circuited, leading to generators which are called Hall generators because *only* Hall current is used.

Analysis of linear generators. The equations of Section 4.3 are used by considering the special geometry and making some assumptions. The first assumption is to consider that the properties of the gas are constant in a local duct cross section horizontal to the axis of the generator. Furthermore, the time variations of the parameters are usually ignored. When this is done, the problem becomes one-dimensional and Eqs. (4.20), (4.21), and (4.22) become, respectively,

$$m = \rho v S, \tag{4.36}$$

where S is the area of the cross section and m is the constant mass flow,

$$\rho v \frac{dv}{dx} + \frac{dp_r}{dx} + JB = 0, \tag{4.37}$$

where the viscosity has been neglected, and

$$\rho v \frac{d}{dx}(C_p T + \tfrac{1}{2}v^2) + EJ = 0, \tag{4.38}$$

where both the viscosity and the thermal conductivity have been neglected. The enthalpy has been replaced by $C_p T$, where C_p is the heat capacitance at constant pressure. There are three equations and eight variables: ρ, v, p_r, T, J, B, E, and S. Five other equations are needed: first, the equation of state. By assuming a perfect gas,

$$p_r = \rho \imath T, \tag{4.39}$$

where $\imath = C_p - C_v = C_v(\gamma - 1)$, where C_v is the heat capacitance at constant volume and γ is the ratio of heat capacitances. Three other equations are given by Ohm's generalized law (Eq. 4.19), the calorific equation (Eq. 4.24) and Maxwell's equation $\nabla \times \mathbf{B} = \mu_0 \mathbf{J}$. To these equations, Eqs. (4.19) and (4.25) are added, and an assumption is made on one of the parameters, leading to five different approaches:

a) Constant velocity analysis, by taking $v = $ constant,
b) Constant temperature analysis, by taking $T = $ constant,
c) Constant pressure analysis by taking $p_r = $ constant,
d) Constant area analysis, by taking $S = $ constant,
e) Constant Mach number analysis, with $v^2/T = $ constant.

If the effect of the different collisions is taken into account, Ohm's generalized law yields, in scalar notation,

$$J_x = \frac{\sigma}{(1 + \beta_e \beta_I)^2 + \beta_e^2} [(1 + \beta_e \beta_I)E_x - \beta_e(E_y - vB)], \qquad (4.40)$$

and

$$J_y = \frac{\sigma}{(1 + \beta_e \beta_I)^2 + \beta_e^2} [(1 + \beta_e \beta_I)(E_y - vB) + \beta_e E_x], \qquad (4.41)$$

where $\beta_I = \omega_I/v_{in}$ and $\beta_e = \omega_e/v_{en}$.

The different types of linear generators listed above can be discussed separately. For continuous electrodes, $E_x = 0$, and Eq. (4.40) becomes

$$J_x = \frac{-\sigma \beta_e(E_y - vB)}{(1 + \beta_e \beta_I)^2 + \beta_e^2}.$$

Under open circuit conditions, $J_y = 0$; therefore $E_y = vB$. Under short circuit conditions $E_y = 0$, and under load conditions, $0 < E_y < vB$ or $0 < \mathcal{K}_F < 1$, where

$$\mathcal{K}_F = E_y/vB, \qquad (4.42)$$

is called the Faraday generator loading factor. Writing J_x and J_y as functions of $\mathcal{K}_F$, one obtains

$$J_x = \frac{\sigma \beta_e vB}{(1 + \beta_e \beta_I)^2 + \beta_e^2} (1 - \mathcal{K}_F),$$

and

$$J_y = \frac{-\sigma(1 + \beta_e \beta_I)vB}{(1 + \beta_e \beta_I)^2 + \beta_e^2} (1 - \mathcal{K}_F).$$

Only J_y is used for power generation in this type of generator, whereas J_x is responsible for the Lorentz force and becomes a source of loss.

For parallel segmented electrodes, $J_x = 0$, Eqs. (4.40) and (4.41) become

$$E_x = \frac{\beta_e vB}{1 + \beta_e \beta_I} (\mathcal{K}_F - 1),$$

and

$$J_y = \frac{\sigma v B}{1 + \beta_e \beta_I} (\mathcal{K}_F - 1).$$

As in the former case, J_y is used for power generation and is equal to zero when $\mathcal{K}_F = 1$ under open circuit conditions.

For the Hall generator, $E_y = 0$, Eqs. (4.40) and (4.41) yield

$$J_x = \frac{\sigma}{(1 + \beta_e \beta_I)^2 + \beta_e^2} [(1 + \beta_e \beta_I)E_x + v B \beta_e], \qquad (4.43)$$

and

$$J_y = \frac{\sigma}{(1 + \beta_e \beta_I)^2 + \beta_e^2} [-v B(1 + \beta_e \beta_I) + \beta_e E_x]. \qquad (4.44)$$

Under open circuit conditions $J_x = 0$ and

$$E_{xo} = -\frac{v B \beta_e}{(1 + \beta_e \beta_I)}.$$

Under short circuit conditions, $E_x = 0$, and

$$J_{xs} = \frac{\sigma v B \beta_e}{(1 + \beta_e \beta_I)^2 + \beta_e^2}.$$

Under load conditions,

$$-\frac{v B \beta_e}{(1 + \beta_e \beta_I)} < E_x < 0,$$

or $0 < \mathcal{K}_H < 1$, where $\mathcal{K}_H$ is called the Hall loading factor. It is equal to

$$\mathcal{K}_H = -\left(\frac{1}{\beta_e} + \beta_I\right)\frac{E_x}{v B}. \qquad (4.45)$$

Equations (4.43) and (4.44) can be written as

$$J_x = \frac{\sigma v B \beta_e}{(1 + \beta_e \beta_I)^2 + \beta_e^2} (1 - \mathcal{K}_H),$$

and

$$J_y = \frac{\sigma v B \beta_e^2 (1 - \mathcal{K}_H)}{(1 + \beta_e \beta_I)[(1 + \beta_e \beta_I)^2 + \beta_e^2]} - \frac{\sigma v B}{(1 + \beta_e \beta_I)}.$$

For series connected electrodes, $E_y = -\alpha E_x$, and Eqs. (4.40) and (4.41) become

$$J_x = \frac{\sigma(1 + \beta_e \beta_I + \alpha \beta_e)}{(1 + \beta_e \beta_I)^2 + \beta_e^2} E_x + \frac{\sigma \beta_e v B}{(1 + \beta_e \beta_I)^2 + \beta_e^2},$$

$$J_y = \frac{-\sigma \alpha(1 + \beta_e \beta_I - \beta_e/\alpha)}{(1 + \beta_e \beta_I)^2 + \beta_e^2} E_x - \frac{\sigma(1 + \beta_e \beta_I)v B}{(1 + \beta_e \beta_I)^2 + \beta_e^2}.$$

The theory of this type of generator has been given by De Montardy [22].

This more general type includes the cases $E_x = 0$, when α goes to infinity, and $E_y = 0$, when α goes to zero.

Numerical example [95]. As an example, consider a constant velocity linear generator using helium with 1 % cesium seed. To Eqs. (4.36), (4.37), (4.38), and (4.39) Ohm's law, $J = \sigma(E + \nu B)$, and Poisson's law, $E = -V/d$, where V is the variable voltage, are added. By assuming further a constant applied magnetic field, this problem can be solved, i.e., the variables ρ, p_r, T, J, E, and S can be found in terms of the coordinate x and the initial values (at $x = 0$). For the chosen working gas,

$$\imath = 1570 \text{ J/kg,}$$
$$\gamma = 5/3,$$
$$\text{Molecular weight} = 5.29,$$

The following values are chosen at the inlet of the duct:

$$\rho_0 = 0.1527 \text{ kg/m}^3,$$
$$p_{r0} = 5.935 \times 10^5 \text{ N/m}^2,$$
$$T_0 = 2475°\text{K,}$$
$$E_0 = 1.04 \text{ kV/m,}$$
$$S_0 = 1328 \text{ m}^2.$$

Further, the generator is desired to produce 500 MW under the following conditions:

$$B = 2 \text{ Tasla,}$$
$$\nu = 1272 \text{ m/sec.}$$

The results of the calculations, reported partially by Way [95], are summarized below for the most important quantities.

At the outlet of the duct,

$$p_{ri} = 2.967 \times 10^5 \text{ N/m}^2,$$
$$T_i = 1982°\text{K,}$$
$$S_i = 1720 \text{ m}^2.$$

The other characteristics of the duct were found to be

$$\text{length} = \ell = 15.44 \text{ m,}$$
$$\text{efficiency} = \eta_i = 81\%,$$
$$\text{output voltage} = V_{out} = 3.275 \text{ kV,}$$
$$\text{conductivity} = \sigma = 18.88 \text{ mho/m.}$$

Vortex generators. In this geometry the gas is introduced tangentially into an outer cylinder and withdrawn along the surface of an inner coaxial cylinder. The two cylinders constitute the electrodes. The magnetic field is in the axial

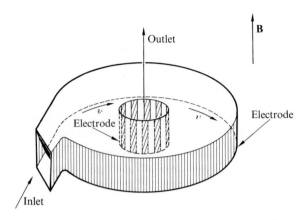

Figure 4.6. A vortex MHD generator. The plasma flows in the azimuthal direction and the electric current flows in the radial direction.

direction, as shown in Fig. 4.6. If the inner diameter is much smaller than the outer, the gas will make several revolutions before leaving the generator. Thus, this geometry permits a long magnetic interaction length. Clearly, this is a variety of the continuous electrode linear generator and the Faraday current is utilized. The Hall current flowing in the tangential direction remains a source of loss.

Radial outflow generator. This (Fig. 4.7) is a variation of the Hall generator in which the gas is injected radially outward from the inner cylinder. In this case, the Hall current flows radially and is the current flowing through the load. Due to the interaction of this current with the magnetic field, the flow will spiral

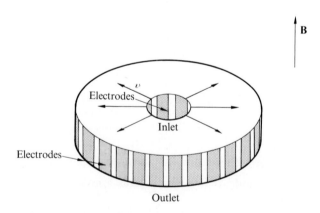

Figure 4.7. Radial outflow MHD generator. Both the plasma and the electric current flow in the radial direction. This is a Hall generator.

outwardly instead of following a radial path. The Hall current then creates a Lorentz force equal to the centrifugal force of the fluid. The Faraday current flows tangentially and is a source of loss.

Annular Hall generator. In this annular geometry (Fig. 4.8), the magnetic field is radial, being produced by a cylindrical permanent magnet. The gas flows in the axial direction. The Faraday current forms a complete loop around the cylindrical magnet. This generator has the advantage of requiring only one annular electrode at each end of the geometry. It has the disadvantages of necessitating a radial magnetic field and of having a large wall area.

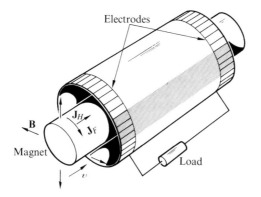

Figure 4.8. Annular Hall generator. The radial magnetic field is produced by a permament magnet.

AC power generation. The generators described above produce direct current. In many applications, and in most central power stations, the use of alternating currents is necessary. If dc power is generated, it could be converted to ac power by the use of an inverter stage, which will complicate the system and make it more expensive.

These two shortcomings can be avoided by ac MHD generators. The possibility of building an ac generator has been considered by many, and several propositions have been presented. Three different generator types can be distinguished: ac generators with varying magnetic field, ac generators with varying gas velocity, and ac generators with varying gas conductivity.

Variable magnetic field. Adequate electrical conductivity of the working gas could be obtained only if the electrons move in a closed path, since the electronic mobility is usually much larger than the ionic mobility. This can be achieved by the use of thermionically emitting cathodes, and collecting anodes in contact with the gas. Because of high surface temperatures, the problem of corrosion becomes extremely important. A cooling system could be used but this will create new problems and a lowering of efficiency. The most attractive solution seems to eliminate the electrodes. AC power could be extracted inductively from closed electron current loops by the use of magnetic field coils in which an alternating magnetic field is generated. The alternating magnetic field should have a phase variation along the duct corresponding to a traveling magnetic wave. This wave interacts with the working plasma which is then decellerated, and its kinetic energy is converted into alternating electrical energy in the field coils. This type of generator has been studied by Jackson and Pierson [43], and Bernstein [3] and his co-workers. The feasibility of such an ac generator remains questionable, its power factor is extremely low, field swinging devices are very costly and extremely heavy, and small scale experiments were unconvincing.

Variable gas velocity. To avoid difficulties of producing an alternating magnetic field, the gas velocity could be varied. However, this velocity variation will certainly lead to fluctuations in pressure and temperature and it is technically difficult to achieve variation over the entire duct length. The use of fissile gases* to produce alternating velocities has been proposed by Colgate and Aamodt [17], but it seems beyond the capacity of present technology.

Variable gas conductivity. Many methods can be used to achieve a variable gas conductivity. If nonthermal ionization is used, a pulsating ionization source, as proposed by Devime [24], could be utilized. It is also possible to introduce charges of fuel and oxygen at discrete intervals to produce hot spots in the working gas as proposed by Thring [90]. This can be used with or without similar pulses of seeding material. The current generated through the electrodes is a modulated direct current, and inverters should be provided to invert the dc component. This will introduce additional expenses. Furthermore, there are many problems of instability which make this generator appear unfeasible.

4.6 Magnetohydrodynamic Materials

The different materials used in MHD generators can be divided into three categories: conducting gases and seedings, materials for plasma containment, and electrode materials.

The working fluid is chosen for its desirable high conductivity. However, other properties should be considered, such as chemical and thermodynamic properties. The problem of chemical stability and corrosion is extremely important. In open-cycle experiments, the problem of the cost of the fuel becomes decisive for any choice of working gas. In these experiments, hydrogen, alcohol, and fuel oil mixed with air or oxygen were used as working fuels. In closed-cycle systems, since the working plasma can be recovered, expensive materials can be

---

* In the paper of Colgate and Aamodt the possibility of producing a *fission* plasma (thus gaseous) as opposed to *fusion* plasma has been mentioned. See reference.

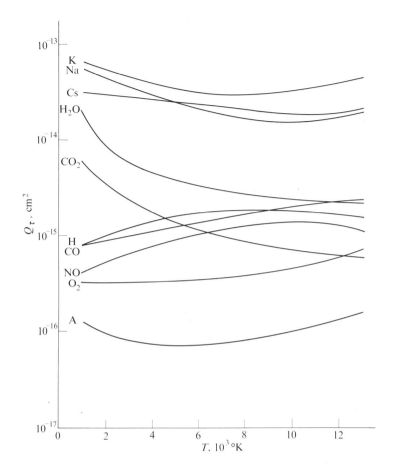

Figure 4.9. Temperature variation of the collision cross sections of electrons with several atoms. The highest values for all temperatures are those of the alkali metal atoms: potassium, sodium, and cesium.

used. To avoid the problem of corrosion, the choice is narrowed to the five noble gases: helium, neon, argon, krypton, and xenon. These are seeded by some low collision cross section metallic materials such as potassium or cesium. The use of pure alkali metals is quite expensive. They have been considered to be used in the form of droplets of liquid, as was proposed by Rowe and Kerrebrock [76], and Smith [81]. The collision cross sections of electrons with several different atoms are plotted in Fig. 4.9 as functions of the temperature, and the ionization potentials V_i of several materials are listed in Table 4.1.

As mentioned in Section 4.2, the conductivity of a plasma increases drastically as the amount of seed is increased. An optimum seeding ratio is reached with

Table 4.1

Ionization potentials of several materials*

Noble gases	V_i, V	Other gases	V_i, V	Alkali metals	V_i, V
Xenon	12.08	NO	9.50	Cesium	3.87
Krypton	13.93	O_2	12.50	Rubidium	4.16
Argon	15.68	H_2O	12.56	Potassium	4.32
Neon	21.47	H	13.53	Sodium	5.12
Helium	24.46	H_2	15.60	Barium	5.19
		CO	14.10	Lithium	5.36
		CO_2	14.40	Aluminum	5.96
		N	14.48	Calcium	6.09
		N_2	15.51	Mercury	10.39

* From Handbook of Chemistry and Physics. 33rd edition,
Chemical Rubber Publishing Co.

only a few percent of the seed. The conductivity of several seeded noble gases is plotted in Fig. 4.10 as a function of the temperature. Fig. 4.11 represents the electrical conductivity of seeded combustion products at various pressures and temperatures. It is clear that the conductivity increases with increasing temperature and decreasing pressure.

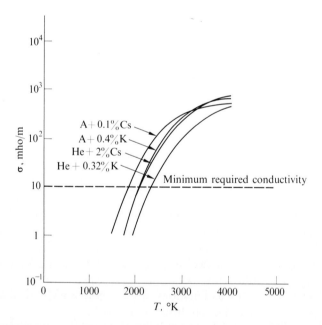

Figure 4.10. Temperature variation of the electrical conductivity of seeded noble gases. The effect of the seed is to increase this conductivity.

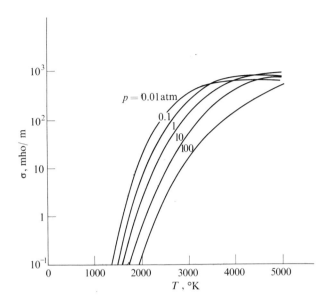

Figure 4.11. Electrical conductivity of seeded combustion gases as a function of temperature and pressure: seed = $1\%K$, cross section $Q_g = 10^{-15}$ cm^2. The conductivity decreases with increasing pressure in the useful temperature range.

In an MHD generator, the conductivity requirements dictate the working temperatures. These are at least 2000°C, and therefore the problem of flow containment may be solved only in one of the following two ways: either the material used for the walls must withstand the hot and corrosive atmosphere of the working plasma or the walls must be cooled to a satisfactory temperature. Furthermore, the wall material should be refractory at such high temperatures. The materials [38] most often preferred are magnesia (MgO) and strontium zirconate (SrO·ZrO$_2$) as tested by Westinghouse [96], although magnesia evaporates and the zirconate is a poor insulator. Other materials are alumina (Al$_2$O$_3$), thorium dioxide (ThO$_2$), hafnium dioxide (HfO$_2$), beryllium oxide (BeO), and boron nitrate (BN). Every one of these materials has some disadvantage. Alumina has a low melting point, ThO$_2$ is radioactive, HfO$_2$ is too expensive, BeO is extremely toxic, and BN oxidizes easily. Ceramic oxides seem to be the only suitable materials for the insulating ducts, even though they are attacked by the alkali metal components. It seems unlikely that any material having a life of the order of months or even weeks at such temperatures will be found. The alternative approach of cooled walls seems attractive. This necessitates the use of materials having good thermal conductivities. Alumina may be considered but, because of its conductivity, serious short circuitry problems are created. To avoid this short circuiting, the walls should be segmented in some way. Several methods of segmentation were proposed by Louis [54], Maycock [56, 57], and their co-workers.

The electrodes, on the other hand, are required to be good electrical conductors at the high working temperatures. The electrodes also serve to make electrical contact with the working plasma. The characteristics of solid-gas contacts are, however, more complicated than those of solid-solid contacts. Furthermore, the electrodes should be good thermionic emitters to avoid a reduction in the conduction current. The most used materials are tungsten, rhenium, tantalum, and carbon, and the carbonates, borates, and nitrates of tantalum, hafnium, niobium (columbium), titanium, and zirconium. At temperatures around 3000°K these materials are oxidized and the same materials used for the walls at lower temperatures have been considered for use as electrodes. The use of cooled electrodes has been proposed and experimented by Maycock [56, 57].

4.7 Production of Magnetic Fields

It can be shown that the output power delivered by an MHD generator is proportional to B^2, whereas the capital and operating-cost dependencies would certainly be less than B^2. Therefore, the magnet should be as large as possible. The economic considerations are very important since the magnet is likely to be one of the most expensive parts of the MHD system. It is then necessary to optimize the production of large magnetic fields. In search of a method of optimization, experiments with several types of magnets have been carried out. Four types of such magnets can be distinguished:

1. Permanent magnets,
2. Water-cooled electromagnets,
3. Cryogenically cooled electromagnets,
4. Superconducting magnets.

For low magnetic fields, iron cores are used to reduce the reluctance of the magnetic circuit. For field strengths larger than 2 Wb/m^2, the effect of the magnetic core becomes negligible, due to the magnetic saturation effect. Most of the designs use air-cooled electromagnets.

Two important considerations determine the magnet design for minimum dissipation of electrical energy. First, the dissipation is proportional to the length of the duct and the resistivity of the magnet windings, but it is independent of the cross sectional area of the duct. Second, the capital cost of the magnet is proportional to the volume of the field windings.

Permanent magnets. These do not consume any electric power, but they do have the disadvantages of being expensive and producing weak magnetic field strengths over large volumes. They are used only for some special applications involving very small scale experiments.

Water-cooled electromagnets. The coils of such magnets are wound of conventional material such as copper and to a lesser extent, aluminum. The flow of water serves to carry away the Joulean heat dissipated in the coils. The dissipation of such magnets can be approximated by $P_m = 400B^2\ell$ kW and the capital cost by $C_m = 0.08B\ell P_{in}$ dollars, where P_{in} is the thermal input to the generator in kilowatts, B is the magnetic flux density in webers per square meter, and ℓ is the length of the duct in meters. The power dissipated can be an important part of the total power output—about 8% for a 500-mW plant, but it decreases for larger plants. This power can, however, be taken directly from the power generated by the MHD generator, making it self-excited, similar to conventional dc machines.

Cryogenically cooled electromagnets. The resistance of pure metals decreases drastically with decreasing temperature and so does the power dissipated in the magnet coils. If the total capital cost of the refrigerator and the magnet coils is smaller than that of the coils and the additional power dissipated at room temperature, then cryogenically cooled electromagnets become advantageous. The total power consumed by the magnet can be reduced by a factor of 10 by operating at very low temperatures; 8°K for sodium coils, 20°K for aluminum coils, and 25°K for copper coils.

Superconducting magnets. There is no Joulean heating in a superconducting magnet. The only energy used is that necessary for the refrigerator to maintain liquid helium temperatures by counterbalancing the heat leakage through the walls. For every superconducting material, there is a critical magnetic field strength above which the material loses its superconductivity. This critical field is usually small, but materials which maintain their superconducting properties at high magnetic field strengths have recently been devised. Niobium-zirconium is such a high-field superconductor. With this material, fields approaching 6 Wb/m² at 200 A/mm² have been used. It also has the advantage of being sufficiently ductile to be made into wires. However, two problems arise with such a material: its characteristics can be changed drastically by any mechanical work, and its cost is extremely high.

4.8 Power Output, Losses, and Efficiency

The power generated by an MHD generator is given by Eq. (4.26), or considering the components of J and E in the x and y directions, by

$$P' = -\sigma(E_x J_x + E_y J_y), \tag{4.46}$$

where J_x and J_y are given by Eqs. (4.40) and (4.41). If

$$\mathscr{C}_1 = \frac{\beta_e}{(1 + \beta_e \beta_I)^2 + \beta_e^2},$$

and

$$\mathscr{C}_2 = \beta_I + 1/\beta_e,$$

then

$$P/P_{max} = -4\mathscr{C}_1\mathscr{C}_2[\mathscr{K}_F^2 + (\mathscr{K}_H/\mathscr{C}_2)^2 - \mathscr{K}_F - (\mathscr{K}_H/\mathscr{C}_2)], \qquad (4.47)$$

where P_{max}, $\mathscr{K}_F$, and $\mathscr{K}_H$ are given by Eqs. (4.33), (4.42), and (4.45), respectively. The different types of generators can now be treated separately.

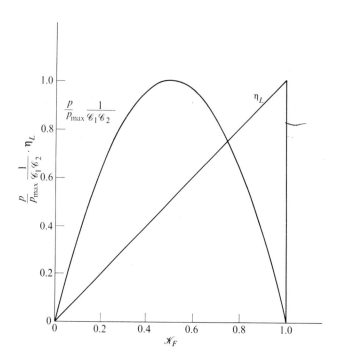

Figure 4.12. Power and efficiency of Faraday generators. Continuous electrodes ($E_x = 0$) and parallel segmented electrodes ($J_x = 0$). The maximum power output is at $\mathscr{K}_F = \frac{1}{2}$.

Continuous electrodes: $E_x = 0$. In this case, $\mathscr{K}_H = 0$ and Eq. (4.47) becomes

$$P/P_{max} = -4\mathscr{C}_1\mathscr{C}_2(\mathscr{K}_F^2 - \mathscr{K}_F). \qquad (4.48)$$

The local efficiency is given by $\eta_L = P/P'$, P' being the braking power density as given by Eq. (4.27). Then $\eta_L = \mathscr{K}_F$. Figure 4.12 shows $(P/P_{max})(\mathscr{C}_1\mathscr{C}_2)^{-1}$ plotted as a function of η_L. It is a parabolic function reaching its maximum value (unity) at $\eta_L = \frac{1}{2}$.

Parallel segmented electrodes: $J_x = 0$. From Eq. (4.46) it is seen that the power density will be the same as in the previous case. It is given by Eq. (4.48) and is illustrated in Fig. 4.12.

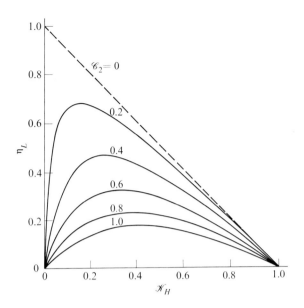

Figure 4.13. Efficiency of Hall generators as a function of the Hall loading factor $\mathscr{K}_H$. The highest efficiencies are for small values of the applied magnetic field and small values of the electronic collision frequency.

The Hall generator: $E_y = 0$. In this case, $\mathscr{K}_F = 0$ and Eq. (4.47) becomes

$$\frac{P}{P_{max}} = -4\frac{\mathscr{C}_1}{\mathscr{C}_2}(\mathscr{K}_H^2 - \mathscr{K}_H),$$

whereas the braking power density, given by Eq. (4.27), is

$$\frac{P_b'}{P_{max}} = 4\frac{\mathscr{C}_1}{-\mathscr{C}_2}(\mathscr{C}_2^2 + \mathscr{K}_H).$$

The local efficiency is

$$\eta_L = \frac{1 - \mathscr{K}_H}{1 + \mathscr{C}_2^2/\mathscr{K}_H}. \qquad (4.49)$$

This local efficiency is plotted in Fig. 4.13 as a function of the loading factor $\mathscr{K}_H$ and with $\mathscr{C}_2$ as parameter. The highest efficiencies in a Hall generator are given by small values of $\mathscr{C}_2$, i.e., small values of B and large values of β_e. In a Hall generator, maximum efficiency is achieved when operation is as close as possible to short circuit, which is the opposite of the Faraday generator case.

Series connected electrodes: $E_y = -\alpha E_x$. In this case, $\mathscr{K}_H/\mathscr{C}_2 = \mathscr{K}_F/\alpha$. Introducing this relation in Eq. (4.48) yields

$$P/P_{max} = 4\mathscr{C}_1\mathscr{C}_2\mathscr{K}_F(1 - \mathscr{K}_F)(\alpha^2 + 1)/\alpha^2.$$

The local efficiency will be

$$\eta_L = \frac{\mathcal{K}_F(1 - \mathcal{K}_F)(\alpha^2 + 1)\mathcal{C}_2}{\alpha[\alpha\mathcal{C}_2 - \mathcal{K}_F(1 + \alpha\mathcal{C}_2)]}. \tag{4.50}$$

Losses in an MHD generator. There are two important sources of loss in an MHD generator: electrical losses and losses at the magnet. The electrical losses are:

1. End losses [7] associated with eddy currents at the inlet and the outlet of the generator,

2. Effects in the vicinity of the segmented electrodes,

3. Instabilities associated with the Lorentz force and fluctuations in the electrical properties of the gas which tend to short out the Hall current.

The current flowing in the generator produces magnetic fields which tend to reduce the applied magnetic field upstream and to increase it downstream. Another cause of end losses is the short circuit current between the electrodes in the plasma at each end of the generator. This effect can be reduced by using adequate aspect ratios d/ℓ (the height divided by the length of the duct). Assuming that the gas properties, velocity, and magnetic field are constant, that the boundary layer is negligible, and that the sides of the duct are straight and parallel, it is found that

$$\eta_{end} = \mathcal{K}_F\left(1 - \frac{\mathcal{K}_F a^*}{1 - \mathcal{K}_F}\right), \tag{4.51}$$

where $a^* = 2d \ln 2/\pi\ell$ is an aspect ratio. The maximum value for η_{end}, after taking the derivative of Eq. (4.51) with respect to $\mathcal{K}_F$, becomes

$$\eta_{e,max} = 1 - 2a^*[(1/a^* + 1)^{1/2} - 1]. \tag{4.52}$$

From this equation it is seen that the aspect ratio a^* (therefore h/ℓ) should be as small as possible. The end losses can be further reduced by several methods such as inserting thin insulating vanes across the duct or extending the magnetic field beyond the generator ends [85]. These losses can be completely eliminated if the working gas does not conduct until it has entered the magnetic field and ceases to conduct before leaving it.

The effects in the vicinity of segmented electrodes have been investigated by mapping the equipotential lines. It was found that possibilities of short circuits between the segmented electrodes are not negligible [85], especially in Hall generators [26].

Flows in MHD generators are usually highly turbulent. This creates fluctuations in the induced magnetic field, which in turn creates circulating currents which are an important source of loss. Also, because the Lorentz force tends to retard the flow, small disturbances may be amplified. Other effects, such as heat transfer and electrical power generation itself, have a stabilizing influence on the unavoidable fluctuations in velocity, density, temperature, and conductivity.

Except in permanent magnets and to a lesser extent superconducting magnets, the power required to generate the magnetic field can become an appreciable part of the total output power.

Experiments in seeded combustion gases have satisfactorily verified the power relations. However, in these experiments, the electrical conductivity and generator dimensions are relatively small, so that the Lorentz force is small compared to the friction pressure drop, making it very difficult to verify the efficiency relations experimentally.

4.9 MHD Power Generation Systems

Apart from the MHD power generator, other apparati are necessary to form the overall MHD system. It is necessary to burn the fuel and the oxidizer, to add the seed, and to make arrangements for exploiting the generated electrical power. The fuel is usually fossil and the oxidizer is air, for obvious economic reasons. For large systems, some precautions should be taken to limit the amount of losses. The air may be enriched with more oxygen, and preheating of the incoming oxidizer becomes necessary to allow thermal ionization. The exhaust temperature is still above the thermal ionization temperature of the gas. Further extraction of the thermal energy is often desired and can be accomplished by the use of steam turbines. Another method can be used by making the system completely closed. The exhaust gas can then be reused and pumped back into the heat exchanger where it is heated, then expanded through a nozzle and passed again through the MHD generator. Nuclear energy supplies the necessary heat. However, due to the high levels of temperatures necessary for thermal ionization, a limit is set by the present development of refractory materials. It was suggested by Elliott [27] that work at lower temperatures can be made possible if two working fluids are used instead of one.

Four types of MHD generation systems can be distinguished:

1. Open cycle systems without recovery,

2. Open cycle systems with recovery,

3. Two-working-fluid systems,

4. Closed cycle systems.

Each of these systems will be discussed separately.

Open cycle system without recovery. In this system, solid or liquid fuel is burned with the oxidizer to which seed is added without preheating. The resulting gas mixture is then accelerated through a nozzle and passed through the generator. It is then exhausted directly into the atmosphere. There is no recovery of the seed material or of the remaining thermal energy in the exhaust gases. For seeded combustion gases to be sufficiently conducting, their temperature must exceed 2000°C at atmospheric pressure, which necessitates the use of pure oxygen or any chemical oxidizer leading to hotter flames. Higher flame temperatures will also lead to a better Carnot efficiency of the overall system. The most used fuels are alcohol, kerosene, methane, hydrogen, and cyanogen. Explosives could be used to generate the combustion gases.

Most of the experiments performed are on open cycle systems [46, 96]* in this country, as well as in Europe [23].† However, such a system may be practical for only small to short duration generators. All the generated power can be available for the load, since the amount of energy required to pump the fuel and oxidizer is negligibly small, but the energy required by the magnet may be large. In the laboratory, efficiencies reaching 15% have been attained for hydrocarbon fuels [46, 96] and much higher figures are predicted for systems using oxycyanogen fuels [79].

Open cycle system with recovery. The system described above is obviously not economical. To be able to use air as an oxidizer, arrangements to compress it to an adequate combustion pressure should be made. To be able to use regular fossil fuels, the inlet oxidizer should be heated to temperatures high enough to permit thermal ionization. A regenerative cycle becomes necessary, as the thermal energy still available in the exhaust gases can be recovered by producing steam for a steam turbine to generate more electrical power. The seed is recovered to be used again. The fuel most often used is coal, but the ashes formed create new difficulties by polluting the seed material. To avoid these difficulties, the ashes are removed by a cyclone burner and electrostatic precipitators are used to remove the seeds. Such a system is illustrated in Fig. 4.14.

Overall efficiencies of about 56% are predicted for these systems, using an enriched air as oxidizer. This efficiency is much higher than the 41% of conventional power plants. The additional cost of the different parts of this system makes the predicted generated electrical power cost about the same as that of steam plants.

Two-working-fluid system. The energy converted in an MHD generator is of two types: the thermal energy and the mechanical energy of the working gas. The first is converted into mechanical energy through expansion of the gas into a nozzle and then, with the second, it is converted into electrical energy by the motion of the conductor gas in a magnetic field. Usually, in a regular MHD generator, the working gas plays this double role. It has been proposed [27] that if these two roles are separated by the use of two different working gases, operation at much lower temperatures will be possible. A gas will be used to transform the thermal energy into mechanical energy and a liquid metal will transform the mechanical energy into electrical energy. This will also lead to much higher efficiencies due to the larger conductivities of the liquid metals. Because of these higher efficiencies, small units can be built for special applications. However, to

---

*   At Westinghouse, Avco-Everett, and General Electric.

†   England, France, Poland, and USSR.

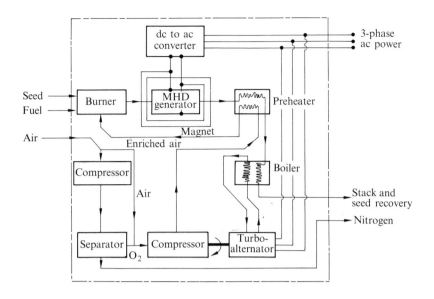

Figure 4.14. MHD open-cycle system with recovery.

use these systems, jet pumps are necessary for the transfer of gas motion to the metal liquid. The use of jet pumps creates new problems which should be attacked. If these problems are solved, ac power could be produced with efficiencies reaching 50%.

Closed cycle system [4, 30, 40]. In this case, the working gas in a heat exchanger is expanded through a nozzle, passed through the generator, cooled, and pumped back through the heat exchanger. Therefore, the working gas is in a completely closed cycle. The heat source should be a nuclear source in a nuclear reactor using an in-pile heat exchanger. A minimum temperature of 2000°C would be required for any appreciable amount of energy to be converted. However, the surface temperature of the fuel elements in present nuclear reactors is not higher than 600°C. High-temperature nuclear reactors should be developed and some form of nonthermal ionization should be used to allow work at lower temperatures. This type of closed-cycle MHD generator attracts the attention of space workers [27, 73], because the pumping power is low and the heat is rejected at a constant temperature. It has also been proposed [28] that liquid metals such as cesium or to a lesser degree potassium could be used as working fluids in a Rankine cycle. However, these liquid metals are highly corrosive at high temperatures. The use of seeded helium has been proposed, along with some kind of magnetically induced ionization. For terrestrial applications, the problem of heat rejection is greatly alleviated and heat can be rejected by conduction through a water flow. Overall thermal efficiencies of about 7% have been predicted for such a system, making it a poor competitor for space applications.

4.10 Economic Aspects and Future Trends [6, 11]

Presently, the most often considered use of MHD generators is as a topping device for conventional steam plants. The fact that there are no moving mechanical parts will make operation at high temperatures feasible. The upper limit temperature in a steam plant is about 750°C, which is far below the temperatures reached by MHD generators (about 2700°C).

To obtain good conducting gases, it is necessary to add cesium or potassium as seed materials and to solve the problem of corrosion. Advances in refractory material techniques are needed. The cost of seeding increases substantially the cost of installed power. Cesium is very expensive, but it is a better seed than the cheaper potassium, due to its lower ionization potential. The degree of ionization of the working gas is not much higher than about 0.01 %, and the bulk of the working gas is composed of neutral atoms which are not affected directly by the magnetic field. Through collisions with the electrons and ions, however, a braking force is created allowing the conversion of the kinetic energy of these neutral atoms into electrical energy.

The cost of the wall material is an important part of the total cost of an MHD generator. Good insulating and refractory materials working for a reasonably long time without deterioration should be found. At the present state of the art, experimental generators operated only for several minutes at the megawatt level and for about 1 hour at about 10 kW. These times are very short and the problem could somehow be alleviated by an increase of the volume to area ratio leading to a decrease in thermal losses to the walls.

The problem of high temperatures could be alleviated by the use of some type of nonthermal ionization. This can also make the possibility of a nuclear reactor–MHD generator coupling feasible, with the advantage of having an entirely static power plant.

In a high power MHD generator, the electronic concentration should be maintained for at least $\frac{1}{10}$ second to allow the gas to pass through the generator. However, a nonequilibrium concentration of electrons dies out quickly through recombination, which is enhanced by a high seed concentration.

The cost of preheaters for the oxidizer becomes very large if the oxidizing air is not adequately enriched in oxygen. The optimum figure is about $N_2 = 2O_2$. Another solution proposed for obtaining high temperatures is by the use of shock waves. (Behind the shock front of the wave passing down the shock tube there is a very high temperature region.)

From the tremendous amount of work done in this field, both theoretically and experimentally, it seems that a fossil fueled MHD topper is the most promising MHD generator and most probably will be the first to be operating on the industrial level. Many problems need to be solved before an MHD power plant becomes competitive, e.g., seed recovery, superconductivity for the magnet, high temperature materials, ash handling, ac power generation, and progress in non-equilibrium ionization techniques.

PROBLEMS

4.1 From Saha's equation, show that for a neutral plasma the electron density is approximately equal to

$$n_e = (\mathcal{K}_1 n_a)^{1/2}[(1 + \tfrac{1}{4}\mathcal{K}_1/n_a)^{1/2} - (\tfrac{1}{4}\mathcal{K}_1/n_a)^{1/2}],$$

where $n_a = n_n + n_e$ and $\mathcal{K}_1 = n_e(n_i/n_n)$.

4.2 The degree of ionization i is defined as the ratio of the number of ions over the total number of neutrals and ions in a gas. Show that for a slightly ionized gas ($i \leqslant 10^{-2}$), the conductivity is approximately equal to $\sigma = 10^{-9}i/\bar{Q}$ in MKSC units.

4.3 Consider a cesium plasma ($V_i = 3.89$ V) of constant density $n_n = 10^{20}$ particles/m³. (a) Find the temperature at which the degree of ionization is equal to that of potassium plasma ($V_i = 4.34$ V) having an electronic temperature equal to 3000°K. (b) Calculate the conductivities of both plasmas. (c) Comment on both results.

4.4 A basic problem in photoionization is the production of light at the required short wavelength and in the desired intensity. The rate of photoionization is

$$(dn_e/dt)_{\text{photo}} = n_i Q_i \Phi_i,$$

where n_i is the number of ionizable atoms, Q_i is their ionization cross section, and Φ_i is the total flux of photons per unit area per unit time with energy greater than the ionization energy. Find the temperature of radiation for a blackbody as a function of the frequency of ionization v_i, the electron density n_e, the number of ionizable atoms n_i, the recombination coefficient α_r, and the ionization cross section. [*Hint:* At equilibrium, the rate of ionization equals the rate of loss by recombination, $\alpha_r n_e^2$.]

4.5 Solve the equations of continuity, motion, and conservation of energy for a unidimensional, uncompressible viscous flow.

4.6 Using the MHD approximations, derive from Maxwell's equations the relation

$$\partial \mathbf{B}/\partial t = \gamma_m \nabla^2 \mathbf{B} + \nabla \times (v \times \mathbf{B}),$$

known as the magnetic transport equation, where γ_m is the magnetic diffusity given by $(\sigma\mu)^{-1}$.

4.7 In Section 4.4, the complex notation has been used to show the operating range of an MHD duct. Following the same method, (a) find the efficiency of an MHD generator, and (b) draw the power characteristic for a type $E_y = 0$ linear generator.

4.8 (a) From the general equations of Section 4.3, deduce the following generalized Ohm's law:

$$\mathbf{J} = \sigma \mathbf{E} - \beta_e \mathbf{J} \times \mathbf{B}/B + \beta_e \beta_I (\mathbf{J} \times \mathbf{B}) \times \mathbf{B}/B^2.$$

(This equation shows that both ions and electrons have appreciable mobilities. It includes the Hall effect as well as the ion slip.) (b) From the above equation, find Eqs. (4.40) and (4.41).

4.9 Consider the MHD generator shown in Fig. 4.15. A conducting fluid is inside the cylindrical channel. (a) Calculate the emf induced between points A and B. (b) What will be the output power of such a generator when a load R_L is connected between A and B?

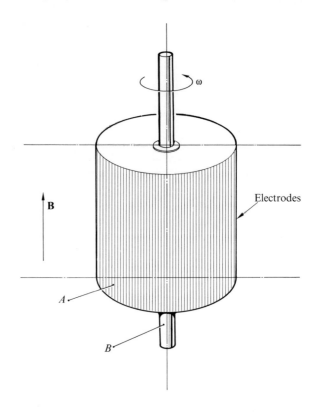

Figure 4.15. Rotating MHD generator.

4.10 Give a complete analysis of a linear MHD generator having a constant rectangular cross section. Do the same analysis for a constant Mach number generator.

4.11 Using your results of Problem 4.10, design a linear constant Mach number MHD generator having the following characteristics:

$$P_{out} = 300 \text{ MW}.$$

Conditions of the plasma at the inlet:

$$v_0 = 700 \text{ m/sec},$$
$$T_0 = 2600°\text{K},$$
$$p_0 = 4.5 \text{ atm}.$$

The applied magnetic field is

$$B = 6 \text{ Wb/m}^2.$$

Use your own judgment for the choice of the type of segmentation. Estimate the amount of heat lost to the wall. Most probably a cooling system would be necessary. What would you propose?

4.12 An example of an ac MHD generator is shown in Fig. 4.16. In this generator, the potential difference in the Faraday circuit is modulated uniformly along the duct at a frequency ω. An ac Hall current is then generated in the axial direction and collected by the Hall electrodes. This current flows in the Hall circuit through the external load. The capacitor in the Hall circuit is chosen for matched conditions at the operating frequency. Assuming that the walls are rigid boundaries, calculate the perturbed values of the Hall and Faraday currents [59].

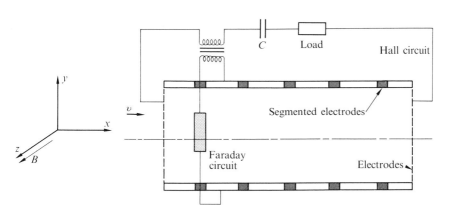

Figure 4.16. AC MHD generator in which the potential difference in the Faraday circuit is modulated uniformly along the duct. After McCune [59].

4.13 Some workers propose liquid metals as working fluids in MHD power generators. (a) Discuss this method and study carefully its pros and cons. (b) Using the magnetic transport equation of Problem 4.6, show that the current density output of a linear Faraday generator having a rectangular cross section and using a liquid metal as the working fluid is given by

$$J(z) = J_0 \exp\left(\frac{v_0}{\gamma} z\right),$$

where the subscript 0 refers to the conditions at the inlet of the duct.

4.14 In the case of series connected electrodes $E_y = -\alpha E_x$. (a) Calculate the value of α for maximum local efficiency. (b) What should be, in this case, the value of $\mathscr{C}_2$ for a loading factor equal to 80%? (c) Is this value of $\mathscr{C}_2$ feasible?

4.15 Equations (4.40) and (4.41) can be written in the form

$$J_x = \sigma^* E_x - tg\phi J_y,$$
$$J_y = \sigma^* E_y - tg\phi J_x.$$

(a) What are the values of σ^* and $tg\phi$ in terms of β_e, β_I, and B? (b) Show that the power output of a segmented generator (no Hall current) is given by

$$P_{\text{out}} = \mathcal{K}_F(1 - \mathcal{K}_F)B^2\sigma^*v^2\mathcal{V},$$

where $\mathcal{V}$ is the volume of the generator. (Some typical values are $\sigma^* = 100$ mhos/m, $v = 3000$ m/sec, $B = 2$ Wb/m^2.)

REFERENCES AND BIBLIOGRAPHY

1. Attard, M. C., *Electron Temperatures in Flame Plasmas*, Proceedings of 2nd International Symposium on MHD Electrical Power Generation, Paris, Vol. 1, p. 21, 1964.

2. Bates, D. R. (ed.), *Atomic and Molecular Processes*, Academic Press, New York, 1962.

3. Bernstein, I. B., et al., *An Electrodeless MHD Generator*, Proceedings of 2nd Symposium on the Engineering Aspects of MHD, Philadelphia, March 1961, p. 255, Columbia University Press, 1962.

4. Bienert, W. B., W. H. Young, and E. N. Zavodny, "Electrical Output from Closed Loop MPD Experiment Using Auxiliary Ionization," *Adv. Energy Conversion*, Vol. 6, No. 1, p. 25, 1966.

5. Blackman, V. H., M. J. Jones, and A. Demitriades, *MHD Power Generation Studies in Rectangular Channels*, Proceedings of 2nd Symposium on Engineering Aspects of MHD, Philadelphia, March 1961, p. 180, Columbia University Press, 1962.

6. Booth, L. A., *Prospects for a 1000 Mw(e) Nuclear Reactor/MHD Power-Plant*, 6th Symposium on Engineering Aspects of MHD, Pittsburgh, March 1965.

7. Boucher, R. A., and D. B. Ames, "End Effect Losses in d.c. Magnetohydrodynamic Generators," *J. Appl. Phys.*, Vol. 32, No. 5, p. 755, 1961.

8. Brederlow, G., R. Hodgson, and W. Riedmuller, *Nonequilibrium Electrical Conductivity and Electron Temperature Measurements in Electric Fields and Crossed Electric and Magnetic Fields*, 6th Symposium on Engineering Aspects of MHD, Pittsburgh, March 1965.

9. Brocher, E. F., "The Constant Velocity MHD Generator with Variable Electrical Conductivity," *J. Aerospace Sci.*, Vol. 29, p. 626, 1962.

10. Brogan, T. R., "MHD Power Generation," *IEEE Spectrum*, Vol. 1, No. 2, p. 58, 1964.

11. Brown, J. W., *Some Aspects of MHD Power Plant Economics*, 3rd Symposium on Engineering Aspects of MHD, Rochester, New York, March 1962.

12. Brown, S. C., *Basic Data of Plasma Physics*, Wiley, New York, 1959.

13. Bürgel, B., "A Graphical Method for the Investigation of MHD Generators," *The Brown Boveri Review*, Vol. 49, No. 12, p. 493, 1962.

14. Byron, S., P. Bortz, and G. Russell, *Electron-Ion Reaction Rate Theory*, 4th Symposium on Engineering Aspects of MHD, University of California, March 1963.

15. Celinski, Z. N., "Electrical Parameters of the d.c. MHD Generator with Arbitrary Connected Electrodes," *Adv. Energy Conversion*, Vol. 6, No. 4, p. 223, 1966.

16. Celinski, Z. N., "Electrical Equivalent Circuits of d.c. MHD Generators," *Adv. Energy Conversion*, Vol. 7, No. 1, p. 67, 1967.

17. Colgate, S. A., and R. L. Aamodt, "Plasma Reactor Promises Direct Electric Power," *Nucleonics*, Vol. 15, No. 8, p. 50, 1957.

18. Compton, K. T., and I. Langmuir, "Electrical Discharges in Gases: Part I, Survey of Fundamental Processes," *Rev. Modern Phys.*, Vol. 2, No. 2, p. 123, 1930.

19. Cook, M. A., R. T. Kayes, and L. L. Udy, "Propagation Characteristics of Detonation-Generated Plasmas," *J. Appl. Phys.*, Vol. 30, No. 12, p. 1881, 1959.

20. Coombe, R. A. (ed.), *Magnetohydrodynamic Generation of Electrical Power*, Chapman and Hall, London, 1964.

21. Crookes, W., "Helium, Argon, and Krypton in the Periodic System," *Roy. Soc. Proc. (London)*, Vol. 63, p. 408, 1898.

22. De Montardy, A., *MHD Generators with Series-Connected Electrodes*, Paper No. 19, 1st International Symposium on MHD Electrical Power Generation, Newcastle/Tyne, England, Sept. 1962.

23. De Montardy, A., and I. Fells, *Experiments and Measurements on Open Cycle Systems*, 3rd International Symposium on MHD Electrical Power Generation, Salzburg, Austria, July 1966. Proceedings Vol. III, p. 543.

24. Devime, R., H. Lecroart, and P. Zettwoog, *Conversion en Veine Inhomogene; Experience de Modulation de Temperature par Effet Joule dans un Gaz de Combustion*, Paper No. 49, 2nd International Symposium on MHD Electrical Power Generation, Paris, July 1964.

25. Dzung, L. S., "The Magnetohydrodynamic Generator with Hall Effect of the Duct Ends," *The Brown Boveri Review*, Vol. 49, No. 6, p. 212, 1962.

26. Dzung, L. S., *Hall Effect and End Loop Losses of MHD Generators*, 1st International Symposium on MHD Electrical Power Generation, Newcastle/Tyne, England, Sept. 1962.

27. Elliott, D. G., "Two-Fluid Magnetohydrodynamic Cycles for Nuclear-Electric Power Conversion," *ARS J.*, Vol. 32, p. 924, 1962.

28. Elliott, D. G., D. J. Cerini, and E. Weinberg, *Investigation of Liquid Metal MHD Power Conversion*, 3rd Biennial Aerospace Power Systems Conference, Philadelphia, Sept. 1964.

29. Elliott, D. G., *d.c. Liquid-Metal Magnetohydrodynamic Power Generation*, 6th Symposium on Engineering Aspects of MHD, Pittsburgh, March 1965.

30. Emmerich, W. S., et al., *A Closed-Loop Device with Cesium-seeded Helium*, 1st International Symposium on MHD Electrical Power Generation, Newcastle/Tyne, England, Sept. 1962.

31. Faraday, M., *Experimental Researches in Electricity*, reprinted from *Phil. Trans.*, 2nd ed. (1831–1836), Taylor, London, 1849.

32. Frost, L. S., "Conductivity of Seeded Atmospheric Pressure Plasmas," *Appl. Phys.*, Vol. 32, No. 10, p. 2029, 1961.

33. Gallant, H., "Development of a Combustion Chamber for MHD Generators," *The Brown Boveri Rev.*, Vol. 51, p. 817, 1964.

34. Gaydon, A. C., and H. G. Wolfhard, "Spectroscopic Studies of Low-Pressure Flames; V. Evidence for Abnormally High Electronic Excitation," *Proc. Roy. Soc. (London)*, Vol. 205, p. 118, 1951.

35. Gunson, W. E., et al., "MHD Power Conversion," *Nucleonics*, Vol. 21, No. 7, p. 43, 1963.

36. Harris, L. P., and J. D. Cobine, "The Significance of the Hall Effect for Three MHD Generator Configurations," *Trans. ASME*, Ser. A., Vol. 83A, p. 392, 1961.

37. Harris, L. P., and C. E. Moore, *Some Electrical Measurements on MHD Channels,* 3rd Symposium on Engineering Aspects of MHD, Rochester, New York, March 1962.

38. Hepworth, M. A., and G. Arthur, *Ceramic Materials for MPD Power Generation,* 1st International Symposium on MHD Electrical Power Generation, Newcastle/Tyne, England, Sept. 1962.

39. Hoffman, M. A., and G. C. Oates, *Electrode Current Distribution in Linear MHD Channel Flows,* 6th Symposium on Engineering Aspects of MHD, Pittsburgh, March 1965.

40. Hundstad, R. L., F. A. Holmes, and T. C. Tsu, *Operating Experience of a Closed-Loop MHD Test Facility,* Paper 105, 2nd International Symposium on MHD Electrical Power Generation, Paris, July 1964.

41. Hurwitz, H., Jr., R. Kilb, and G. W. Sutton, "Influence of Tensor Conductivity on Current Distribution in a MHD Generator," *J. Appl. Phys.,* Vol. 32, p. 205, 1961.

42. Hurwitz, H., Jr., G. W. Sutton, and S. Tamor, "Electron Heating in Magnetohydrodynamic Power Generation," *ARS J.,* Vol. 32, p. 1237, 1962.

43. Jackson, W. D., and E. S. Pierson, *Operating Characteristics of the MHD Induction Generator,* Paper 26, 1st International Symposium on MHD Electrical Power Generation, Newcastle/Tyne, England, Sept. 1962.

44. Jones, M. S., Jr., et al., *Large Scale Explosively Driven MHD Generator Experiments,* 6th Symposium on Engineering Aspects of MHD, Pittsburgh, March 1965.

45. Käch, J., "Die Auflösung Von Gewöhnlichen und Partiellen Differentialgleichungen Mittels der ein- und Mehridimensionalen Laplace-Transformation fur Beschränkte Gebiete," *Arch. Elekt. Ubertr.,* Vol. 15, No. 1, p. 39, 1961.

46. Karlovitz, B., and D. Halasz, *History of the K & H Generator and Conclusions Drawn from the Experimental Results,* 3rd Symposium on Engineering Aspects of MHD, Rochester, New York, March 1962.

47. Kerrebrock, J. L., *Conduction in Gases with Elevated Electron Temperatures,* 2nd Symposium on Engineering Aspects of MHD, Philadelphia, 1961.

48. Kerrebrock, J. L., *Segmented Electrode Losses in MHD Generators with Nonequilibrium Ionization,* 6th Symposium on Engineering Aspects of MHD, Pittsburgh, March 1965.

49. Kerrebrock, J. L., "MHD Generators with Nonequilibrium Ionization," *AIAA J.,* Vol. 3, p. 591, 1965.

50. Lapp, M., and J. A. Rich, "Electrical Conductivities in Seeded Flame Plasmas in Strong Electric Fields," *Phys. Fluids,* Vol. 6, No. 6, p. 806, 1963.

51. Lin, S. C., E. L. Resler, and A. R. Kantrowitz, "Electrical Conductivity of Highly Ionized Argon Produced by Shock Waves," *J. Appl. Phys.,* Vol. 28, p. 95, 1955.

52. Lindley, B. C., *Some Economic and Design Considerations of Large-Scale MHD Generators,* 1st International Symposium on MHD Electrical Power Generation, Newcastle/Tyne, England, Sept. 1962.

53. Louis, J. F., J. Lothrop, and T. R. Brogan, "Fluid Dynamic Studies and MHD Generators," *Phys. Fluids,* Vol. 7, p. 362, 1964.

54. Louis, J. F., G. Gal, and P. R. Blackburn, *Detailed Theoretical and Experimental Study on a Large MHD Generator,* 5th Symposium on Engineering Aspects of MHD, Cambridge, Mass., 1964.

55. Massey, H. S., and E. H. Burhap, *Electronic and Ionic Impact Phenomena*, Oxford University Press, New York, 1962.

56. Maycock, J., J. A. Noe, and D. T. Swift-Hook, "Permanent Electrodes for Magneto-hydrodynamic Power Generation," *Nature*, Vol. 193, No. 6814, p. 467, 1962.

57. Maycock, J., D. T. Swift-Hook, and J. K. Wright, "Permanent Insulating Duct Walls for MHD Power Generation," *Nature*, Vol. 196, No. 4851, p. 260, 1962.

58. McCune, J. E., *Wave Growth and Instability in Partially Ionized Gases*, 2nd Symposium on MHD Electrical Power Generation, Paris, July 1964.

59. McCune, J. E., "Linear Theory of a MHD Oscillator," *Adv. Energy Conversion*, Vol. 5, No. 3, p. 221, 1965.

60. McCune, J. E., *Non-Linear Effects of Fluctuations on MHD Performance*, 6th Symposium on Engineering Aspects of MHD, Pittsburgh, March 1965.

61. McDaniel, E., *Collision Phenomena in Ionized Gases*, Wiley, New York, 1964.

62. McNab, F. R., and R. Brown, *Electrical Conductivity Experiments in an MPD Generator*, 6th Symposium on Engineering Aspects of MHD, Pittsburgh, March 1965.

63. Milde, H., and H. Gallant, *Electrodeless Conductivity Measurements within the Reaction Zone of a Combustion Chamber*, International Symposium on MHD Power Generation, Salzburg, Austria, July 1966.

64. Mullaney, G. J., and N. R. Dibelius, "Small MHD Power Generator Using Combustion Gases as an Energy Source," *ARS J.*, Vol. 31, p. 555, 1961.

65. Neuringer, J. L., *Optimum Power Generation Using a Plasma as a Working Fluid*, 3rd Biennial Gas Dynamics Symposium, Northwestern University Press, Evanston, Ill., p. 1953, 1966.

66. Olson, R. A., and E. C. Lary, "Conductivity Probe Measurements in Flames," *AIAA J.*, Vol. 1, p. 2513, 1963.

67. Pipkin, A. C., "Electrical Conductivity of Partially Ionized Gases," *Phys. Fluids*, Vol. 4, No. 1, p. 154, 1961.

68. Pitkin, E. T., "Optimum Radiator Temperature for Space Power Systems," *ARS J.*, Vol. 29, p. 596, 1959.

69. Podolsky, B., and A. Sherman, "The Influence of Tensor Conductivity on End Currents in Crossed Field MHD Channels with Skewed Electrodes," *J. Appl. Phys.*, Vol. 33, p. 1414, 1962.

70. Rosa, R. J., and A. Kantrowitz, "MHD Energy Conversion Techniques," in *Direct Conversion of Heat to Electricity*, J. Kaye and J. A. Welsh (eds.), Wiley, New York, 1960.

71. Rosa, R. J., "Nuclear Fueled MHD Power Plants," *IRE Intern. Conv. Record*, Vol. 8, pt. 9, p. 72, 1961.

72. Rosa, R. J., "The Hall End Ion Slip Effects in a Non-uniform Gas," *Phys. Fluids*, Vol. 5, No. 7, p. 1081, 1962.

73. Rosa, R. J., "Power Conversion," in *Plasma Physics in Theory and Application*, W. B. Kunkel (ed.), McGraw-Hill, New York, 1966.

74. Rosner, M., and L. S. Dzung, *Efficiency of Large-Scale Open-Loop MHD Power Generation*, 6th Symposium on Engineering Aspects of MHD, Pittsburgh, March 1965.

75. Rosner, M., *The Oil Fired MHD Power Plant*, 3rd International Symposium on MHD Electrical Power Generation, Salzburg, Austria, July 1966.

76. Rowe, A. W., and J. L. Kerrebrock, "Nonequilibrium Electric Conductivity of Two-Phase Metal Vapors," *AIAA J.*, Vol. 3, p. 361, 1965.

77. Saha, M. N., "Ionization in the Solar Chromosphere," *Phil. Mag.*, Vol. 40, No. 238, p. 472, 1920.

78. Shair, F. H., and A. Sherman, *Electron Beam Preionization in an MHD Generator*, 6th Symposium on Engineering Aspects of MHD, Pittsburgh, March 1965.

79. Sherman, A., *A High Performance Short Time Duration MHD Generator System*, ARS Space Power Conference, Sept. 1962, Preprint 2558-62.

80. Shkarofsky, I. P., "Values of the Transport Coefficients in a Plasma for any Degree of Ionization Based on a Maxwellian Distribution," *Can. J. Phys.*, Vol. 29, p. 1619, 1961.

81. Smith, J. M., "Nonequilibrium Ionization in Wet Alkali Metal Vapors," *AIAA J.*, Vol. 3, p. 648, 1965.

82. Sporn, P., and A. Kantrowitz, "MHD—Future Power Processes?," *Power*, Vol. 103, No. 11, p. 62, 1959.

83. Sternglass, E. J., et al., *MHD Power Generation by Nonthermal Ionization and Its Application to Nuclear Energy Conversion*, 3rd Symposium on Engineering Aspects of MHD, Rochester, New York, March 1962.

84. Sutton, G. W., "Hall Effect in a Lorentz Gas," *Phys. Fluids*, Vol. 4, No. 10, p. 1273, 1961.

85. Sutton, G. W., H. Hurwitz, Jr., and H. Poritsky, Jr., "Electrical and Pressure Losses in a MHD Channel due to End Current Loops," *Trans. AIEE Communications and Electronics*, Vol. 80, p. 687, 1962.

86. Sutton, G. W., and F. Robben, *Preliminary Experiments on MHD Channel Flow with Slightly Ionized Gases*, Proceedings of Symposium on Electromagnetics and Fluid Dynamics of Gaseous Plasma, Brooklyn Polytechnic Press, New York, 1962, p. 307.

87. Sutton, G. W., "End Losses in MHD Channels with Tensor Electrical Conductivity and Segmented Electrodes," *J. Appl. Phys.*, Vol. 34, p. 396, 1963.

88. Sutton, G. W., and E. Witalis, *Linearized Analysis of MHD Generator Flow Stability*, 2nd International Symposium on MHD Electrical Power Generation, Paris, July 1964.

89. Swift-Hook, D. T., "MHD Generation," in *Direct Generation of Electricity*, K. H. Spring (ed.), Academic Press, New York, 1965.

90. Thring, M. W., in *Advances in MHD*, I. A. McGrath, R. G. Siddall, and M. W. Thring (eds.), Pergamon Press, New York, 1963, p. 3.

91. Tsu, T. C., *Performance Prediction of MHD Generators over a Wide Range of Magnetic Fields*, Paper 104. 2nd International Symposium on MHD Electrical Power Generation, Paris, July 1964.

92. Tsu, T. C., "MHD Power Generators in Central Stations," *IEEE Spectrum*, Vol. 4, No. 6, p. 59, 1967.

93. Velikhov, E. P., *Hall Instability of Current-Carrying Slightly Ionized Plasmas*, 1st International Symposium on MHD Electrical Power Generation, Newcastle/Tyne, England, Sept. 1962.

94. Von Engel, A., and J. R. Cozens, "Origin of Excessive Ionization in Flames," *Nature*, Vol. 202, No. 4931, p. 480, 1964.

95. Way, S., *MHD Power Generation*, Westinghouse Research Laboratories. Scientific Paper 6-40599-2.P2, April 1960.

96. Way, S., et al., "Experiments with MHD Power Generation," *Trans. ASME J. Eng. Power*, Vol. 83A, p. 397, 1961.

97. Wilson, G. W., and D. C. Roberts, *Superconducting Magnet for MPD Power Generation*, 1st International Symposium on MHD Electrical Power Generation, Newcastle/Tyne, England, Sept. 1962.

98. Witalis, E. A., "Analysis of Linear MHD Power Generators," *Plasma Physics*, Vol. 7, p. 455, 1965.

99. Woodson, H. H., "A.C. Power Generation with Transverse Current MHD Conduction Machines," *IEEE Trans. PAS*, Vol. 86, p. 1066, 1965.

100. Wright, J. K., et al., *Some Factors Influencing the Design of Open-Cycle Fossil-Fuel MHD Generators for the Electricity Supply Industry*, 1st International Symposium on MHD Electrical Power Generation, Newcastle/Tyne, England, Sept. 1962.

101. Zauderer, B., *Shock Tube Studies of Magnetically Induced Ionization*, 6th Symposium on Engineering Aspects of MHD, Pittsburgh, March 1965.

5 THERMOELECTRIC POWER GENERATION

5.1 Introduction

Thermoelectric power generation is based on the Seebeck effect, which was noted by Seebeck [84] in the 1820's. This effect can be stated as follows: when in a circuit constituted of two different metals, there is a difference in temperature between the two metals, a voltage is developed.

The first such junctions were made up of bismuth and antimony and their efficiencies were around 1 %. Work in the field of thermoelectricity stopped during the nineteenth century and the first half of the twentieth century. In the 1950's [24, 37, 45] the subject was revived as a result of the surge in understanding of semiconductor theory and materials.

Many advantages already appear in the use of thermoelectric generators. First, since there are no moving parts, the hot source of the thermodynamic cycle can be made much hotter than the 600°C ceiling of conventional generators. The mechanical stresses are eliminated and the efficiency can be made much higher than the 41 % maximum of steam turbine generating plants. The problem is reduced to finding materials with acceptable thermoelectric properties at such high temperatures.

The absence of moving parts means absence of noise. In some applications this is of the utmost importance. It also means greater compactness.

Thermoelectric devices can be used as "toppers" or "tailers" in conjunction with other methods of power generation.

The disadvantage of thermoelectric generators is that they produce direct currents. For some applications, an ac to dc converter stage becomes necessary.

5.2 Thermoelectric Effects

Seebeck effect. When a difference in temperature between the two ends of a metal or a semiconductor occurs, electrons at the hotter end have a higher kinetic energy and therefore higher speed than those at the colder end. This will lead both

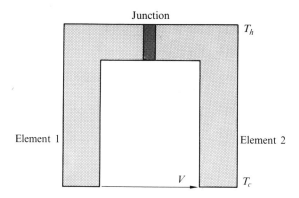

Figure 5.1. An elementary thermoelement. The two elements are usually made up of two different types of semiconductors.

to a transfer of energy (thermal conductivity) and to a diffusion of electrons toward the colder end. Due to this motion of electrons, an emf will be established between the two ends, as shown in Fig. 5.1. This emf will increase until the net flow of electrons ceases completely. This voltage is called *Seebeck voltage*. If V is the open circuit voltage thus developed, and if T_h and T_c are the temperatures of the hot source and the cold source, respectively, then for small temperature differences,

$$V = \Xi(T_h - T_c), \tag{5.1}$$

where Ξ is the Seebeck coefficient for the junctions between the two metals; Ξ is a function of the material and of temperature (see Section 5.8). For a given material, Ξ has usually a maximum value, e.g., $\Xi = 230 \ \mu V/°K$ at 380°K for a p-type 30% Bi_2Te_3–70% Sb_2Te_3 semiconductor with an excess of 2% tellurium [43]. One of the highest values measured is $\Xi = 400 \ \mu V/°K$ for p-type lead telluride with 0.1 atomic percent sodium [32].

Peltier effect. In the 1830's, Peltier [94] observed an effect complementary to the Seebeck effect. The Peltier effect can be described as follows: heat is absorbed or generated whenever a current passes through a junction between different materials, depending on the direction of the current. If $\mathscr{Q}$ is the reversible heat generated and I is the current passing through the junction, then

$$\mathscr{Q} = \Pi I, \tag{5.2}$$

where Π is the Peltier coefficient.

Thomson effect. In the 1860's, William Thomson (Lord Kelvin) [93, 94] proved the relationship between the Seebeck and Peltier effects. He also showed that when a current passes through a single homogeneous conductor along which a temperature gradient exists, heating or cooling will occur, depending on the direction of the current. This is known as the Thomson effect. If P_h is the reversible heat rate generated by the Thomson effect, then

$$P_h = \tau I(T_h - T_c),\qquad(5.3)$$

where τ is the Thomson coefficient. Thomson showed also that

$$\Xi T = \Pi,$$

and

$$T\frac{d\Xi}{dT} = \tau_1 - \tau_2,$$

where T is the absolute temperature, and the subscripts 1 and 2 refer to the two different materials forming the junction.

5.3 Classical Thermodynamic
Analysis of the Thermoelectric Effects [46]

Each of the three effects described above is reversible. In a more general notation, Eqs. (5.1), (5.2), and (5.3) can be written as

$$\Xi_{12} = \lim_{\Delta T \to 0} \frac{\Delta V_{12}}{\Delta T}.$$

Thus

$$\Xi_{12} = \frac{dV_{12}}{dT},\qquad(5.4)$$

where Ξ_{12}, the Peltier coefficient of junction (1, 2), is a function of the temperature T,

$$\Pi_{12} = \mathscr{Q}_{12}/I,\qquad(5.5)$$

where Π_{12}, the Peltier coefficient of the junction, is a function of the temperature, and

$$\tau_1 = \lim_{\Delta T \to 0} \frac{\Delta(P_{h1}/I)}{\Delta T},\qquad(5.6)$$

where τ_1 is the Thomson coefficient of material 1.

In addition to these three thermoelectric effects, two other thermodynamic processes occur in the conductor. These are Joulean heating and heat conduction, both being irreversible:

$$\mathscr{Q} = RI^2$$

and

$$q = -\kappa\,\nabla T,\qquad(5.7)$$

where q is the heat flux vector and κ is the thermal conductivity.

The tensor properties of the thermal conductivity led to the law of the symmetry of the coefficients of coupled irreversible phenomena. This led to the development of the science of *irreversible thermodynamics* [18, 22].

It is possible to use the first and second laws of thermodynamics by assuming that the heat interactions which occur with the reversible thermoelectric effects can be separated from those occurring with the irreversible Joulean heating and heat conduction processes.

This assumption is valid only because the ratio of reversible to irreversible effects is proportional to the current and can be neglected for very small currents.

The first law of thermodynamics states that $dU = \delta\mathcal{Q} + \delta W$. Since temperatures and voltages are to be constants in time, $dU = 0$, and

$$\delta\mathcal{Q} = -\delta W. \qquad (5.8)$$

The work δW done by the system is $\delta W = IV_{12}\,dt$ and the heat produced is $\delta\mathcal{Q} =$ Peltier heat + Thomson heat. Thus,

$$\delta\mathcal{Q} = \left\{\Pi_{12c} - \Pi_{12h} + \int_{T_c}^{T_h}(\tau_1 - \tau_2)\,dt\right\}I\,dt,$$

where c stands for cold and h for hot. From Eq. (5.8),

$$\Pi_{12c} - \Pi_{12h} + \int_{T_c}^{T_h}(\tau_1 - \tau_2)\,dT = -V_{12}.$$

After taking the derivative with respect to T and taking the limit as T_c approaches T_h,

$$\frac{dV_{12}}{dT} = \frac{d\Pi_{12}}{dT} - (\tau_1 - \tau_2),$$

and, introducing the Seebeck coefficient of Eq. (5.4),

$$\Xi_{12} = \frac{d\Pi_{12}}{dT} - (\tau_1 - \tau_2). \qquad (5.9)$$

The second law of thermodynamics gives

$$d\mathcal{S} = \delta\mathcal{Q}/T, \qquad (5.10)$$

or

$$d\mathcal{S} = \left\{\frac{\Pi_{12c}}{T_c} - \frac{\Pi_{12h}}{T_h} + \int_{T_c}^{T_h}(\tau_1 - \tau_2)\frac{dT}{T}\right\}I\,dt, \qquad (5.11)$$

where $\mathcal{S}$ is the entropy. Since all effects are assumed reversible, $d\mathcal{S} = 0$. Thus, after differentiation by dT and taking the limit as T_c approaches T_h,

$$\frac{d}{dT}\left(\frac{\Pi_{12}}{T}\right) = \frac{(\tau_1 - \tau_2)}{T},$$

or

$$\frac{d\Pi_{12}}{dT} = (\tau_1 - \tau_2) + \frac{\Pi_{12}}{T}. \qquad (5.12)$$

Comparison of Eqs. (5.9) and (5.10) yields

$$\Xi_{12} = \Pi_{12}/T \tag{5.13}$$

and

$$T\frac{d\Xi_{12}}{dT} = (\tau_1 - \tau_2). \tag{5.14}$$

These are the results which were proved experimentally in the 1850's by Lord Kelvin.

5.4 Thermoelectric Coefficients in Semiconductor Theory

The thermoelectric effects of Seebeck, Peltier, and Thomson are related to the atomic structures of given materials. This is also true for the thermal and electrical conductivities. For an explanation of the physical nature of these effects in a semiconductor, return to Chapter 2. The electrical conductivity of a semiconductor is given by $\sigma = e(n_e \ell_e + n_h \ell_h)$, where e is the electronic charge, n_e and n_h are the number densities of the electrons and holes, respectively, and ℓ_e and ℓ_h are their mobilities. To obtain a more accurate expression for the electrical conductivity, it is necessary to take into consideration the velocity distribution f_e of the electrons. This can be obtained from Boltzmann's equation:

$$\frac{\mathbf{F}}{M} \cdot \nabla_u f_e + v \cdot \nabla f_e = -\left.\frac{\partial f_e}{\partial t}\right|_{\text{ext forces}} = \left.\frac{\partial f_e}{\partial t}\right|_{\text{coll}}, \tag{5.15}$$

where $\mathbf{F}$ is the force acting on the particle, $\nabla_u f_e$ is the gradient of the distribution in the hodographic space,* v is the velocity of the electrons, and $\partial f_e/\partial t|_{\text{coll}}$ is the collisional term which can be represented by a relaxation time ℓ according to

$$\left.\frac{\partial f_e}{\partial t}\right|_{\text{coll}} = \frac{f_o - f_e}{\ell}, \tag{5.16}$$

where f_o is the local isotropic distribution function. When only an electric field is present in the x-direction and the system is near equilibrium, ∂f_e may be replaced by ∂f_o, obtaining, from Eqs. (5.15) and (5.16), for the distribution function,

$$f_e = f_o + \frac{\ell e E}{M} \cdot \frac{\partial f_o}{\partial u} - \ell u \frac{\partial f_o}{\partial x}, \tag{5.17}$$

where u, v, and w are the components of the velocity in the x-, y-, and z-directions, respectively. When f_o is taken as the Fermi-Dirac distribution of the electronic energies, Eq. (5.17) becomes, after some rearrangement,

$$f = f_o + \frac{\ell}{Mu}\left[eE + T\frac{d}{dT}\left(\frac{\mathscr{E}_F}{T}\right)\frac{\partial T}{\partial x} + \frac{\mathscr{E}}{T} \cdot \frac{\partial T}{\partial x}\right],$$

---

* Space in which the three coordinates are the velocities of the particles in the x-, y-, and z-directions, respectively.

since

$$\mathscr{E} = \tfrac{1}{2}Mv,$$

$$\frac{\partial f_o}{\partial T} = \frac{\partial f_o}{\partial T} \cdot \frac{\partial T}{\partial x},$$

$$\frac{\partial f_o}{\partial T} = T\frac{d}{dT}\left(\frac{\mathscr{E} - \mathscr{E}_F}{T}\right)\frac{\partial f_o}{\partial \mathscr{E}}$$

where $\mathscr{E}_F$ is the Fermi energy, and

$$f_o = \frac{1}{1 + \exp\,(\mathscr{E} - \mathscr{E}_F)/kT}.$$

The electrical conductivity for the electrons is given by

$$\sigma = \frac{ne}{E}\int\!\!\!\int\!\!\!\int_{-\infty}^{+\infty} f_e u\;du\;dv\;dw.$$

After using the necessary averages,

$$\sigma = \frac{ne}{M}\left[e + \frac{T}{E}\cdot\frac{\partial}{\partial x}\left(\frac{\mathscr{E}_F}{T}\right)\right]\langle t\rangle + \frac{n}{MT}\cdot\frac{e}{E}\langle t\mathscr{E}\rangle\frac{dT}{dx}.$$

For an open circuit ($\sigma = 0$), the current density J will be equal to zero; therefore the electric field due to the temperature gradient will be equal to

$$E = -\frac{1}{e}\left[\frac{1}{T}\cdot\frac{dT}{dx}\frac{\langle t\mathscr{E}\rangle}{\langle t\rangle} + T\frac{\partial}{\partial x}\left(\frac{\mathscr{E}_F}{T}\right)\right],$$

or

$$E = \frac{T}{e}\cdot\frac{d}{dT}\left(\frac{\langle t\mathscr{E}\rangle - \mathscr{E}_F\langle t\rangle}{\langle t\rangle T}\right)\frac{dT}{dx}.$$

The Seebeck effect was defined by Eq. (5.4) and, since $dV_{12} = E\,dx$, then

$$\Xi_{12} = E\frac{dx}{dT} = \frac{T}{e}\cdot\frac{d}{dT}\left(\frac{\langle t\mathscr{E}\rangle - \mathscr{E}_F\langle t\rangle}{\langle t\rangle T}\right).$$

Therefore,

$$\Xi_{12} = \frac{\mathscr{E}_F\langle t\rangle - \langle t\mathscr{E}\rangle}{e\langle t\rangle T}. \tag{5.18}$$

From Eq. (5.14), for the Thomson coefficient,

$$(\tau_1 - \tau_2) = T\frac{d\Xi_{12}}{dT}.$$

Thus

$$(\tau_1 - \tau_2) = T\frac{d}{dT}\left(\frac{\mathscr{E}_F\langle t\rangle - \langle t\mathscr{E}\rangle}{e\langle t\rangle T}\right). \tag{5.19}$$

Peltier's coefficient is given by Eq. (5.13). Thus

$$\Pi_{12} = \frac{\mathscr{E}_F}{e} - \frac{\langle t\mathscr{E} \rangle}{e\langle t \rangle}.$$

(5.20)

The values of $\mathscr{E}_F$, $\langle t \rangle$, and $\langle t\mathscr{E} \rangle$ are characteristics of the different types of semiconductors.

5.5 General Equations

A more general description of the thermoelectric effects can be given by the *irreversible thermodynamics theory*. This has the important limitation of being able to cover small departures from equilibrium. But, since most engineering applications use materials in local thermochemical quasi-equilibrium, the irreversible thermodynamic theory is extremely satisfactory. The thermodynamic properties describing these irreversible effects are the thermal conductivity κ, the viscosity v, the electrical conductivity σ, and the diffusion coefficient $\mathscr{D}$, as used in the following linear laws:

Fourier's law of thermal conduction given by Eq. (5.7); Newton's law of viscosity,

$$\psi = \tfrac{2}{3}v \, \nabla^2 v,$$

(5.21)

where ψ is the pressure gradient and v is the velocity of the particle; Ohm's law of electrical conduction,

$$\mathbf{J} = -\sigma \, \nabla V,$$

(5.22)

where $\mathbf{J}$ is the electrical current density and V is the potential; and Fick's law of diffusion,

$$\mathbf{\Gamma} = -\mathscr{D} \, \nabla n,$$

(5.23)

where $\mathbf{\Gamma}$ is the random current density and n is the number density of the particles.

The theory of irreversible thermodynamics will relate all the effects dealing with these coefficients—thermoelectricity and thermomagnetism, among others.

To these equations are added the equation of continuity,

$$\partial \rho_e / \partial t + \nabla \cdot \mathbf{J} = 0,$$

(5.24)

the first and second laws of thermodynamics,

$$dU = \partial \mathcal{Q} + \delta W,$$
$$d\mathscr{S} = \delta \mathcal{Q}(1/T_c - 1/T_h),$$

(5.25)

the equation of motion,

$$\rho \, Dv/Dt = -\nabla p_r + \psi,$$

(5.26)

and, finally, the equation of conservation of energy, which in its simplified form is given by

$$\mathbf{E} \cdot \mathbf{J} + \nabla \cdot (\kappa \nabla T) = 0.$$

Since the Thomson coefficient is a function of temperature,

$$\mathbf{E} \cdot \mathbf{J} = J^2/\sigma - \tau \mathbf{J} \cdot \nabla T,$$

and the equation of conservation of energy becomes

$$\nabla \cdot (\kappa \nabla T) - \tau \mathbf{J} \cdot \nabla T + J^2/\sigma = 0. \qquad (5.27)$$

5.6 Thermoelectric Devices [12, 13]

The electric power output of a thermoelectric cell can be obtained from Poynting's law,

$$P_{\text{out}} = -\mathbf{J} \cdot \mathbf{E}. \qquad (5.28)$$

The heat input is given by

$$\mathscr{Q} = I\Xi T_h + K \Delta T - \tfrac{1}{2}RI^2, \qquad (5.29)$$

where K is the thermal conductance and R is the resistance of the thermocouple. The power output can also be stated as

$$P_{\text{out}} = R_L I^2, \qquad (5.30)$$

where R_L is the resistance of the load. Therefore, the quantity of heat which is not converted is given by $\mathscr{Q}_L = -P_{\text{out}} + \mathscr{Q}$. Thus

$$\mathscr{Q}_L = -(R_L + \tfrac{1}{2}R)I^2 + K \Delta T + I\Xi T_h. \qquad (5.31)$$

There are two different important uses of the thermoelectric cell. When $\mathscr{Q}_L$ is positive, heat is converted into electricity, and the thermoelectric cell works as a thermoelectric power generator. When $\mathscr{Q}_L$ is negative, heat is transported from a low-temperature reservoir to one of higher temperature. This is possible only by the absorption of electric power. In this case, the cell works as a thermoelectric refrigerator.

For thermoelectric refrigeration, Eq. (5.29) becomes

$$\mathscr{Q}_r = -\tfrac{1}{2}RI^2 - K \Delta T + I\Xi T_c, \qquad (5.32)$$

where the first term on the right-hand side represents the Joulean heating caused by the current flowing through the cell, the second term is the rate of heat conduction to the cold junction because of the temperature gradient, and the third term is the Peltier cooling at the cold junction.

The power input to the refrigerator is

$$P_{\text{in}} = RI^2 + I\Xi T,$$

where the first term represents the heat produced by the Joulean effect of the current in both legs of the junction and the second term is the Seebeck input power.

Many thermoelectric refrigerators have been built for commercial use in the U.S. [11, 37, 39] and in the Soviet Union [56]. One of these devices is a thermoelectric detector cooler [36, 72]* used in many spot cooling applications in electronic equipment. A baby bottle cooler-warmer [60]* which can work either as a refrigerator or as a heater has also been produced. The unit works at 30°C ambient temperature and can either cool the bottle to a temperature of 4.5°C in 1 hour or heat it to a temperature of 38°C. The same unit can act as a heat regulator. A thermoelectric hostess cart [60] is an interesting commercial unit which works simultaneously as both heater and refrigerator. The unit has a 0.0566-m³ cooling chamber located at the bottom of the cart, maintained at 1°C in an ambient temperature of 24°C. On the top of the cart is a 0.0368-m³ heating chamber (oven) which uses the heat rejected by the refrigerator. All these commercial units can be operated by residential line electric power or by individual batteries. Other similar devices have been built for laboratory [36, 65] and medical [47] applications.

Thermoelectric refrigeration has one of its most important uses in military applications. In this case, the thermoelectric module should be designed to withstand shocks and other stresses. The U.S. Navy† has been developing a thermoelectric air-conditioner for submarines [4]. This unit is designed to produce cold water at about 10°C with an inlet temperature of 13°C. The sea, at about 20°C, is used as the heat sink. The design form was a modified shell and a tube heat exchanger; the thermoelectric couples were built inside the tube in which the chilled water flows, while the sea water flows on the shell side. Each unit contains 30 identical subunits and has a capacity of 10 tons of refrigeration. Some difficulties arose, however, from electrical and thermal insulation problems.

Another example is a Peltier temperature control device for use with micro miniature electronics.‡ A similar thermoelectric controlled temperature chamber has been developed by Kirby [57] and his co-workers. The chamber of this device has a volume of 0.014 m³ and was designed to be maintained at any preset temperature between 0 and 100°C.

Thermoelectric refrigerating units for frozen food storage rooms aboard submarines have been developed [75]. One of these units has a capacity of 8500 Btu/hr at rated conditions of −18°C for air and +13°C for water.

A host of small thermoelectric cooling and heating units, such as a number of ice makers, water coolers, and small refrigerators, have been marketed in several countries in the last few years. Because of their present low efficiencies, these thermoelectric devices are competitive with the corresponding conventional devices only under certain conditions. The Navy is interested in silent operation,

---

*   Westinghouse.

†   Bureau of Ships.

‡   Built by the Diamond Ordnance Fuze Labs., Washington, D.C.

and the consumer market is limited to small units. It is hoped that with progress in thermoelectric technology, new markets will be open, especially in the household refrigeration field.

5.7 Different Types of Thermoelectric Generators

There are three ways of classifying the thermoelectric generators based on (1) the number of thermoelement stages used and the modes of staging, (2) the source of heat utilized, and (3) the types of materials used for the thermoelements. The most important types can thus be listed as follows:

1. Staging classification
   a) Single-stage generators
   b) Multistage generators
      · Segmented generators
      · Cascaded generators

2. Fuel classification
   a) Fossil fueled generators
   b) Solar powered generators
   c) Radioisotope powered generators
   d) Nuclear powered generators

3. Material classification
   a) Metallic thermoelements
   b) Semiconductor elements
   c) High-temperature thermoelements
   d) Low-temperature thermoelements

Single-stage series-coupled generator. The simplest thermoelectric generator is shown in Fig. 5.2. This thermoelectric cell consists of two different materials, n-type and p-type semiconductor elements. These join at the heat source through a third material to form the thermoelectric junction and at the heat sink through the load. The circuit is thus closed and generation of electrical energy is possible.

If it is assumed that there is no heat transfer between the source and the sink, and that the electrical conductivity σ, the thermal conductivity κ, and the Seebeck coefficient Ξ of the material are independent of temperature, then Eq. (5.27) becomes (since $\tau = 0$, and for $\Xi = $ constant), for a linear configuration,

$$\kappa \, d^2 T/dx^2 + J^2/\sigma = 0. \tag{5.33}$$

It is possible to solve Eq. (5.32) by using the following boundary conditions:

$$T = T_h \quad \text{at} \quad x = 0,$$
$$T = T_c \quad \text{at} \quad x = \ell,$$

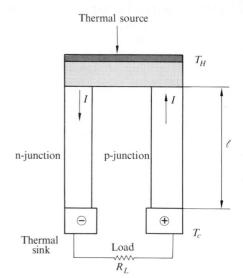

Figure 5.2. An elementary thermoelectric generator. Current flows in the direction of the heat flow in the n-junction and opposite to it in the p-region.

where ℓ is the total length of the thermoelement. Therefore,

$$\frac{T(x)}{T_h} = 1 - \frac{x}{\ell}\frac{\Delta T}{T_h} + \frac{1}{2}\frac{J^2}{\sigma\kappa T_h}\left(\frac{x}{\ell}\right)\left(1 - \frac{x}{\ell}\right),\tag{5.34}$$

where $\Delta T = T_h - T_c$. This equation can be represented by part of a parabola, as shown in Fig. 5.3.

As shown by Eq. (5.34), the heats involved are of two types: heat due to the Peltier effect and heat due to thermal conduction.

The assumption of constant σ, κ, and Ξ is not always acceptable. If these characteristics vary with temperature, Eq. (5.27) becomes, for a linear configuration,

$$\frac{d}{dx}\left(\kappa\frac{dT}{dx}\right) - \tau J\frac{dT}{dx} + \frac{J^2}{\sigma} = 0.\tag{5.35}$$

This is a nonlinear differential equation, since κ, τ, and σ are functions of temperature.

Several elementary thermoelectric generators can be arranged to be in parallel thermally and in series electrically. This type of single-stage thermoelectric device is shown in Fig. 5.4. The total output voltage of the N thermoelements is here equal to N times the output voltage of one element, whereas the current is the same as for a single thermoelement. Therefore, the total power output and the total heat input will be equal to N times the output and input of a single thermoelement, respectively, the efficiency remaining unchanged.

Multistage generators. The properties of the thermoelectric materials vary considerably with temperature, and, since there are large temperature differences between the heat source and the sink, it becomes necessary to use different

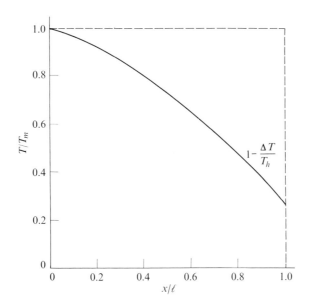

Figure 5.3. Temperature distribution in a thermoelement.

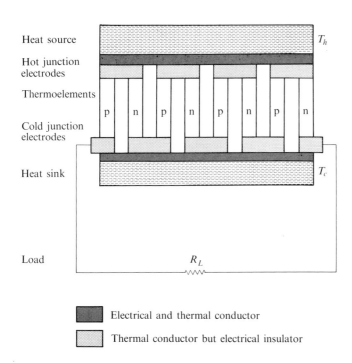

Figure 5.4. A single-stage series generator.

materials for different temperatures. This can be made possible in a multistage generator. There are two important methods of staging: segmenting or cascading (or both).

An example of a segmented generator is shown in Fig. 5.5. In this generator, the legs of the thermoelement are placed electrically and thermally in series. By choosing the proper material for different stages, the efficiency of this element is greatly improved. Several elements of this type can be placed in series (Fig. 5.4) or in parallel.

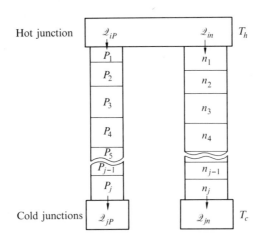

Figure 5.5. Segmented thermoelectric generator. Each stage material of the generator is chosen for its best thermoelectric properties at the local stage temperature.

An example of a cascaded generator is shown in Fig. 5.6. Cascaded generators are thermally in series, i.e., the heat rejected by one stage is the heat input to the following stage. This is shown in Fig. 5.7, where

$$Q_N - Q_{N-1} = P_N. \tag{5.36}$$

The thermal efficiency of the Nth stage is

$$\eta_{T(N)} = P_N/Q_N. \tag{5.37}$$

Combining Eqs. (5.36) and (5.37) yields

$$Q_N(1 - \eta_{T(N)}) = Q_{N-1}, \tag{5.38}$$

and similarly

$$Q_{N-1}(1 - \eta_{T(N-1)}) = Q_{N-2}. \tag{5.39}$$

.

.

.

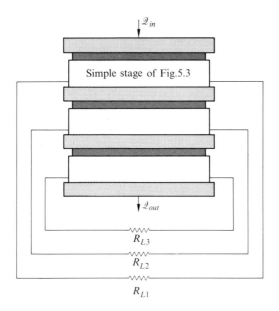

Figure 5.6. Cascaded thermoelectric generator.

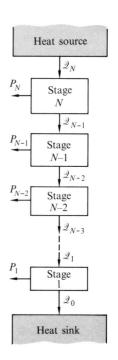

Figure 5.7. Flow of energies in a cascaded thermoelectric generator.

The overall thermal efficiency of the generator is

$$\eta_T = \frac{\sum_{j=1}^{N} P_j}{\mathcal{Q}_N},$$ (5.40)

and since

$$\mathcal{Q}_N - \mathcal{Q}_0 = \sum_{j=1}^{N} P_j,$$

then

$$\mathcal{Q}_0 = \mathcal{Q}_N(1 - \eta_T).$$ (5.41)

From Eqs. (5.38) and (5.39), in general,

$$\mathcal{Q}_0 = \mathcal{Q}_N(1 - \eta_{T(N)})(1 - \eta_{T(N-1)}) \cdots (1 - \eta_{T(1)}).$$ (5.42)

Combining Eqs. (5.41) and (5.42) and rearranging yields

$$\eta_T = 1 - \prod_{j=1}^{N}(1 - \eta_{T(j)}).$$ (5.43)

Fuel classification. In fossil fueled generators, the thermoelement assemblies are usually grouped around an aluminum core combustion chamber acting as the heat source. The heat is obtained by burning a fuel such as propane in an oxidizer such as air. Heat transfer from the burner system to the combustion chamber is obtained mainly by thermal radiation and then by conduction through the thermoelements to the outer heat exchanger fins, from which heat is rejected by free convection. The spacing and the size of the fins are fixed to permit maximum heat dissipation and the length of the thermoelements is limited by heat transfer considerations. Figure 5.8 illustrates this type of generator.

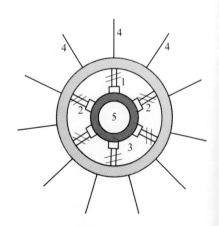

1. Springs
2. Thermoelectric element
3. Core
4. Heat exchanger fins
5. Combustion chamber

Figure 5.8. Fossil fueled thermoelectric generator.

Solar powered generators are considered for use in satellites with power levels around 100 mW. At this level, solar batteries based on the photovoltaic effect become more competitive and are most often used. In solar thermoelectric systems, an oriented mirror is utilized to concentrate the solar radiation on a thermal storage material such as lithium hydride. The thermal energy stored in this material is then transferred to a radiator through the thermoelements.

Isotopes are produced by reactor irradiation of some heavy elements and can be either α- or β-emitters. There are several hundred radioisotopes, but only a few are useful for thermoelectric power generation. A useful radioisotope should be processed to a utilizable form with relatively low cost and its radiations should be easily attenuable* to avoid harmful irradiation. Its life should be relatively long and its specific power (in kilowatts per kilogram) should be relatively high. The design of the generator itself is not very different from that of a fossil fueled device, except that the combustion chamber is replaced by a nuclear source and special attention is given to protection of operating personnel from harmful radiation by shielding the overall structure of the thermoelectric generator.

It would be most interesting if heat could be converted directly into electricity in a nuclear reactor. Theoretically, this can be done by making the reactor work in an "in-pile" by using the fuel elements as the heat source. The hot junctions of the thermoelements should, then, be in direct thermal contact with the fuel, leading to higher efficiency and minimum size for the nuclear plant. New problems are avoidably created by this method. For instance, it is not yet clear how important is the radiation influence on the thermoelectric coefficients and on the properties of the semiconductor in general. This subject will be treated in more detail in the sections on thermoelectric systems. Examples will be given and difficulties will be outlined.

5.8 Thermoelectric Materials [1, 2, 3, 9, 18, 26]

Many materials are in competition for use in thermoelectric applications. Not only should a useful material have acceptable thermoelectric characteristics in the temperature range of interest, but other considerations, such as thermal and electrical conductivities and mechanical characteristics, influence their choice by an engineer.

Figure of merit. Equation (5.29) gives the amount of heat supplied to the hot junction, whereas the power output is given by Eq. (5.30). The voltage output is given as a function of the Seebeck effect by

$$V = \Xi \, \Delta T = (R + R_L)I. \qquad (5.44)$$

---

* Attenuation means here the decrease in the amount of emitted particles with increasing distance.

Taking into consideration R and introducing Eq. (5.44) into Eq. (5.30) yields

$$P_{\text{out}} = \frac{R_L(\Xi\,\Delta T)^2}{(R + R_L)^2},$$ (5.45)

and, since the overall efficiency is $\eta = P_{\text{out}}/\mathcal{Q}$, then

$$\eta = \frac{\Delta T}{\frac{3}{2}T_h + \frac{1}{2}T_c + 2(R + R_L)K/\Xi^2}.$$ (5.46)

If the resistance of the load is

$$R_L = m_L R,$$

the maximum value of the efficiency will occur when $d\eta/dm_L = 0$. This will give $m_L = (1 + Z\bar{T})^{1/2}$, where $\bar{T} = \frac{1}{2}(T_h + T_c)$ and

$$Z = \Xi^2/KR,$$ (5.47)

and, after some mathematical rearrangements,

$$\eta_{\text{max}} = \gamma_m \frac{\Delta T}{T_h},$$ (5.48)

where

$$\gamma_m = \frac{(1 + Z\bar{T})^{1/2} - 1}{(1 + Z\bar{T})^{1/2} + T_c/T_h}.$$

Note that the characteristics of the thermoelement appear only in the quantity Z. Therefore, Z is called the *figure of merit of the junction*. For high efficiency, it is clear that $Z\bar{T}$ should be as large as possible.

Equation (5.47) is valid for a junction made up of two different materials. Therefore,

$$\Xi_{12} = (\Xi_1 - \Xi_2),$$

$$R = \frac{\ell_1}{S_1\sigma_1} + \frac{\ell_2}{S_2\sigma_2},$$

$$K = \frac{S_1\kappa_1}{\ell_1} + \frac{S_2\kappa_2}{\ell_2}.$$

The best performance can be obtained when RK is minimum, i.e., for

$$\frac{\ell_1 S_2}{\ell_2 S_1} = \left(\frac{\sigma_1\kappa_1}{\sigma_2\kappa_2}\right)^{1/2}.$$

The figure of merit Z_{12} becomes

$$Z_{12} = \left[\frac{\Xi_1 - \Xi_2}{(\kappa_1/\sigma_1)^{1/2} + (\kappa_2/\sigma_2)^{1/2}}\right]^2.$$

The Seebeck effect can be positive or negative, and the maximum value is obtained for $(\Xi_1 - \Xi_2) = 2\Xi_1$. Therefore $Z = \Xi^2\sigma/\kappa$ can be taken as the *figure of merit of the thermoelectric material*, or, introducing the temperature,

$$ZT = \Xi^2 \sigma T / \kappa \qquad (5.49)$$

should be, in general, as high as possible.

Metallic materials. When mechanical considerations such as ductility become important to obtain the highest reliability possible, or when it is necessary to obtain high thermal conductivities, the use of metallic thermoelements becomes almost imperative. The Seebeck coefficient of metallic elements is, however, low.

Table 5.1

Metallic thermoelectric materials

Metal	$\mu V/°K$	ZT^*
Bismuth	-75	0.208
Nickel	-18	0.112×10^{-1}
Copper	$+2.5$	0.231×10^{-3}

* Calculated by the author.

In a metal, the relation between the electrical and thermal conductivities is given by the Wiedemann-Franz law:

$$\kappa / \sigma = L k^2 T / e^2.$$

Therefore

$$ZT = \Xi^2 e^2 / L k^2,$$

where $L k^2 / e^2$ is the Lorentz number. For a metal, $L = 3.3$; therefore ZT is independent of temperature and is directly proportional to the square of the Seebeck coefficient. Since $e^2 / L k^2 = 0.37 \times 10^8$, then $ZT = 0.37 \times 10^8 \Xi^2$.

The most used metals, with their characteristics, are listed in Table 5.1.

Semiconductors. The theory of the thermoelectric properties of semiconductors is still in its infancy and most of the results are obtained from empirical experimentations. A certain level of theoretical knowledge should be reached so as to control the properties of semiconductors. Thus, new materials could be fabricated with the best factor ZT possible.

Extrinsic* semiconductors are the most promising materials in thermoelectric power generation. Today the most used material in this category is lead telluride [31, 32]† (PbTe). The values of ZT for various levels of doping of this material

---

*   The definitions of "intrinsic" and "extrinsic" are identical to those given in Chapter 2. As mentioned above, both lead and tellurium are soluble in the lead telluride compound. This solubility is a function of the temperature. It disappears completely at temperatures higher than 900°K, thus transforming the extrinsic semiconductor into an intrinsic one. The same explanation can be given for the other compounds.

†   Manufactured by the Minnesota Mining and Manufacturing Co.

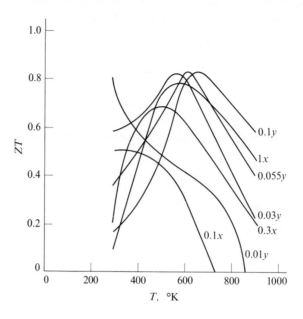

Figure 5.9. Figures of merit of several lead telluride materials. p-type semiconductor with percentage of sodium as parameter (%x). n-type semiconductor with percentage of lead diiodide as parameter (%y).

are shown in Fig. 5.9 for n- and p-types PbTe. This is an intermetallic compound which melts at 922°C and contains around 61.9% Pb and 38.1% Te in its pure form. Both lead (Pb) and tellurium (Te) are soluble to a very small extent in PbTe. An excess of Te leads to a p-type semiconductor, while an excess of Pb leads to an n-type material. Dopants are used to improve the characteristics of the material at a desired temperature. These are mostly lead diiodide (PbI_2), bismuth tritelluride (tetradymite Bi_2Te_3), tantalum tritelluride ($TaTe_3$), etc., for the n-type material and sodium (Na) among others for the p-type.

As shown in Fig. 5.9, PbTe is a low temperature material at least when it is extrinsic (i.e., at temperatures lower than 900°K where ZT has the highest values).

Other materials, such as binary compounds made from bismuth (Bi), antimony (Sb), tellurium (Te), and selenium (Se) have large figures of merit at low temperatures. Most of them are used either in refrigeration applications or in the cold parts of the thermoelectric legs. They may be n- or p-type semiconductors. Figure 5.10 illustrates the properties of such materials. These materials are extrinsic in the lower temperature range.

At temperatures higher than 900°K, lead telluride and the other compounds become intrinsic.* At first sight, an intrinsic semiconductor seems unsuitable

---

* Manufactured by the Minnesota Mining and Manufacturing Co.

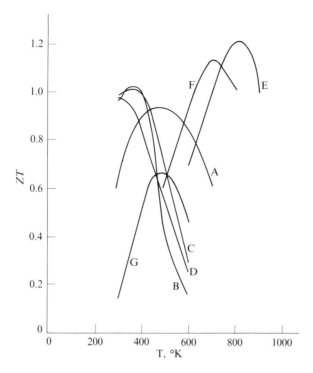

Figure 5.10. Figures of merit of some low and medium temperature thermoelectric materials. Curve A: n-type, 0.75 Bi_2Te_3, 0.25 Bi_2Se_3; curve B: p-type, 0.30 Bi_2Te_3, 0.70 Sb_2Te_3(Te); curve C: p-type, 0.25 Bi_2Te_3, 0.75 Sb_2Te_3(Te); curve D: p-type, 0.25 Bi_2Te_3, 0.75 Sb_2Te_3(Se); curve E: p-type, 0.95 GeTe, 0.05 Bi_2Te_3; curve F: p-type, 0.90 GeTe, 0.10 AgSbTe; curve G: ZnSb.

for thermoelectric applications, due to the low value of Ξ and the large value of the thermal conductivity κ. However, if the electron mobility is much larger than the mobility of the holes, the characteristics of the material become attractive.

Silicon-germanium (Si-Ge) alloys are the best materials in the high temperature range. They also have good mechanical properties: high mechanical strength, low coefficient of expansion, and light weight. The III–V compounds* are also good high temperature thermoelectric materials. These are indium antimonide (InSb), indium arsenide (InAs), and gallium arsenide (GaAs) among others. Figure 5.11 illustrates the properties of some high temperature materials. Note that at a very high temperature, thermoelectric conversion becomes more efficient, at least at the present state of the art.

The product ZT rarely reaches unity. This leads to a stage efficiency of about 17% of the Carnot cycle efficiency, which was found to be the maximum value attainable. The theoretical explanation of this limit is not yet clear.

---

* Compounds made up of materials from Group III and Group V of the periodic table.

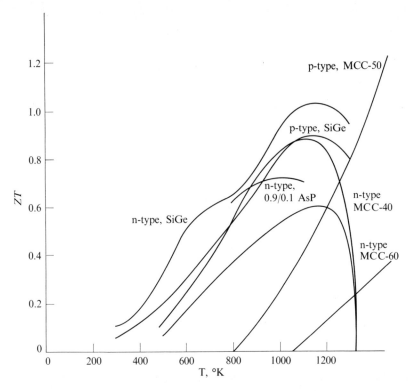

Figure 5.11. Figures of merit of some high-temperature thermoelectric materials. (MCC-40, MCC-50, and MCC-60 are trademarks for materials produced by Monsanto Research Co.)

Silicon-germanium thermoelements [22, 23] were used in SNAP-10A and proved very efficient [97]. On the other hand, Monsanto Research Co. produced materials [52] having extremely good properties at high temperatures, as shown in Fig. 5.11. The latter are called MCC-40, 50, and 60.

All the materials described above are broad-band materials. A material is called narrow band when there is very little overlap of the electron orbits between the nearest similar ions. The transport mechanism in this case is quite different from that in broad-band materials. An example of a narrow-band material is nickel oxide (NiO).

5.9 Power Output, Losses, and Efficiency

Losses. The losses of a thermoelectric generator can be considered as the part of the heat which has not been converted into electricity through the Seebeck effect. It is a function of the heat conduction and the Seebeck effect. The amount of this lost heat is given by Eq. (5.31).

Electrical power output and efficiency [14]. The electrical power output is given by Eq. (5.28) and the efficiency of a thermoelement is given by Eq. (5.48), where $\Delta T/T_h$ = Carnot efficiency. Equation (5.48) gives the maximum efficiency possible. The optimum power for this efficiency will then be, from Eq. (5.45),

$$P_{opt} = \frac{(1 + Z\bar{T})^{1/2}(\Xi\,\Delta T)^2}{R[1 + (1 + Z\bar{T})^{1/2}]},$$

and the optimum current is

$$I_{opt} = \frac{\Xi\,\Delta T}{R[1 + (1 + Z\bar{T})^{1/2}]}.$$

In many applications, obtaining the maximum power possible is most important. This will be found at the usual matched impedance condition, $R = R_L$. Thus

$$P_{max} = \tfrac{1}{2}(\Xi\,\Delta T)^2/R_L.$$

In this case the thermal efficiency will be, from Eq. (5.46),

$$\eta_T = \frac{\Delta T}{\tfrac{3}{2}T_h + \tfrac{1}{2}T_c + 4KR/\Xi^2},$$

or

$$\eta_T = \frac{\eta_c}{1 + \bar{T}/T_h + 4/ZT_h}. \tag{5.50}$$

Efficiency of a cascaded generator. The total efficiency of a cascaded generator is given by Eq. (5.43). When an infinitesimal generator similar to the one shown in Fig. 5.4 is considered, the maximum efficiency will occur when T_h approaches T_c:

$$d\eta = \frac{(1 + ZT)^{1/2} - 1}{(1 + ZT)^{1/2} + 1}\frac{dT}{T}. \tag{5.51}$$

If

$$\eta_0(T) = \frac{(1 + ZT)^{1/2} - 1}{(1 + ZT)^{1/2} + 1},$$

where $ZT = f(T)$, then $d\eta = \eta_0(T)\,dT/T$. From Eq. (5.40),

$$\eta = 1 - \mathcal{Q}_0/\mathcal{Q}_N.$$

On the other hand, the thermal efficiency for the infinitesimal generator can be stated as $d\eta = d\mathcal{Q}/\mathcal{Q}$. Thus

$$d\mathcal{Q}/\mathcal{Q} = \eta_0(T)\,dT/T,$$

and

$$\ln(\mathcal{Q}_0/\mathcal{Q}_N) = -\int_{T_c}^{T_h}\eta_0(T)\,dT/T. \tag{5.52}$$

Finally

$$\eta_T = 1 - \exp\left[-\int_{T_c}^{T_h} \eta_0(T)\,dT/T\right]. \tag{5.53}$$

The integration of Eq. (5.52) may be readily carried out for a metal since, in that case, $\eta_0(T)$ is constant; and graphically for a semiconductor. Ure and Heikes [95] have shown that η_T is not greatly increased by staging.

5.10 Thermoelectric Systems [17]

The first work on thermoelectric generators was done with the interesting idea of transforming solar energy into electricity, as explained by Maria Telkes [90]. Due to the problems of solar concentration, Telkes' work quickly ran into difficulties. By the end of the 1950's, work started in both Europe [56] and the United States [33] on fossil fueled thermoelectric generators. This work was linked with the space program and use in remote terrestrial areas. More recently the use of radioisotopes and nuclear reactors has been considered.

Apart from the theoretical problem of understanding the microscopic behavior of the thermoelectric effect, many technical problems are associated with thermoelectric power generation systems. One very important problem is obtaining satisfactory hot junction bonds. Due to the differential expansion between the thermoelements and the electrodes, thermal stresses which tend to destroy the bonds are created. Furthermore, apart from the necessity of choosing the right solder, it is important that the chemical characteristics of the junction not be altered by the medium in which it works, often making it necessary to encapsulate the entire junction in an inert gas. A similar problem is the chemical instability of the thermoelements. It is of great necessity that the characteristics of their materials remain approximately constant during the lifetime of the generator. Other problems are related to heat transfer considerations and to the effect of high temperatures on the chemical and metallurgical characteristics of the thermoelements.

Fossil fueled systems. The thermoelectric generator built [6] for the U.S. Coast Guard to be used on buoys or remote shore locations is an example of such a system. The 10-W generator, working as a power source for aids to navigation, has a 10-year life span, during which it needs minimum servicing and maintenance. The thermoelectric generator charges a storage battery through a voltage regulator. The load on the storage battery is a flashing lamp. The fuel is propane, stored in a fuel supply tank (able to hold enough propane for 3 years of continuous operation) and injected into the combustion chamber of the generator through a pressure regulator.

The thermoelectric arms, made from n- and p-type lead telluride, are arranged around the central combustion chamber in a cylindrical configuration. The elements are segmented, and there are 102 couples. The length of the legs is about 20 mm and their diameter is 4 mm for the n-legs and 5 mm for the p-legs.

The total cross section of the thermoelectric elements is 32% of the area of the combustion chamber, leading to a relatively large leg packing density.

A thermoelectric efficiency of 8.5% was measured for this generator with non-optimum loading and a nominal output power of 15 W with the generator evacuated. This efficiency fell to about 7.4% when the generator was filled with a gas mixture (0.95 argon and 0.05 hydrogen) at atmospheric pressure to avoid oxidation of the elements. This loss in efficiency is due to the increased thermal conductivity of the insulation.

The flow rate of the propane is about 368 g/hr. Since the energy input per unit weight of propane is 690 W-hr/kg, the total chemical power input is 254 W. With a burner efficiency of 67%, the overall efficiency is then reduced to approximately 5%. It should be noted, however, that the outdoor, unattended performance of this generator is considerably lower than might be suggested by these values.

Many technical details were studied to give this system the maximum possible reliability and lifetime. For instance, the entire active portion of the generator was hermetically sealed after being filled with the nonoxidizing argon-hydrogen mixture. To provide enough space for thermal expansion and possible sublimation, as well as to give the best resistance to vibration and shocks, each thermoelectric arm was independently spring loaded. Lead telluride (PbTe) has a good thermal efficiency in the 610 to 670°C range, but to increase the life expectancy of the generator, operation is at the lower temperature of 500°C at the hot junction. Special consideration was also given to the electrical contacts. At the hot end of the elements, the joints are made by pressure contact, and at the cold end by the use of pretinned copper shoes soldered to the pure tin ultrasonically coated surface of the elements. The thermal losses are minimized by the use of good thermal insulators (such as diatomaceous earth).

Fossil fueled thermoelectric generators of the high-temperature type are currently built.* These generators use silicon-germanium thermoelements and work in a wide temperature range between 650 and 1000°C. A low temperature methane fueled thermoelectric generator is also being produced.† In this device, the thermoelectric material is bismuth telluride. This generator is built mainly for experimental purposes in American and Canadian universities.

Solar fueled systems [10, 33, 34]. A typical solar fueled thermoelectric converter has been developed for space application [87].‡ This system competes favorably with photovoltaic converters and has a better resistance to radiation damage. One aim of the project was to obtain low specific weight and volume at low cost, whereas efficiency was a secondary consideration.

The system consists of the thermoelectric converter, a concave mirror to collect and concentrate solar energy, a focal target to absorb radiation and deliver

---

*   RCA at Harrison, N.J.

†   General Instrument Co. Thermoelectric Division, Newark, N.J.

‡   Boeing Airplane Co., Seattle, Washington.

it in the form of heat to the hot junction of the thermoelectric converter, a radiator acting as the heat sink, thermal and electrical connections, and the frame structure on which all the elements are mounted. The thermoelectric elements and the paraboloidal metallic reflectors are joined through an insulator. The reflectors are rectangular, their back is blackened, and their function is multiple: they work as collectors, radiators, and electrical and thermal connectors, and they form the frame structure. The metal is aluminum, thus allowing an integrated low weight system. The collector-radiator configuration is made up of 54-cm^2 units. The thermoelectric material is lead telluride, used in thermoelements 6.3 mm in length and 2.5 mm in diameter. The heat path is not longer than 20 mm. Heat is radiated from both sides of the radiator in such a way as to keep the temperature of the sink at 0°C at sea level and 64°C in space. The generator is of the cascaded type, where lead telluride is used up to the maximum temperature of 300°C and chrome and gold-nickel wires are used for higher temperatures (up to 800°C).

To minimize the packaging volume, spacing of the thermoelements was limited to 2.5 mm. The radiators and collectors were made identical by using multipurpose metallic mirrors.

Because there is only one thermocouple assembly at each collector focal point, insulation separating focal targets from thermocouples and neighboring thermocouples is no longer needed. This leads to a relatively large power to weight ratio.

The complete unit has been designed to produce 90 mW. Its total weight is only 2.7 g, leading to a specific weight of 60 g/W. The specific volume of the system is equal to 226 m^3/W and its specific area is equal to 576 m^2/W. The overall efficiency of the projected system is equal to 2% of the sea level radiation of 64.6 mW/cm^2.

Radioisotope fueled systems [16, 19]. Isotopes of an element are atoms having the same number of protons and electrons but different numbers of neutrons. Consequently, their chemical characteristics are identical and their difference is only in weight. When an isotope is unstable, it decays and emits radiation; it is then called a *radioisotope*. The radiation emitted is of three types: alpha, beta, and gamma. The rate of decay varies from one radioisotope to another and it is usually characterized by the *half-life*, which is the time required for half of the unstable nuclei in a pure sample to decay. This half-life varies between values shorter than 12 msec for nitrogen-12 to values larger than 16 million years for iodine-129.* The kinetic energy of the decay particles can be transformed into heat energy, which, in turn, can be used to power a thermoelectric generator.

The first radioisotope thermoelectric generator was tested in 1959. It was SNAP-3 [7] which used polonium-210 as fuel and produced 2.5 W of electricity. SNAP-3 generators were built to power satellites, moon probes, navigational aids,

---

* The half-lives of helium-5 and beryllium-8 are shorter than 12 msec, whereas the half-life of rubidium-87 is larger than 16 million years.

and weather stations. Polonium was chosen because of its availability and its safety. Its half-life is 138 days and its initial power density of 141 W/g is one of the highest of all radioisotopes. Its major radiation is α-particles and its melting point is near 254°C. SNAP-3 weighed 1.8 kg and had a cylindrical form 12 cm in diameter and about 14 cm in height.

SNAP-9 [64] was roughly spherical with a diameter of 13.5 cm. The radioisotope was in a central chamber and the outer shell was made of white coated copper to avoid heat absorption from the sun. The radioisotope used was plutonium-238, which has a half-life of 89 years. Plutonium-238 emits α-particles, melts at 640°C, and has a rather low initial power density of 0.55 W/g. SNAP-9 produced 2.7 W of electricity and was the first thermoelectric space generator.

SNAP-9A [64], which powered two satellites in 1963, was a bigger but similar version of SNAP-9. Its power output was 25 W and its lifetime was 5 years. The generator weighed about 14.5 kg and was cylindrical in form, about 51 cm in diameter, including the radiating fins, and 24 cm in height.

SNAP-11 [64] is being developed, in connection with NASA's Surveyor lunar landing missions, to work in conjunction with photovoltaic generators. The fuel used is curium-242, which has a half-life of 162 days, a power density of 121 W/g, a melting point at 950°C, and emits α-particles. SNAP-11 weighs 13.6 kg and produces 25 W in lunar nights and 21 W under solar radiation.*

SNAP-27 [32] has a power output of 50 W and uses plutonium-238. It is under development in conjunction with the Apollo program, and is designed to power the Apollo lunar surface experiment packages which will transmit measurements automatically to the earth. The total weight of the generator is only 13.6 kg and its dimensions are 46 cm in diameter and 46 cm in height.

SNAP-19 [64] is similar to SNAP-9A and serves to power the Nimbus-B spacecraft, which orbits in low altitudes for atmospheric measurements. In this satellite, two 30-W thermoelectric generators work in connection with solar batteries.

The Axel Heiberg Island in northern Canada was the site of the first weather station to be powered by a thermoelectric generator [64] (1961). It is an unmanned station which collects and relays information on wind, barometric pressure, and temperature. The radioisotope powered generator is made up of 60 pairs of lead telluride legs arranged around a cylindrical strontium-90 source. Strontium-90 has a 28-year half-life and a 0.93 W/g power density; it melts at 770°C and emits, mainly, β-particles. The generator produces 5 W of electricity.

The SNAP-7 [64] series of generators is similar to the Axel Heiberg type, differing from it only in weight and power output. SNAP-7A powers a navigational buoy, B a fixed navigational light, C a weather station in Antarctica, D a floating

---

* During lunar nights—which last 14 earth days—the efficiency of this generator increases because of the larger temperature difference between the source and the sink of heat of the thermoelectric elements, thus explaining the higher efficiency during the night in spite of the loss of solar energy to the photovoltaic cells.

weather station, E an ocean bottom beacon, and F an offshore oil rig in the Gulf of Mexico.

SNAP-15A [64] is a tiny generator for military applications. It weighs less than $\frac{1}{2}$ kg and produces 1 mW of electric energy for at least 10 years. The fuel used is plutonium-238, and the thermoelements are metallic. It can be a very reliable lightweight source of electricity for communication use.

Many other radioisotope powered thermoelectric generators are under study for space and terrestrial applications. Many of these are designed to use strontium-90 as fuel. The latter is a common by-product of nuclear reactors; it is cheap and the only difficulty is related to the safety requirements.

Nuclear fueled systems [5, 8]. Two important nuclear fueled systems using thermoelectric power generation were built for space applications. These are SNAP-10 and SNAP-10A.

In the SNAP-10 [34] system, two thermoelectric generators are mounted directly on two cylindrical reactors. In each unit, heat is transferred directly from the reactor to the thermoelements through conduction, whereas the waste heat is dissipated by an outer radiator. The power delivered by the unit tested at Atomics International, was 127 W of electricity with 2.15% efficiency. The hot junction of the thermoelements was at 614°C and the cold junction at 364°C. If this temperature were lowered to 30°C, the efficiency of the generator would be raised to about 16.6%.

Each unit was made up of 384 thermoelements, about 12.7 mm in diameter and 6.3 mm in length, the p-leg being made of germanium-bismuth telluride and the n-leg of lead telluride. The weight of the two units totaled about 63.5 kg, including about 18 kg for the braid spring assemblies which were used to force the straps against the insulation and to reinforce the resistance of the system to stresses and shocks.

A revision of SNAP-10, SNAP-10A [99], was completed and launched in orbit in April 1965 from Vandenberg Air Force Base, California. This 500-W generator was used to help power the ion-propulsion unit of the satellite as well as its telemetry. A power absorber was used as the heat sink. The reactor worked satisfactorily for 43 days. In SNAP-10A, heat from the nuclear reactor was carried by a flow of liquid NaK past the thermoelectric couples. The thermoelements were made of Si-Ge materials and worked between 473°C at the hot junction formed by the liquid metal pipe, and 321°C at the cold junction. The overall efficiency of this system was not higher than 1.6%, with a specific mass of 908 kg/kW.

It seems that for high power levels, thermionic conversion is more promising for nuclear power applications in space. However, with advances in high-temperature thermoelement research, it is hoped that thermoelectricity will be competitive up to the 20-kW level. This explains plans to improve the technology of the SNAP-2 and 10 systems by including thermoelectricity in their conversion stages in the 1970's.

5.11 Applications and Future Trends

From a modest beginning in the early 1950's, thermoelectric power generation is now becoming a primary source of electric energy in many applications. Further developments in material technology and simplicity will surely widen the range of these applications.

Because of the high cost of thermoelectric materials, thermoelectric generators are not yet used widely in commercial applications. Whenever standby auxiliary power is needed, fossil fueled thermoelectric generators are a handy, reliable, and simple source of electrical energy. An example is the kerosene fueled generators built in the Soviet Union to be used by peasants in the remote stretches of Siberia [56]; other examples can be cited [23].

Thermocouple powered devices are now used more and more widely in the control field, mainly in the gas and steel industries. Thermoelectric generators power the safety interlock with the gas valve in gas furnaces, remote communication sites, cathodic protection of gas and oil pipelines, and blowers in gas furnaces.

Thermoelectricity has an even greater prospect for space applications [49, 58, 62], since development of strontium fueled generators is expected to reduce the overall costs of the special systems. Units for manned and unmanned missions requiring outputs in the vicinity of 5 kW are being studied. These systems will be powered by strontium-90, plutonium-238, and polonium-210, and are proposed for the extended Apollo, Manned Orbiting Laboratories and Manned Mars missions. Research is being conducted in many American and foreign laboratories to provide fuels with higher melting points and higher power densities, to improve the cost and the availability of the isotopes, and to solve the many technological problems of high-temperature thermal and electrical losses.

In the nuclear field, research is directed at building small nuclear reactors and the use of adequate coolants. As for solar energy, the problem remains one of concentration and storage of the incoming radiation from the sun.

The efficiency of the working thermoelectric generators increased from about 6% in 1956 to 14% in 1967. Currently, in the low-power range (between 1 and 1000 W), thermoelectric power generators using gas as a fuel are more economical than any internal combustion engine. If this trend continues, the market for thermoelectric devices will widen rapidly.

PROBLEMS

5.1 From the viewpoint of thermodynamics, thermoelectricity can be considered as the interaction between the electric current density given by Ohm's law (Eq. 5.22) and the heat flux given by Eq. (5.7). A change in electrostatic potential V will affect both the electric and the thermal properties of the material. This can be summarized by the following equations:

$$\mathbf{2} = L_{11}\, \nabla T + L_{12}\, \nabla V,$$
$$\mathbf{J} = L_{21}\, \nabla T + L_{22}\, \nabla V,$$

where the L's are known as the kinetic coefficients. From these two equations derive Kelvin's relation of Eq. (5.14).

5.2 Consider the differential thermoelectric element shown in Fig. 5.12. Derive the differential equation for the temperature distribution in such an element. [*Hint:* Use the equation of conservation of energy.]

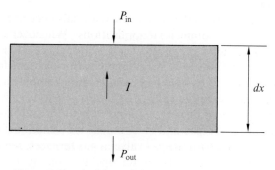

Figure 5.12. A differential thermoelectric element.

5.3 The Seebeck coefficient of a certain thermoelectric material is given by the relation

$$\Xi(T) = \mathscr{A}_1 T^2 + \mathscr{A}_2 T + \mathscr{A}_3$$

in a certain temperature range. $\mathscr{A}_1$, $\mathscr{A}_2$, and $\mathscr{A}_3$ are known constants. Calculate the Peltier and Thomson coefficients of this material as a function of temperature.

5.4 Assuming that in Problem 5.3

$$\mathscr{A}_1 = -0.2 \times 10^{-8} \text{ V} \times {}^\circ\text{K},$$
$$\mathscr{A}_2 = 1.6 \times 10^{-6} \text{ V},$$
$$\mathscr{A}_3 = -1.4 \times 10^{-4} \text{ V}/{}^\circ\text{K},$$

calculate the amount of heat absorbed and generated at the junctions of a thermoelectric refrigerator using the above material as a leg. The other leg is constituted by a material having a Seebeck coefficient given by

$$\Xi_2(T) = -\Xi_1(T).$$

The refrigerator operates between $T_h = 400^\circ\text{K}$ and $T_c = 300^\circ\text{K}$. The input current is $I = 20$ A. (The above material is a p-type 25% Bi_2Te_3–75% Sb_2Te_3 with 1.75% excess Se.)

5.5 By assuming that in Eq. (5.35) κ, σ, and τ are linear functions of temperature, solve Eq. (5.35) for a linear thermoelement.

5.6 Solve Eq. (5.33) for a cylindrical leg of radius r and length ℓ.

5.7 Show that the thermal power input in a pn thermoelectric junction is given by the following relation for an area ratio and a load adjusted for a maximum power output:

$$\mathcal{Q} = \frac{S_t}{\ell} \Delta T \left[\frac{\Xi^2 \bar{\sigma}}{32} (\Delta T + 4\bar{T}) + \bar{\kappa} \right],$$

where $S_t = S_n + S_p$ = total surface area, $\bar{\sigma} = \frac{1}{2}(\sigma_n + \sigma_p)$ = average electrical conductivity,

$\bar{\kappa} = \frac{1}{2}(\kappa_n + \kappa_p) =$ average thermal conductivity, $\bar{T} = \frac{1}{2}(T_h + T_c) =$ average temperature, and $S_t/\ell =$ shape factor.

5.8 Consider a two-stage generator in which the first stage has N_1 couples and operates between T_h and T_0 and the second stage has N_2 couples and operates between T_0 and T_c. The couples are thermally in parallel and electrically in series. Find the ratio N_2/N_1 in the case where the currents in the stages are such that $I_1 = I_2 = I$, where I is the current corresponding to maximum efficiency.

5.9 Design an elementary pn thermoelectric generator operating between 10 and 90°C. Consider a maximum efficiency design. The materials used are lead telluride:

n-type:

$$\Xi_n = -180\ \mu V/°K,$$
$$\sigma_n = 1000\ \text{mho/cm},$$
$$Z_n = 0.4 \times 10^{-3}\ 1/°K;$$

p-type:

$$\Xi_p = 220\ \mu V/°K,$$
$$\sigma_p = 900\ \text{mho/cm},$$
$$Z_p = 1.7 \times 10^{-3}\ 1/°K.$$

Calculate (a) the power output per unit area, (b) the thermal efficiency, (c) the required heat input per unit area, and (d) the output voltage.

5.10 Answer the questions of Problem 5.9 and use the same materials, but consider a maximum power output design.

5.11 From the description of the fossil fueled generator mentioned in Section 5.10, prove that the efficiency is about 8.5% and that the output power is 15 W.

5.12 SNAP-9 uses plutonium-238 as fuel and produces 2.7 W. From the data given in Section (5.10), can you make an educated "guess" as to what type of thermoelectric material is used?

5.13 The SNAP-10 generator is described in Section 5.10. It was mentioned that if the cold junction temperature were lowered from 364°C to 30°C, the efficiency of the generator would have risen from 2.1% to 16.6%. Prove this statement mathematically (use the data given in the text).

5.14 The SNAP-3 generator uses p- and n-doped lead telluride. The specifications of the thermoelectric elements are

$$S_n/S_p = 0.89,$$
$$\ell = 2.5\ \text{cm},$$

with the n-leg segmented and the p-leg not segmented. It is fueled by Po^{210}. Calculate its total power output and its efficiency. [*Answer:* $P_{out} = 5\ W, \eta = 5.5\%$.] Comment on the eventual discrepancy of your results.

5.15 Design a thermoelectric generator fueled by strontium-90 and producing 50 W of electric power. The device is desired to operate at maximum efficiency at the South Pole.

REFERENCES AND BIBLIOGRAPHY

1. Abeles, B., and R. W. Cohen, "Ge-S: Thermoelectric Power Generator," *J. Appl. Phys.*, Vol. 35, p. 247, 1964.

2. Abowitz, G., et al., "Thin Film Thermoelectrics," *Semiconductor Prod.*, Vol. 8, No. 2, p. 18, 1965.

3. Amith, A., "Seebeck Coefficient in n-Type Germanium-Silicon Alloys 'Competition' Region," *Phys. Rev.*, Vol. 139, Series A, p. 1624, 1965.

4. Andersen, J. R., "Thermoelectric Air Conditioner for Submarines," *RCA Rev.*, Vol. 22, p. 292, 1961.

5. Anderson, G. M., "Nuclear Reactor Systems," *Astronautics and Aerospace Engineering*, May 1963.

6. Bargen, D. W., "Thermoelectric Generators for Battery-Charging Applications," *Adv. Energy Conversion*, Vol. 3, No. 2, p. 507, 1963.

7. Barmat, M., G. M. Anderson, and E. W. Bollmeir, "Electricity from Radionuclides and Thermoelectric Conversion," *Nucleonics*, Vol. 17, p. 166, May 1959.

8. Barmat, M., "Direct Conversion Applied to Nuclear Heat Sources," Chap. 18, in *Thermoelectric Materials and Devices*, I. B. Cadoff and E. Miller (eds.), Reinhold, New York, 1960.

9. Bauerle, J. E., P. H. Sutter, and R. W. Ure, Jr., "Measurements of Properties of Thermoelectric Materials," in *Thermoelectricity Science and Engineering*, R. R. Heikes and R. W. Ure, Jr. (eds.), Interscience, New York, 1961.

10. Baum, V. A., "The Conversion of Solar Energy into Electricity," *Sol. Energy*, Vol. 7, p. 180, 1963.

11. Boffi, V. C., V. G. Molinari, and D. E. Parks, "Thermoelectric Theory of a Plasma Diode," *Adv. Energy Conversion*, Vol. 7, No. 1, p. 1, 1967.

12. Burshteyn, A. I., *Semiconductor Thermoelectric Devices*, Temple Press, London, 1964.

13. Cadoff, I. B., and E. Miller, *Thermoelectric Materials and Devices*, Reinhold, New York, 1960.

14. Cohen, R. W., and B. Abeles, "Efficiency Calculations of Thermoelectric Generators with Temperature Varying Parameters," *J. Appl. Phys.*, Vol. 34, p. 1687, 1963.

15. Conwell, E. M., and J. Zucker, "Thermoelectric Effect of Hot Carriers," *J. Appl. Phys.*, Vol. 36, No. 7, p. 2192, 1965.

16. Corliss, W. H., and D. L. Harvey, *Radioisotope Power Generation*, Prentice-Hall, Englewood Cliffs, N.J., 1964.

17. Corry, T. M., and G. Spira, "Thermoelectric Generator Design, Performance and Application," *IRE Transactions on Military Electronics*, MIL-6, p. 34, 1962.

18. Damchask, Ya. I., V. Y. Prokhorenko, and N. M. Klym, "Structure and Thermoelectric Properties of the Bismuth-Cadmium System in the Solid and Liquid States," *Fiz. Tverdogo Tela*, Vol. 7, No. 7, p. 1974.

19. Defrane, A., *Underwater Thermoelectric Reactor Plant*, AIAA 2nd Annual Meeting and Technical Demonstration, San Francisco, July 26–29, 1955.

20. DeGrott, S. R., *Thermodynamics of Irreversible Processes*, Interscience, New York, 1951.

21. DeHass, E., *Radioactive Isotope Fueled Thermo-Electric Generators for Space Missions*, Thesis, Technische Hogeschool, Eindhoven, Holland, 1964.

22. Dismukes, J. P., et al., "Thermal and Electrical Properties of Heavily Doped Ge-Si Alloys Up to 1300°K," *J. Appl. Phys.*, Vol. 35, p. 28–99, 1964.

23. Dismukes, J. P., and L. Ekstrom, "Homogeneous Solidifications of Ge-Si Alloys," *Am. Inst. Mining Engrs. Met. Soc. Trans.*, Vol. 233, No. 4, p. 672, 1965.

24. Domenicali, C. A., "Irreversible Thermodynamics of Thermoelectricity," *Rev. Mod. Phys.*, Vol. 26, p. 237, 1954.

25. Egli, P. H. (ed.), *Thermoelectricity*, Wiley, New York, 1960.

26. Egli, P. H., *Thermoelectric Materials The Present and the Potential*, AIEE Proceedings, Pacific Energy Conversion Conference, 1962.

27. Eichhorn, R. L., "Experimental Cooling Devices," in *Thermoelectric Materials and Devices*, I. B. Cadoff and E. Miller (eds.), Reinhold, New York, 1960.

28. Freedman, S. I., "Heat Transfer Considerations in Space Power Supplies," Chap. 15 in *Developments in Heat Transfer*, W. M. Rohenow (ed.), M.I.T. Press, Cambridge, Mass., 1964.

29. Freedman, S. I., "Thermoelectric Power Generation," in *Direct Energy Conversion*, G. W. Sutton (ed.), McGraw-Hill, New York, 1966.

30. Freud, P. J., and G. M. Rothberg, "Thermoelectric Power of Germanium Effect of Temperature-Dependent Energy Levels," *Phys. Rev.*, Vol. 140, Series A, p. 1007, 1965.

31. Fritts, R. W., "Design Parameters for Optimizing the Efficiency of Thermoelectric Generators Using p-Type and n-Type Telluride," *Trans. AIEE Communications and Electronics*, Vol. 78, p. 817, 1960.

32. Fritts, R. W., "Lead Telluride Alloys and Junctions," Chap. 10, in *Thermoelectric Materials and Devices*, I. R. Cadoff and E. Miller (eds.), Reinhold, New York, 1960.

33. Fritts, R. W., "The Development of Thermoelectric Power Generators," *Proc. IEEE*, Vol. 51, No. 5, p. 713, 1963.

34. Fuschillo, N., R. Gibson, F. K. Eggleston, and J. Epstein, "Flat Plats Solar Thermoelectric Generator for Near-Earth Orbits," *Adv. Energy Conversion*, Vol. 6, p. 103, 1966.

35. Fuschillo, N., and R. Gibson, "Germanium-Silicon, Lead Telluride and Bismuth Telluride Alloy Solar Thermoelectric Generators for Venus and Mercury Probes," *Adv. Energy Conversion*, Vol. 7, No. 1, p. 47, 1967.

36. Garachk, V. K., and V. A. Nayer, "Semiconductor Thermoelectric Coolers of Transistors," *Izv. Vuzov. Instr. Bldg.*, Vol. 8, p. 229, 1965.

37. Goldsmid, H. J., and R. W. Douglas, "The Use of Semiconductors in Thermoelectric Refrigeration," *Brit. J. Appl. Phys.*, Vol. 5, p. 386, 1954.

38. Goldsmid, H. J., *Applications of Thermoelectricity*, Methuen, London, 1960.

39. Goldsmid, H. J., *Thermoelectric Refrigeration*, Plenum Press, New York, 1964.

40. Goldsmid, H. J., "Design of a Thermoelectric Cooling Unit for Non-Ideal Operating Conditions," *Solid-State Electron.*, Vol. 8, No. 8, p. 109, 1965.

41. Goldsmid, H. J., and C. B. Thomas, "Comparison of Peltier-Seebeck and Nernst-Ettinghausen Energy Converters in Intermediate Magnetic Fields," *Adv. Energy Conversion*, Vol. 7, No. 1, p. 33, 1967.

42. Gray, P. E., *The Dynamic Behavior of Thermoelectric Devices*, MIT Press, Cambridge, Mass., 1960.

43. Green, W. B. (ed.), *Thermoelectric Handbook*, Westinghouse Elec. Corp., Youngswood, Pa., 1962.

44. Gus'kov, Yu. K., V. P. Pashchenko, and Y. Y. Sibir, "Study of the Operation of the Thermoelectron Converter with Various Metallic-Film Cathodes," *Izv. Acad. Nauk SSSR, Ser. Fiz.*, Vol. 28, No. 9, p. 33, 1964.

45. Harman, T. C., "Multiple Stage Thermoelectric Generation of Power," *J. Appl. Phys.*, Vol. 29, p. 1471, 1958.

46. Hatsopoulos, G. N., and J. H. Keenan, "Thermodynamics of Thermoelectricity," Chap. 15 in *Direct Conversion of Heat to Electricity*, J. Kaye and J. A. Welsh (eds.), Wiley, New York, 1960.

47. Hayward, J. N., et al., "Peltier Biothermodes," *Am. I. Medical Electr.*, Vol. 4, p. 11, 1965.

48. Hazard, H. R., *Multifuel Thermal Energy Converters*, 16th Power Sources Conference Proceedings, May 1962, PSC Publications Committee, Red Bank, N.J., 1962, p. 124.

49. Hedley, W. H., and H. R. Stroup, *High-Temperature Thermoelectric Systems for Space Power at Multi-KW(e) Levels*, American Nuclear Society 11th Annual Meeting, Gatlinburg, Tenn., June 20–24, 1965.

50. Hedgcock, F. T., and D. P. Mathur, "Low-Temperature Thermoelectric Power of Heavily Doped n-Type Germanium," *Can. J. Phys.*, Vol. 43, No. 11, p. 2008, 1965.

51. Heikes, R. R., and R. W. Ure, Jr., *Thermoelectricity Science and Engineering*, Interscience, New York, 1961.

52. Henderson, C. M., et al., *High Temperature Thermoelectric Generator*, ASD-TDR-62-896, Oct. 1963.

53. Henderson, C. M., and C. W. Glassburn, "High Temperature Thermoelectric Research," *IEEE Trans. On Aerospace*, No. 2, April 1964.

54. Hisakado, T., "Experimental Studies on the Thermoelectric Cooling," *Osaka Univ. Mem. Eng.*, Vol. 6, No. 12, p. 35, 1964.

55. Issi, J. P., and J. M. Streydio, "Evaluation of the Diffusion Thermoelectric Power of Bismuth," *Adv. Energy Conversion*, Vol. 7, p. 99, 1967.

56. Joffe, A. F., *Semiconductor Thermoelements and Thermoelectric Cooling*, Infosearch, London, 1957.

57. Kirby, G., M. Norwood, and R. Marlow, "Construction and Evaluation of a Thermoelectric Controlled Temperature Chamber," *Adv. Energy Conversion*, Vol. 2, p. 265, 1962.

58. Klem, R. L., and W. J. Helwig, *Thermoelectric Technology for Space-Power Systems*, AIAA 2nd Annual Meeting and Technical Demonstration, San Francisco, July 26–29, 1965.

59. Kolbikov, L. O., "Thermoelectric Thermometer," *Byulleten Izobretenii i Towarnykh Znakov*, No. 6, 1960 (Russian).

60. Lackey, R. S., et al., "Applications of Thermoelectric Cooling and Heating," *Refrig. Eng.*, Vol. 66, p. 31, Dec. 1958.

61. Lye, R. G., "The Thermoelectric Power of Titanium Carbide," *J. Phys. & Chem. Solids*, Vol. 26, No. 2, p. 407, 1965.

62. Mackay, D. B., *Design of Space Powerplants*, Prentice-Hall, Englewood Cliffs, N.J., 1963.

63. Mayev, S. A., and I. O. Stakhanov, "Application of the Grad Method of the Design of the Thermoelectron Energy Converter," *Izv. Akad. Nauk SSSR, Ser. Fiz.*, Vol. 28, No. 9, p. 30, 1964 (Russian).

64. Mead, R., and W. R. Corliss, *Power from Radioisotopes*, U.S. Atomic Energy Commission, Division of Technical Information, Series on Understanding of the Atom, 1964.

65. Merritts, T. D., *Thermoelectric Temperature Control for Quartz Crystals and Crystal Oscillators*, Westinghouse Elect. Co., Semiconductor Division, Quarterly Progress Rept. No. 3, Jan. 1964.

66. Millionshehikov, M. D., et al., *High Temperature Direct Conversion Reactor Romashka*, 3rd United Nation International Conference on the Peaceful Uses of Atomic Energy, Geneva, 1964.

67. Morgulis, N. D., "Le Convertisseur Thermoelectronique," *Collection Espace et Electronique*, Gauthier-Villars, Paris, 1965.

68. Mortlock, A. J., "Experiments with a Thermoelectric Heat Pump," *Am. J. Phys.*, Vol. 35, No. 10, p. 813, 1965.

69. Moumouni, A., "Contribution à l'Analyse du Fonctionnement d'un Refrigerateur Thermoélectrique Alimenté par une Pile Thermoélectrique," *Adv. Energy Conversion*, Vol. 7, No. 1, p. 53, 1967.

70. Naumer, D. A., and J. L. McCabria, *Solar Thermoelectric Power Conversion Coupled with Thermal Storage for Orbital Space Applications*, 3rd Biennial Aerospace Power System Conference, Philadelphia, Sept. 1964.

71. Neild, A. B., Jr., "Portable Thermoelectric Generators," *Trans. ASE*, Paper No. 645A, 1963.

72. Paris, J. P., and V. F. Damme, "Thermoelectric Cooler for Spectral Cells," *J. Sci. Instr.*, Vol. 36, No. 7, p. 1058, 1965.

73. Perron, J. C., "Pouvoirs Thermoelectriques des Alliages Liquides Selenium-Tellure," *Acad. Sci. Paris. Compt. Rend.*, Vol. 260, p. 5760, 1965.

74. Petruzzella, N., and R. C. Nelson, "Thermoelectric Power of Some Organic Photo-Conducting Dyes," *J. Chem. Phys.*, Vol. 42, p. 3922, 1965.

75. Phillips, A. F., *Thermoelectric Air Conditioning and Refrigeration for Submarines*, Institut International du Froid, Annex to the Bulletin, Commissions 2, 3, 4, 6A, Washington, D.C., 1962.

76. Phillips, L. S., "The Measurement of Thermoelectric Properties at High Temperatures," *J. Sci. Instr.*, Vol. 42, No. 4, p. 209, 1965.

77. Raag, V., *Silicon-Germanium Thermocouple Development*, 17th Power Sources Conference Proceedings, May 1963, PSC Publications Committee, Red Bank, N.J., 1963.

78. Richards, J. D., *Materials Selection Criteria for Thermoelectric Power Generation*, AIEE Proceedings of Pacific Energy Conversion Conference, 1962.

79. Rosi, F. D., E. F. Hockings, and N. E. Lindenblad, "Semiconducting Materials for Thermoelectric Power Generation," *RCA Rev.*, Vol. 22, No. 3, p. 82, 1961.

80. Russcher, G. E., "Analysis of Thermoelectric Materials for the High Temperature

Direct Conversion of Nuclear Energy," *Nuclear Struc. Eng.*, Vol. 2, No. 10, p. 341, 1965.

81. Russell, A. W., and J. H. Goldsmid, "Thickness-Dependence of the Thermoelectric Properties of a Polycrystalline Bismuth Telluride Alloy," *Adv. Energy Conversion*, Vol. 7, p. 99, 1967.

82. Schulman, F., *Isotopes and Isotope Thermoelectric Generators*, Space Power Systems Adv. Technology Conference Proceedings, Cleveland, Ohio, August 1966, NASA SP-131, p. 73.

83. Scott, W. C., and F. Schulman, "Space Electrical Power," *Astronautics and Aerospace Engineering*, May 1963.

84. Seebeck, T. J., *Abhdlg. Kgl. Akad. Wiss.*, Berlin, 1822/23.

85. Shalyt, S. S., and P. V. Tamarin, "Thermal Conductivity and Thermoelectric Power of InSb at Low Temperatures," *Fiz. Tverdogo Tela*, Vol. 6, No. 8, p. 2327, 1964 (Russian).

86. Shaw, D. T., *A Non-Linear Analysis of the Transient Behavior of Thermoelectric Generators with Temperature Dependent Physical Parameters*, Conference 654-83, American Nuclear Society Meeting, San Francisco, Nov.–Dec. 1964.

87. Shligtig, R. C., "Solar Thermoelectric Converter with Small Multipurpose Collectors," *Adv. Energy Conversion*, Vol. 2, p. 299, 1962.

88. Swanson, B. W., and E. V. Somers, "Optimization of a Conventional-Fuel-Fired Thermoelectric Generator," *J. Heat Transfer*, Vol. 81, p. 245, 1959.

89. Takahaski, T., "Thermoelectric Effect on Ice," *J. Atmospheric Sci.*, Vol. 3, No. 1, p. 74, 1966.

90. Telkes, M., "Solar Thermoelectric Generators," *J. Appl. Phys.*, Vol. 25, p. 765, 1954.

91. Thiele, A. W., and M. G. Coombs, "SNAP Thermoelectric Systems," in *Space Power Systems*, N. W. Snyder (ed.), Academic Press, New York, 1961.

92. Thomson, J. M., "Isotope-Powered Thermoelectric Generator for Ripple I," *Electr. Ltrs.*, Vol. 1, No. 6, p. 93, 1965.

93. Thomson, W. T., *Trans. Edinburgh Soc.*, Vol. 21, p. 153, 1857.

94. Thomson, W. T., *Collected Papers*, Vol. 1, Cambridge University Press, Cambridge, 1882, p. 232.

95. Ure, R. W., Jr., and R. R. Heikes, "Theoretical Calculations of Device Performance," Chap. 15 in *Thermoelectricity Science and Engineering*, R. R. Heikes and R. W. Ure, Jr. (eds.), Interscience, New York, 1961.

96. Van Heyst, H. P., and T. M. Cunningham, *Si-Ge Thermoelectric Power Modules*, 18th Power Sources Conference Proceedings, May 1964, PSC Publications Committee, Red Bank, N.J., 1964, p. 130.

97. Vorinin, A. N., et al., *Radioisotope Fueled Thermoelectric Generators*, 3rd United Nations International Conference on the Peaceful Uses of Atomic Energy, Geneva, 1964.

98. White, D. C., B. D. Wedlock, and J. Blair, *Recent Advances in Thermal Energy Conversion*, 15th Annual Power Sources Conference Proceedings, May 1961, PSC Publications Committee, Red Bank, N.J., 1961.

99. Wilson, R. F., J. E. Brunings, and G. S. Budney, "SNAP-10A-Prologue to Flight," *Nucleonics*, Vol. 22, No. 6, p. 44, 1964.

100. Wright, D. A., "New Ways in Thermoelectricity," *Brit. J. Appl. Phys.*, Vol. 15, p. 217, 1964.

101. Wright, D. A., "Thermoelectric Generation," in *Direct Generation of Electricity*, K. H. Spring (ed.), Academic Press, New York, 1965.

102. Ybarrondo, L. J., and J. E. Sunderland, "Influence of Spatially Dependent Properties on the Performance of a Thermoelectric Heat Pump," *Adv. Energy Conversion*, Vol. 5, No. 12, p. 383, 1965.

6 THERMIONIC POWER GENERATION

6.1 Introduction

A thermionic power generator converts heat into electricity by using the effect of *thermionic emission*. Thermionic emission is the release of electrons from a hot metallic surface. The heat supplies energy to the electrons of the metal, allowing them to escape from the surface and migrate to a colder material. The hot material is called the *emitter* and the cold material is called the *collector*. If an external load is placed between the emitter and the collector, an electric current is obtained. It was known before the nineteenth century that a negatively charged metallic body loses its charge more rapidly when heated. However, the thermionic emission of electrons was discovered much later by Edison [23], who explained it clearly in an 1883 patent. The effect discovered by Edison and known as the *Edison effect* can be described as follows: When a galvanometer connects a plate to the positive end of a battery-heated filament, a current is observed to flow through the galvanometer. Between the plate and the filament was a vacuum.

This effect was translated by Fleming [28] into a thermionic diode rectifier in 1904. Eleven years later, Schlichter [80], at Gottingen University, hinted at the possibility of using the Edison effect to convert heat into electricity. In spite of the tremendous amount of work done in the field of plasmas during the first half of the twentieth century (e.g., by Langmuir [18, Chap. IV] in the 1920's and 1930's), the possibility of a thermionic converter has been seriously considered only since the early 1950's, and then simultaneously in several countries: by Morgulis [66] (USSR), by Champeix [17] (France), and by Hatsopoulos [35] (USA), among others [39, 47, 67]. Since then, theoretical and experimental work in thermionic conversion has progressed at a rapid pace.

Although the efficiencies reached by experimental thermionic converters do not exceed 12%, thermionic conversion promises to be the first direct energy method to reach the efficiencies of present-day power stations. This will require a high-temperature heat source which could be a nuclear reactor. As mentioned in the chapter on MHD power generation, high-temperature nuclear reactors are still in the initial steps of development.

6.2 Thermionic Emission

The free electrons of the emitter have an average energy related to the Fermi level of the emitter (cathode). For these electrons to be able to escape the cathode, the amount of heat energy supplied to them should be at least equal to the cathode work function $\mathscr{E}_c$. The emitted electrons will travel with a certain average velocity toward the collector (anode), with small loss of energy. At the anode, the electrons are absorbed, giving up an energy $\mathscr{E}_a$ in the form of heat, thus reaching the Fermi level of the anode. Since $\mathscr{E}_a$ is smaller than $\mathscr{E}_c$, as shown in Fig. 6.1(a), the amount of energy $eV_L = \mathscr{E}_c = \mathscr{E}_a$ can be transformed into electrical energy to the external load connected between the cathode and the anode, where V_L is the voltage across the external load.

When electrons are emitted in a moderate number, they will be retarded by the repulsion of the electrons already existing in the interelectrode space and will tend to form a negative space charge at the surface of the cathode, creating an energy barrier $\mathscr{E}_m$, as illustrated in Fig. 6.1(b). By increasing the temperature of the emitter, the number of emitted electrons can be increased, leading to a higher space charge potential barrier, and most of the electrons will be reflected toward

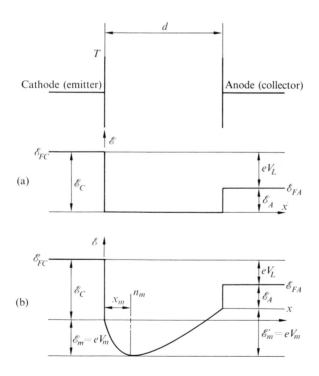

Figure 6.1. Energy diagram of a thermionic cell: (a) with no losses, (b) with a space charge barrier (losses).

the cathode. Only those electrons which have high energies will reach the anode and do useful work. Therefore, the space charge should be removed or neutralized if any appreciable amount of electrons is to reach the anode. Another problem is the existence of accelerating or decelerating gradients in the interelectrode space, leading to lower efficiencies because of the unmatched load conditions.

In 1903, Richardson [79] gave the first complete theoretical analysis of thermionic emission in vacuum. The mathematical analysis was later given by Langmuir [52] in detail. He assumed that electrons are emitted from the cathode with a Maxwellian velocity distribution and that after emission they are accelerated toward the anode by an applied electric field; the effect of collisions is neglected. Boltzmann's equation states that*

$$v \cdot \nabla f_e + \frac{e}{M} \nabla V \cdot \nabla_v f_e = \frac{\partial f_e}{\partial t}, \tag{6.1}$$

where f_e is the velocity distribution function, V is the electric potential, $\nabla_v f_e$ is the gradient of the distribution in hodographic space, $\partial f_e/\partial t$ is the collisional term, and v is the electronic velocity. Using a one-dimensional approach and neglecting collisions, Eq. (6.1) becomes

$$v \frac{\partial f_e}{\partial x} + \frac{e}{M} \cdot \frac{dV}{dx} \cdot \frac{\partial f_e}{\partial v} = 0.$$

The variables x and v are separable, and the general solution is of the form

$$f_e(x, v) = \sum_{k=0}^{\infty} \mathscr{A}_k \exp \left\{ \mathscr{B}_k \left[\frac{eV(x)}{M} - \tfrac{1}{2} v^2 \right] \right\}, \tag{6.2}$$

where $\mathscr{A}_k$ and $\mathscr{B}_k$ are constants determined by the boundary conditions on $f_e(x, v)$. $f_e(x, v)$ is such that

$$\int_{-\infty}^{+\infty} v f_e(x, v) \, dv = 0$$

and the Maxwellian distribution must be truncated† in velocity whenever a net current exists. Thus, Langmuir assumed a truncated Maxwellian distribution for the velocity of the electrons in the emitter, since the net current is often smaller than the saturation current. From Eq. (6.2),

$$f_e(x, v) = \mathscr{A}_0 \exp \left\{ \mathscr{B}_0 \left[\frac{eV(x)}{M} - \tfrac{1}{2} v^2 \right] \right\} \quad \text{for} \quad v \geq |v_\ell|, \quad \text{and}$$

$$\tag{6.3}$$

$$f_e(x, v) = 0 \quad \text{for} \quad v < |v_\ell|,$$

---

* See Sections 2.26 and 5.4.

† In the presence of a net current, the distribution function is 0 for $v < |v_\ell|$ and Maxwellian for $v \geq |v_\ell|$. The overall distribution function is then said to be truncated in velocity.

where

$$\mathscr{A}_0 = 2n_m\left(\frac{M}{2\pi kT_c}\right)^{1/2} \exp\left(-\frac{\mathscr{E}_a}{kT_c}\right),$$

$$\mathscr{B}_0 = M/kT_c \tag{6.4}$$

$$v_\ell = \left\{\frac{2e}{M}[V(x) - V_m]\right\}^{1/2},$$

where v_ℓ is the velocity necessary for an electron to cross a potential barrier equal to $[V(x) - V_m]$, T_c is the cathode temperature, and n_m is the particle density at the maximum potential barrier at distance x_m as shown in Fig. 6.1(b). The emitter saturation current J_s can now be calculated; it is given by

$$J_s = -e\int_0^\infty v f_e(0, v)\, dv, \tag{6.5}$$

and the net current density J is

$$J = -e\int_{-v_\ell}^\infty v f_e(x, v)\, dv. \tag{6.6}$$

Introducing Eq. (6.3) into Eqs. (6.5) and (6.6) yields

$$J_s = -en_m\left(\frac{2kT_c}{\pi M}\right)^{1/2} \exp\left(-\frac{eV_m}{kT_c}\right),$$

and

$$J = -en_m(2kT_c/\pi M)^{1/2}.$$

The magnitude of the emitter saturation current is

$$J_s = \mathscr{A}T_c^{1/2} \exp\left(-eV_m/kT_c\right), \tag{6.7}$$

where

$$\mathscr{A} = en_m(2k/\pi M)^{1/2}.$$

Equation (6.7), known as Richardson's equation, is valid only for vacuum converters, where the effect of collisions is negligible.

6.3 Thermodynamic Analysis of Thermionic Emission [36]

Richardson [79] and Dushman [22], among others [42], used the thermodynamic theory for the calculation of the electron current emitted by a heated solid material. In this case, it is assumed that an electron gas is in equilibrium with a hot conductor. The repulsion forces between the electrons of the gas are neglected, and it is assumed that the system is in equilibrium.

In a heat engine operating between the temperatures T and $T + dT$, the Carnot efficiency is given by

$$\eta_c = dT/T. \tag{6.8}$$

This can also be expressed as the ratio of the output work over the amount of heat supplied to the engine,

$$\eta_c = \frac{\Delta \mathscr{V} \, dp_r}{\Delta \mathscr{Q}},$$ (6.9)

where p_r is the pressure, $\Delta \mathscr{V}$ is the increase in volume of the system due to the evaporation of 1 mole of electrons, and $\Delta \mathscr{Q}$ is the latent heat of evaporation of this same amount of electrons. Equating Eq. (6.9) to Eq. (6.8), one obtains the Clausius-Clapeyron equation:

$$\Delta \mathscr{Q} = \Delta \mathscr{V} T \, dp_r/dT.$$ (6.10)

If a perfect electron gas is assumed, then, for the equation of state, $\Delta \mathscr{V} = \dot{\mathscr{R}}T/p_r$, and, putting this in Eq. (6.10),

$$\frac{dp_r}{p_r} = \frac{\Delta \mathscr{Q}}{\dot{\mathscr{R}}} \cdot \frac{dT}{T^2},$$ (6.11)

where $\dot{\mathscr{R}}$ is the gas constant. Introducing C_p, the heat capacitance at constant pressure for the gas and $\mathscr{Q}_0$, the molecular latent heat at absolute zero, yields

$$\Delta \mathscr{Q} - \mathscr{Q}_0 = \int_0^T C_p \, dT.$$ (6.12)

By applying the principle of equipartition of energy states, the total energy of N molecules, each having an average energy $\mathscr{E}_N$, is found to be $\mathscr{E}_i = \frac{1}{2} n_f n' \dot{\mathscr{R}} T$, where $\mathscr{E}_i = N \mathscr{E}_N$ is the internal energy of the gas, n_f is the number of degrees of freedom, and n' is the number of moles. The specific heat at constant volume is $C_v = \partial \mathscr{E}_i'/\partial T|_\mathscr{V}$ where $\mathscr{E}'$ is the internal energy per unit mole; thus $C_v = \frac{1}{2} n_f \dot{\mathscr{R}}$ and, since $C_p = \dot{\mathscr{R}} + C_v$, then $C_p = (\frac{1}{2} n_f + 1)\dot{\mathscr{R}}$. For a monoatomic gas, $n_f = 3$; thus $C_p = \frac{5}{2} \dot{\mathscr{R}}$. Introducing this equation into Eq. (6.12) yields

$$\Delta \mathscr{Q} = \mathscr{Q}_0 + \frac{5}{2} \dot{\mathscr{R}} T.$$ (6.13)

Introducing Eq. (6.13) into Eq. (6.11) and integrating yields

$$p_r = \mathscr{A} T^{5/2} \exp\left(- \mathscr{Q}_0/\dot{\mathscr{R}} T\right),$$ (6.14)

where $\mathscr{A}$ is a constant of integration. The average current density for electrons striking a unit area and coming from different directions with different speeds is

$$J = \frac{1}{4} n e \langle v \rangle.$$ (6.15)

From the equation of state,

$$n = p_r/kT,$$ (6.16)

and for a Maxwellian distribution of the electronic velocities,

$$\langle v \rangle = \left(\frac{8kT}{\pi M}\right)^{1/2}.$$ (6.17)

Taking into consideration Eqs. (6.14), (6.15), (6.16), and (6.17), one has

$$J_s = \mathscr{A}' T^2 \exp\left(- \mathscr{E}/kT\right),$$ (6.18)

where

$$\mathscr{A}' = \frac{(1 - \mathscr{C}_r)e}{(2\pi Mk)^{1/2}},$$

where $\mathscr{C}_r$ is called the mean reflection coefficient. Equation (6.18) yields a saturation current corresponding to Eq. (6.7). This equation is known as the Richardson-Dushman equation for saturation current density of electrons from a solid. The same equation can be obtained by using quantum mechanics, as will be shown in the next section.

6.4 Fermi-Dirac Analysis of Thermionic Emission

In 1926, Fermi [27] and Dirac [21] proved that the distribution of electronic velocities and, consequently, energies, is not Maxwellian. From statistical mechanics they showed that the distribution is

$$f_e(\mathscr{E}) = \frac{4\pi(2M)^{3/2}}{h} \frac{\mathscr{E}^{1/2}}{1 + \exp\left[(\mathscr{E} - \mathscr{E}_F)/kT\right]}. \tag{6.19}$$

The total number of electrons per unit volume is

$$n = \int_0^\infty f_e(\mathscr{E}) \, d\mathscr{E} \tag{6.20}$$

At absolute zero, $f_e(\mathscr{E})$ is proportional to $\mathscr{E}^{1/2}$ for $\mathscr{E} < \mathscr{E}_F$ (the Fermi level) and is equal to zero for $\mathscr{E} \geqslant \mathscr{E}_F$, as can be seen from Eq. (6.19). Figure 6.2(a) shows that all the levels of energy up to the Fermi level $\mathscr{E}_F$ are filled. For higher temperatures, some electronic energies exceed $\mathscr{E}_F$ and, if the thermal energy given to an electron is higher than the work function $\mathscr{E}$ of the metal, the electron will be able to escape the metal's surface.

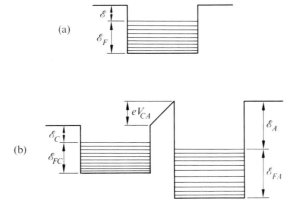

Figure 6.2. (a) Energy levels in a metal at absolute zero. (b) The contact potential V_{CA} as difference of work potentials between two different surfaces.

In the interior of a metal, the bound electrons exist in discrete energy levels, whereas the free electrons fill a continuous region of energies, as seen from Eq. (6.19). Inside the metal, the conduction electrons move in a practically equipotential region due to the presence of the ions. This balance is broken at the surface, where an image force acts on the electron by creating a potential energy barrier. The energy necessary for the electron to cross this barrier is called the *work function* $\mathscr{E}$. It is equal to $\mathscr{E} = (\mathscr{E}_{in} - \mathscr{E}_F)$ where $\mathscr{E}_{in}$ is the internal potential energy. The image force can be given by Coulomb's law applied to an electron of charge e at a distance x from the metal's surface and its image inside the metal; thus

$$F(x) = \frac{e^2}{16\pi\varepsilon_0 x^2},$$

and the potential of the electron is

$$V(x) = -\int_{r_D}^{x} \frac{F(x)}{e}\, dx$$

for $x > r_D$. Coulomb's law is not valid for distances smaller than the Debye radius r_D for which a constant force is postulated by Schottky [81]; thus

$$V_1(x) = -\frac{ex}{16\pi\varepsilon_0 r_D^2} \qquad \text{for} \quad x < r_D,$$

$$V_2(x) = \frac{ex}{16\pi\varepsilon_0}\left(\frac{1}{x} - \frac{1}{r_D}\right) \qquad \text{for} \quad x > r_D.$$

The total work function is given by

$$\mathscr{E} = eV_1(r_D) + eV_2(\infty),$$

or

$$\mathscr{E} = \frac{-e^2}{8\pi\varepsilon_0 r_D}.$$

Figure 6.2(b) shows two different metallic plane surfaces with Fermi levels $\mathscr{E}_{Fc}$ and $\mathscr{E}_{Fa}$ and work functions $\mathscr{E}_c$ and $\mathscr{E}_a$. At equilibrium, the total current flowing from one surface to another is zero, on the average, and a potential difference appears between the surfaces of the two metals. It is called the *contact potential* and is given by

$$eV_{ac} = \mathscr{E}_c - \mathscr{E}_a. \tag{6.21}$$

Now the saturated current emitted from the metallic surface and received at the collector can be calculated. The current density due to electrons having velocities between $v_\perp$ and $v_\perp + dv_\perp$ perpendicular to the metallic surface is given by $dJ_s = ev_\perp \, dn_\perp$, where $dn_\perp$ can be obtained from Eq. (6.20) by using the transformation

$$4\pi v^2 \, dv = dv_\perp \, dv_y \, dv_z.$$

Thus

$$dn_\perp = \frac{2M^3}{h^3} dv_\perp \int_{-\infty}^{+\infty} \int_{-\infty}^{+\infty} \exp\left(\frac{\mathscr{E}_F}{kT}\right) \exp\left[-\frac{1}{2}\frac{M}{kT}(v_\perp^2 + v_y^2 + v_z^2)\right] dv_y\, dv_z,$$

(6.22)

where 1 has been neglected in front of the exponential in the denominator of Eq. (6.19), which is valid for the temperatures in practical thermionic converters. Performing the integrations of Eq. (6.22) yields

$$dn_\perp = \frac{4\pi}{h^3} M^2 kT \exp\left(\frac{\mathscr{E}_F}{kT}\right) \exp\left(-\frac{1}{2}\frac{Mv_\perp^2}{kT}\right) dv_\perp,$$

and the saturation current density J_s will be

$$J_s = \frac{4\pi e M^2 kT}{h^3} kT \exp\left(\frac{\mathscr{E}_F}{kT}\right) \int_v^\infty v_\perp \exp\left(-\frac{1}{2}\frac{Mv_\perp^2}{kT}\right) dv_\perp,$$

(6.23)

where $v = (2\mathscr{E}/M)^{1/2}$. After integrating Eq. (6.23),

$$J_s = \left(\frac{4\pi Mek^2}{h^3}\right) T^2 \exp\left(-\frac{\mathscr{E}}{kT}\right),$$

or, taking into consideration the mean reflection coefficient $\mathscr{C}$,

$$J_s = \mathscr{A}'' T^2 \exp\left(-\mathscr{E}/kT\right),$$

(6.24)

where

$$\mathscr{A}'' = (1 - \mathscr{C}_r) M 4\pi ek^2/h^3.$$

Most of the time the reflection coefficient is smaller than 0.1 and it is often neglected. With $\mathscr{C}_r = 0.1$, $\mathscr{A}'' = 1.08 \times 10^6$ A/m²-°K².

Richardson plots represent $\ln (J_s/T^2)$ versus $1/T$. As can be seen from Eq. (6.24), these plots are linear, according to

$$\ln (J_s/T^2) = 13.9 - 1.16 \times 10^4 V/T.$$

(6.25)

This equation is plotted in Fig. 6.3 with the work potential as parameter. In the experimental Richardson plots [45], it is found that the work potential V ($\mathscr{E} = eV$) is not independent of temperature, but is rather a linear function of it, such as $V = V_0 + \mathscr{C}_\alpha T$, where $\mathscr{C}_\alpha$ is a constant depending on the type of emitter, V is called the effective work function, and V_0 is the Richardson work function. Equation (6.25) can now be written as

$$\ln (J_s/T^2) = (13.9 - 1.16 \times 10^4 \mathscr{C}_\alpha) - 1.16 \times 10^4 V_0/T.$$

The performance characteristic. If space charge effects are neglected, all the electrons emitted from the cathode will reach the anode if the output voltage is equal to or smaller than the contact potential given by Eq. (6.21). The converter

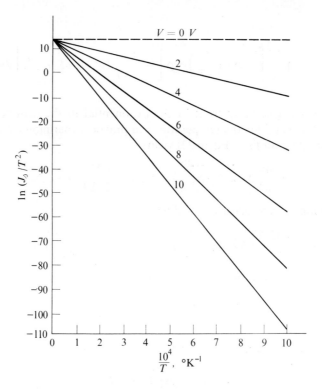

Figure 6.3. The Richardson plots with the work potential as parameter, the effective work function being neglected.

acts as a constant current generator capable of a maximum power density equal to

$$P_m = J_s V_{ac}. \tag{6.26}$$

For an output voltage larger than the contact potential V_{ac}, the variation of the current will be exponential. For the performance equation,

$$J = J_s = \text{constant} \quad \text{for} \quad 0 \leqslant |V| \leqslant V_{ac},$$

and

$$J = J_s \exp\left[-\frac{e}{kT}(|V| - V_{ac})\right] \quad \text{for} \quad |V| > V_{ac}. \tag{6.27}$$

This equation is plotted in Fig. 6.4.

6.5 Different Types of Thermionic Converters [2, 5, 40, 48, 53]

No thermionic converter can operate properly unless the space charge effects are sufficiently reduced in some way. To understand the choice of different types of energy converters, it will be necessary to determine the different considerations in designing the thermionic engine. This can be done by calculating the efficiency. Assuming that the space charge is nullified, the peak power density is given by Eq. (6.26). The most important loss is by radiation. The neighboring regions of the cathode having good absorptive qualities, the net rate of radiant loss density $\mathcal{Q}_r$ can be obtained from

$$\mathcal{Q}_r = \mathcal{B}_1(\Theta_c T_c^4 - \Theta_N T_N^4),$$

where $\mathcal{B}_1$ is Stefan's constant $\mathcal{B}_1 = 5.67 \times 10^{-8}$ W/m²-°K⁴, T_c and T_N are the temperatures of the cathode and its neighboring space, respectively, and Θ_c and Θ_N are their emissivities. Since the temperature of the cathode,

$$\mathcal{Q}_r = \mathcal{B}_1 \Theta_c T_c^4, \tag{6.28}$$

the total input power density is equal to the sum of the heat cooling the cathode $(J_s V_c)$, the heat loss by radiation $(\mathcal{Q}_r)$, and the kinetic energy $(\mathcal{Q}_k)$ of the electrons emitted from the cathode surface. This kinetic energy is equal to $\mathcal{Q}_k = 2kT_c J_s/e$. The loss of heat by conduction and all other losses being neglected,

$$\mathcal{Q}_{in} = J_s V_v + \mathcal{B}_1 \Theta_c T_c^4 + \frac{2k}{e} T_c J_s. \tag{6.29}$$

The efficiency is, by definition, equal to the power output divided by the power input:

$$\eta = \frac{1 - V_a/V_c}{1 + 2(kT_c/eV_c) + (\mathcal{B}_1 \Theta_c T_c^4/J_s V_c)}.$$

Putting $V_T = kT_c/e$ and replacing J_s by its value given by Eq. (6.24), the efficiency

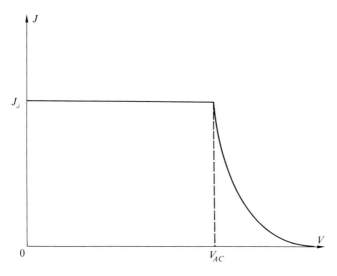

Figure 6.4. Performance characteristic of a thermionic converter. The space charge is neglected.

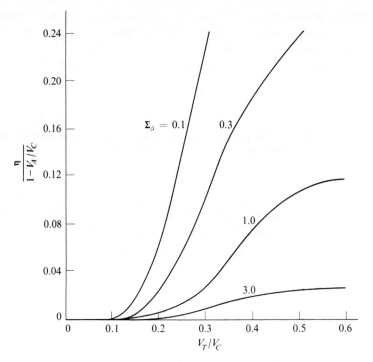

Figure 6.5. Effect of the ratio V_T/V_C on the thermionic efficiency for low values of V_T/V_C. Parameter $\Sigma_s = \mathscr{B}_1 \Theta_c T_c^2 / \mathscr{A}'' V_C$.

is

$$\eta = \frac{1 - V_a/V_c}{1 + 2(V_T/V_c) + (\mathscr{B}_1 \Theta_c T_c^2 / \mathscr{A}'' V_c) \exp (V_c/V_T)}. \tag{6.30}$$

It is seen from Eq. (6.30) that there are three considerations in designing a thermionic converter:

1. To reduce space charge,
2. To increase the ratio V_T/V_c,
3. To decrease the ratio V_a/V_c.

A plot of $\eta/(1 - V_a/V_c)$ is shown in Fig. 6.5, where $\Sigma_s = \mathscr{B}_1 \Theta_c T_c^2 / \mathscr{A}'' V_c$ is a parameter. From this figure, note that η drops sharply with decreasing V_T/V_c.

To accomplish the three objectives mentioned above, four types of thermionic converters are considered:

1. High-vacuum, closely spaced thermionic converters,
2. Low-pressure thermionic converters,
3. High-pressure thermionic converters,
4. Arc mode thermionic converters.

High-vacuum thermionic converters. In this converter, the anode-cathode distance is very small, often smaller than 10 microns. The use of closely spaced electrodes is one of the most practical methods of reducing the space charge barrier, as will be discussed later. The collisionless theory of Section 6.2 leads to the value of the saturation current in this type of converter, as expressed by Eq. (6.7). To calculate the electron density of the interelectrode space as a function of the interelectrode distance, it is necessary to solve the integral

$$n(x) = \int_{-v_\ell}^{\infty} f_e(x, v)\, dv, \tag{6.31}$$

where v_ℓ and $f_e(x, v)$ are defined by Eqs. (6.4) and (6.3), respectively. Integration of Eq. (6.31) yields

$$n(x)/n_m = \exp(\psi_0)[1 \pm \mathrm{erf}\,(\psi_0)^{1/2}], \tag{6.32}$$

where n_m is the space charge at the minimum potential (Fig. 6.1b), ψ_0 is defined by

$$\psi_0 = \frac{1}{2}\frac{Mv_\ell^2}{kT_c},$$

and the error function is defined by

$$\mathrm{erf}\,(\psi_0)^{1/2} = \frac{2}{\sqrt{\pi}} \int_0^{\sqrt{\psi_0}} \exp(-\psi_0^2)\, d\psi_0.$$

From Poisson's law, $\nabla^2 V(x) = en(x)/\varepsilon_0$ or, in dimensionless variables, and taking into consideration Eq. (6.4),

$$2\frac{d^2\psi_0}{d\xi_0^2} = [1 \pm \mathrm{erf}\,(\psi_0)^{1/2}]\exp(\psi_0), \tag{6.33}$$

where

$$\xi_0 = e\left(\frac{2n_m}{\varepsilon_0 kT_c}\right)^{1/2}(x - x_m),$$

or

$$\xi_0 = \left(\frac{2\pi Me^2}{k^3\varepsilon_0^2}\right)\frac{|J|^{1/2}}{T^{3/4}}(x - x_m).$$

The solution of Eq. (6.33) is given by

$$\xi_0 = \int_0^{\psi_0} \frac{dx}{[\exp(x) - 1 \pm \mathrm{erf}\,(x)^{1/2} \pm 2(x/\pi)^{1/2}]^{1/2}}, \tag{6.34}$$

where x is an arbitrary variable. Langmuir plotted ψ_0 versus ξ_0 as given by the above equations. This plot is shown in Fig. 6.6, where $\xi_0 = 0$ represents the point of minimum potential, positive ξ_0 represents the space between the potential minimum and the anode, and negative ξ_0 represents the space between the potential minimum and the cathode. Curve A represents the case where collisions are neglected, as given by Eqs. (6.32), (6.33), and (6.34), whereas curve B represents

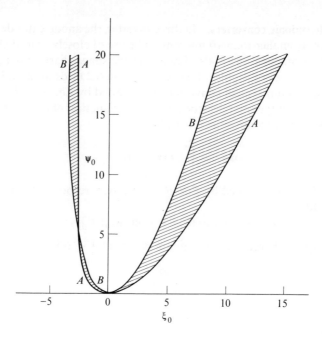

Figure 6.6. Effect of the distance between the electrodes on the space charge. Curve A: collisions neglected; curve B: electron-electron collisions considered with a Maxwellian distribution in the mass [8]. The actual $\psi_0(\xi_0)$ function should be situated in the shaded area.

the case where electron-electron collisions are considered by assuming that they are numerous enough to lead to an isotropic electronic distribution (Maxwellian distribution) in the center of mass [8]. Most of the experimental results are between these two theoretical extreme cases. From the $\psi_0(\xi_0)$ characteristic, J can be easily calculated as a function of the voltage and with the interelectrode distance as a parameter, for given saturated current density J_s, cathode temperature T_c, and contact potential V_{ac}. The total output voltage will be

$$V_L = V_{ac} + (V_m - V'_m).$$

Vacuum converters always operate in the space-charge-limited mode and the output current is never equal to the saturation current. This output current rarely exceeds 10% of the saturation current in most of the experiments.

Because of the high temperatures, the close spacing of the electrodes becomes a difficult problem. Furthermore, the evaporated tungsten coating of the anode increases its potential, limiting the prospects of such converters to very small devices, not exceeding 10 W power output.

Low-pressure thermionic converters. Because of the difficulties of close spacing in vacuum thermionic converters, another way of reducing space charge effects becomes necessary at higher power densities. As was proposed by Wilson [95, 96], the negative space charge between the electrodes can be neutralized by the presence of positive ions within the interelectrode space. Such positive ions can be generated in appreciable quantity through both surface and volume ionization if the gas inside the thermionic converter has an ionization potential smaller than the work function of the emitter. Cesium has the lowest ionization potential and is the most used gas. A cesium atom striking the emitter surface will be ionized by giving one of the valence electrons to the cathode. Another quality of the cesium is its low work function. Cesium condenses on the relatively cold anode surface, lowering the apparent work function of the anode.

The pressure inside the converter determines the amount of cesium vapor. It was shown experimentally by Taylor and Langmuir [86] that the cesium vapor pressure p_{cS} in torrs is related to the cesium pool temperature T_{cS} by the relation

$$p_{cS} = 121.2 T_{cS}^{-1.35} \exp\left(-1755/T_{cS}\right).$$

In the low-pressure converter, the mean free path is much larger than the interelectrode spacing, for cesium pressures of the order of magnitude of 10^{-4} mm of mercury.

The performance characteristic of the low pressure converter is similar to the one given in Fig. 6.4. The potential distribution in the interelectrode space can be calculated by considering an isothermal metallic enclosure containing neutral cesium atoms, cesium ions, and electrons. The enclosure is assumed to consist of two parallel plates of infinite extent (to neglect end effects), separated by a distance d. In equilibrium the particle velocity distribution function becomes Maxwellian. The equations of conservation of momentum are given by Boltzmann's transfer equation: for the neutral atoms, having a density n_n,

$$dn_n/dx = 0;$$

for the ions, having a density n_i,

$$\frac{kT}{e} \cdot \frac{dn_i}{dx} + n_i \frac{dV}{dx} = 0;$$

and for the electrons, having a density n_e,

$$\frac{kT}{e} \cdot \frac{dn_e}{dx} - n_e \frac{dV}{dx} = 0.$$

Poisson's law states

$$d^2V/dx^2 = e(n_e - n_i)/\varepsilon_0.$$

These equations can be solved for $V(x, n_e/n_i)$ by taking the appropriate boundary conditions. The solutions, as found by Auer [3] and Goldstein [31],

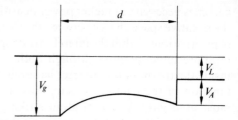

Figure 6.7. Potential distribution in a low-pressure thermionic conversion.

are similar to the distribution shown in Fig. 6.7. It was shown that for low-pressure converters, the Debye radius is much smaller than the interelectrode space, suggesting that Coulomb interactions* are dominant.

In low-pressure thermionic converters, the effect of the ions is taken into account by using Boltzmann's equation for the electrons and for the ions. If the collisions are neglected as was done by Auer [3], the Boltzmann equations become

$$v_e \frac{\partial f_e}{\partial x} + \frac{e}{M} \cdot \frac{dV}{dx} \frac{\partial f_e}{\partial v_e} = 0,$$

and

$$v_i \frac{\partial f_i}{\partial x} - \frac{e}{M} \cdot \frac{dV}{dx} \frac{\partial f_i}{\partial v_i} = 0,$$

where f_i is the ionic velocity distribution function, and f_e is the electronic velocity distribution function. The ionic and electronic densities are, respectively,

$$n_i(x) = \int_{-\infty}^{+\infty} f_i(x, v_i) \, dv_i,$$

and

$$n_e(x) = \int_{-\infty}^{+\infty} f_e(x, v_e) \, dv_e.$$

They are related to the voltage by Poisson's law;

$$d^2 V/dx^2 = e(n_e - n_i)/\varepsilon_0. \tag{6.35}$$

By assuming that the ions and the electrons leave the emitter with Maxwellian distribution functions of their velocities, it can be found that the electronic saturation current of the emitter is

$$J_{ecS} = -en_{em}\left(\frac{2kT_c}{\pi M_e}\right)^{1/2} \exp\left(-\frac{eV_m}{kT_c}\right),$$

and the net electron current density is

$$J_e = J_{ecS} \exp\left(eV_m/kT_c\right).$$

---

*   Coulomb interactions are interactions in which the forces involved are functions of the reverse of the square of the distance between the interacting particles.

For the ions,

$$J_{icS} = en_{ic}\left(\frac{2kT_c}{\pi M_i}\right)^{1/2} \quad \text{for} \quad V_a \leqslant 0,$$

and

$$J_{iaS} = en_{ia}\left(\frac{2kT_a}{\pi M_i}\right)^{1/2}\left[1 + \text{erf}\left(\frac{eV_a}{kT_c}\right)^{1/2}\right]^{-1} \quad \text{for} \quad V_a > 0.$$

The net ion current density is, for $V_a \leqslant 0$, $J_i = J_{iaS}$ and for $V_a > 0$,

$$J_i = J_{iaS} \exp\left(-eV_a/kT_c\right).$$

The performance characteristic has been calculated by Goldstein [31], using Eq. (6.34). The result of a numerical calculation is shown in Fig. 6.8 with $\mathscr{A}$ as parameter; $\mathscr{A}$ is defined by the ratio

$$\mathscr{A} = (M_i/M_e)^{1/2}(-J_{iaS}/J_{eaS}).$$

In this figure the output current density is equal to $J = J_e + J_i$. This collisionless analysis also assumes that the saturation current of the anode is negligible. This assumption is no longer valid if the anode radiates heat to the emitter.

Because low-pressure converters require low emitter work functions, most of the experimental work was done for $\mathscr{A} \geqslant 1$, as shown by Hernqvist [41] and Steele [84]. It can be easily seen from their experiments that the characteristics

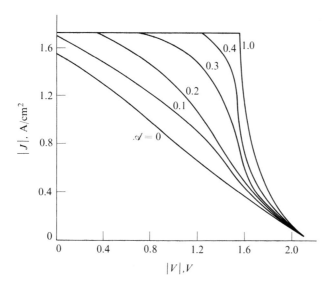

Figure 6.8. The performance characteristic of a low-pressure thermionic converter (collisionless theory): $T_C = 2000°\text{K}$, $V_C = 3.35$ V, $V_A = 1.80$ V, and $d = 25.4$ microns. (Adapted from Blue and Ingold [8].)

of the converter are greatly limited by the lower temperatures of the cesium vapor, leading to an increase of the limiting ion current.

High-pressure thermionic converters. The power output of thermionic converters is a function of the cesium pressure, as shown in Fig. 6.9. At low pressure, almost all the incident cesium ions are ionized at the emitter. The density of the ions increases with increasing pressure, leading to a decrease in the space charge. The power density increases until a plateau is reached where all the space charge has been neutralized, at a cesium pressure of about 10^{-5} mm Hg. When the pressure is further increased, the mean free path becomes larger and the probability of collision increases accordingly, leading to a drop in the output power at a pressure of about 10^{-1} mm Hg. At still higher pressures, cesium is deposited on the anode surface, decreasing its work function, and the conductivity of the cesium plasma is increased through volume ionization. The output power increases again, suggesting the use of a high-pressure converter.

For the theoretical calculation of the performance characteristic, collisions should be taken into consideration in Boltzmann's equation. These are of two types: elastic collisions and inelastic collisions. Each individual particle current is assumed to be independent of position. If the velocity of the center of mass of the particles is neglected and if all the temperatures of the different particles are assumed to be equal, the diffusion equations for the ions and electrons can be written for the steady state. To these equations, the momentum equation is added

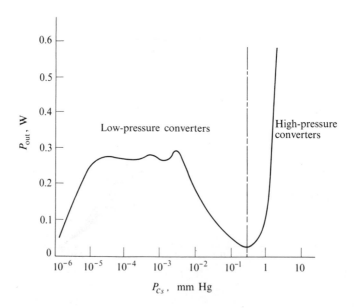

Figure 6.9. Experimental variation of power output with cesium pressure: $T = 2523°K$, $d = 1$ mm. (Adapted from Wright and Harrowell [98].)

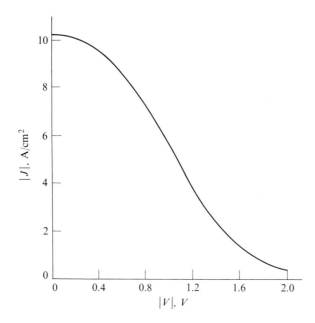

Figure 6.10. Performance characteristic of a high-pressure thermionic converter: $T_C = 2000°K$, $V_C = 2.72$ V, $V_A = 1.70$ V, $p_{CS} = 10$ mm Hg, and $d = 18$ microns. (From Warner and Hansen [88].)

as well as Poisson's law and the energy equation, which is

$$\frac{d\mathcal{Q}}{dx} + J\frac{dV}{dx} = 0.$$

These equations have been studied by Hernqvist [39]* and his co-workers. A typical performance characteristic is shown in Fig. 6.10, as reported by Warner and Hansen [88]. For more detail, papers written by Hatsopoulos [38] and Hansen and Warner [33] should be consulted.

Arc mode thermionic converters [83]. The high-pressure converter is characterized by an apparent saturation much smaller than the saturation current at the emitters. In this case, ion generation is due mainly to the resonance ionization at the cathode, which has a temperature in the 1500 to 1800°C range.

Higher efficiencies can be obtained in the arc mode of operation (often called the ignited mode) at low cathode temperatures (in the 1200 to 1500°C range). The most important mechanism of ion generation is impact ionization. The ignited converter becomes an externally heated hot cathode arc discharge, the cathode

---

* General Electric Vallecitos Atomic Laboratory.

becomes space charge limited, and the performance characteristic rises to much higher current densities (relative to the unignited mode). As compared to the idealized space-charge-neutralized converter of Fig. 6.4, there is a shift in the output voltage due to the loss of power necessary for the ionization process. In the simplest case, when this internal voltage drop V_d is independent of current, the performance characteristic of the arc mode can be obtained just by shifting the exponential by V_d toward zero. Experimentally, there is found a deviation from the exponential characteristic at both high and low current densities. This discrepancy is due primarily to the potential distribution in the interelectrode space. The converter arc discharge is composed of a Langmuir double sheath at the cathode and an anode sheath, the rest of the discharge being made up of a glow plasma characterized by a space constant potential and a high electronic temperature.

With the statement that the power input in the discharge is equal to the power delivered to the anode by the hot plasma electrons, it can be found that

$$V = (V_{c,\text{eff}} - V_a) - \frac{2k}{e}(T_e - T_c),$$

where $V_{c,\text{eff}} = V_c$ for $J = J_s$ and

$$V_{c,\text{eff}} = \frac{kT_e}{e} \ln (\mathscr{B}T_e^2/J) \qquad \text{for} \quad J < J_s,$$

with $\mathscr{B}$ being the electron emission constant, and the losses being neglected. This result suggests that for small currents, the electronic temperature T_e decreases with the increasing current, reaches a minimum, and then increases.

6.6 Thermionic Materials

The necessary materials for thermionic power generation are those for the cathode, the anode, the filling gases, and the envelope.

Cathodes [32]. A good cathode material should have an adequate emission capability at the operating temperature and an acceptable life span. It should have a low thermal emission and high thermal and electrical conductivities. Other factors such as chemical stability and price are important for the choice of a material.

Some metals which make good cathodes are tungsten (W), molybdenum (Mo), and tantalum (Ta). The use of these materials is not practical because of their short life due to high evaporation at the working temperatures. To avoid this evaporation, a partial monolayer of positive cesium ions is laid by condensation on the metal, leading to the elimination of the evaporation problem and to a higher emission at temperatures under 2000°C. This can be a good arrangement for high-pressure converters, where the monolayer on the emitter is maintained by a rate of condensation of the cesium on the cathode which offsets the rate of evaporation of the layer. For low-pressure and vacuum converters this rate is very small and a film of thorium instead of cesium is placed on the metal. The thorium film has a very low evaporation rate. Often thoriated metals are used, leading to a higher lifetime, since the evaporated thorium atoms are replaced by a diffusion of thorium from inside the cathode to its surface.

Table 6.1

Properties of some thermionic emitters [34, 44, 70]*

Material	V_c, V	J_c, A/cm^2	$V_c J_c$, W/cm^2	T_c, °K	Evaporation rate, mg/m^2/hr
Cs-W	2.00	50	100	1500	—
LaB$_6^{51}$	2.90	10	29.00	1900	0.3000
UC^{99}	3.16	5	15.80	2000	0.7200
BaO·SrO	1.55	10	15.50	1100	0.0009
W (impregnated)	2.12	6	12.72	1410	0.0058
Lemmens cathode	2.15	5	10.75	1400	0.0720

* Average values are taken.

Some nonmetallic materials which are also used are barium oxides and uranium carbides, and these are often mixed with strontium and calcium oxides. The characteristics of several materials are given in Table 6.1. From this table it can be seen that cesiated tungsten gives the best performance. Figure 6.11 shows Richardson curves of a tungsten filament in equilibrium with the cesium vapor at the filament temperature T_c. The rate $\dot{n}_{cs}$ of arrival of cesium on the tungsten surface per square centimeter per second is taken as a parameter, the diagonal lines representing the fraction of filament surface covered by cesium. Figure 6.12 shows the same for several cesiated materials when the cesium arrival flux is equal to 10^{16} atoms/cm^2·sec, as reported by Wilson [94].

Anodes [7]. For a material to be used as an anode, it should have a low work function, low thermal emissivity, and low electrical resistivity. It should have good chemical and mechanical characteristics. Cesium has one of the lowest work functions and often is condensed on copper, nickel, or silver in such a way that the final work potential of the anode becomes that of cesium (about 1.8 V).

Cesium layers on refractory metals may lead to lower work potentials; however, the problem of thermal emissivity becomes extremely important. Cesium layers on tungsten [85], antimony, and other materials were investigated [34, 44] with relative success. Layers of other alkali metals [97] on refractory materials have also been studied. The alkali metals used were sodium and potassium, while many refractory metals such as tantalum and molybdenum were considered. Finally, as for the cathode, the use of strontium and barium oxides was considered, but the problem of anode poisoning* should be solved. As for the evaporation of the anodic material, the problem is less acute than it is in the case of the cathode, because of the lower temperature of the collector.

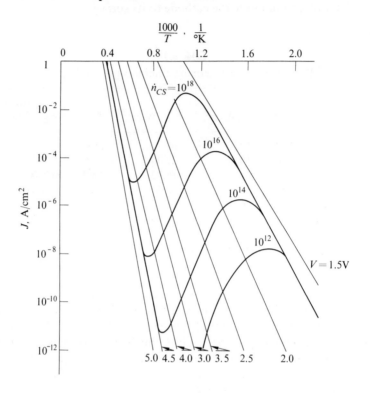

Figure 6.11. Electron emission from cesiated tungsten. These Richardson plots should be compared with the theoretical ones of Fig. 6.3. (Adapted from Wilson [94].)

Neutralization gases. A neutralization gas should have a low ionization potential, and for this reason cesium is the most used material, followed by the other alkali metals. Because of the corrosive nature of these metal vapors, workers [23, 30, 49]† investigated the use of mercury, cesium fluoride, barium, and some noble gases such as argon, xenon, and helium. These gases led to a loss in efficiency because of their high ionization potentials. To avoid this loss, mixtures of alkali and mercury vapors have been used for experiments, but with limited success.

---

* An anode is said to be poisoned when atoms evaporating from the cathode deposit on the anode surface and increase its work function.

† See also papers in Report on Thermionic Conversion Specialists Conference, Cleveland, Ohio, 1964.

Envelopes. The envelope of a thermionic device usually absorbs only a small quantity of the cathodic heat. This material should have a high thermal conductivity in order to avoid an accumulation of heat energy on the envelope. Most often glass is used, but there are instances where, because of the high operating temperatures, metallic or ceramic materials are considered. The envelope material should have a low thermal expansion to resist thermal shocks, and it should have a low electrical conductivity to avoid electrical losses.

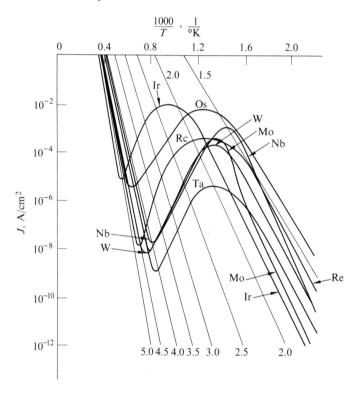

Figure 6.12. Electron emission from cesiated refractory metals for a cesium arrival flux of 10^6 atoms/cm^2 sec. (Adapted from Wilson [94].)

The corrosive effect of the alkali metals, the mechanical characteristics, and the price should also be taken into consideration for any choice of material.

6.7 Space Charge Neutralization

Several methods are used to reduce the space charge in the interelectrode space. Two of them, described in Section 6.5, are space charge reduction by the use of a small interelectrode space and space charge reduction by the introduction of positive ions. The second method can be divided into three categories of space charge neutralization:

1. By surface ionization,
2. By volume ionization,
3. By the use of auxiliary electrodes.

The number n_i of cesium ions emitted from an emitting solid surface per unit area per unit time is given by the Saha-Langmuir equation

$$\frac{n_i}{n_n} = \left\{ 1 + 2 \exp \left[\frac{e(V_i - V_c)}{kT_c} \right] \right\}^{-1},$$

where n_n is the arrival rate of the cesium atoms impinging on the surface and V_i is the ionization potential of cesium. When the density of the emitted ions equals the density of emitted electrons, the space charge becomes completely neutralized. This happens when

$$\frac{J_s}{Sen_i} = \left(\frac{M_i}{M_e} \right)^{1/2},$$

where S is the area of the emitter, M_i is the mass of the ions, and M_e is the mass of the electrons.

When the emitter work potential V_c becomes smaller than the minimum value for which space charge neutralization by surface ionization alone is possible, a double performance characteristic appears for the converter. That means that new ions are generated in an electron potential well through volume ionization in a certain region of the gas. The energy absorbed to produce the ions results in a potential drop V_0, the value of which is

$$V_0 = 2k(T_e + T_\varepsilon) + \frac{V_i J_i + V_a J_a}{J_e},$$

where the subscripts e, ε, i, and a refer to the electrons, electrons in the ion generation region, ions, and excited atoms, respectively. The ion current from the ion generation region is given by

$$J_i = \mathscr{B} m_a n_\varepsilon d' \exp(-eV_i/kT_\varepsilon),$$

where d' is the width of the ionization region and $\mathscr{B}$ is a quantity given by

$$\mathscr{B} = \left(\frac{2}{\pi M_e} \right)^{1/2} (kT_\varepsilon)^{3/2} \mathscr{C}_\delta \left(2 + \frac{eV_i}{kT_\varepsilon} \right),$$

with the assumption that the ionization cross section Q_i is proportional to the electron energy above ionization potential, as $Q_i = \mathscr{C}_\delta(V - V_i)$ with $\mathscr{C}_\delta$ as constant of proportionality.

The use of auxiliary electrodes has been considered by Johnson and Webster [47]. In this method, the space charge is neutralized by the gating action of positive ion sheaths which surround auxiliary grids introduced between the

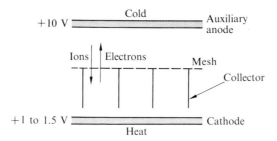

Figure 6.13. Space charge neutralization by the use of auxiliary electrodes. Gabor's generator.

cathode and the anode. This method is not very promising and other versions of the same method have been proposed. One of them is the device proposed by Gabor [30], the features of which are shown in Fig. 6.13. In this device, the inter-electrode space is separated into two regions by the presence of a mesh at 0 V. The region between the mesh and the auxiliary anode contains a luminous ionizing discharge; that in the cells formed by collectors in the space between the mesh and the cathode contains a dark conducting plasma. The electrons flow from the auxiliary anode to the cathode; the ions flow in the opposite direction.

Another method of space charge neutralization is by the use of combined electric and magnetic fields. These so-called magnetic triodes have been studied by Welsh, Hatsopoulos, and Kaye [92] and others [82, 100]; their principle is illustrated in Fig. 6.14. In this device, the electrons are accelerated by a potential of about 100 V to a third electrode and are diverted by a transverse magnetic field to the collector. This has the advantage of increasing the possibilities of heat shielding and of generating alternating energy by modulating the magnetic field.

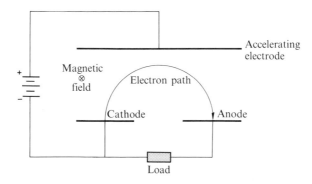

Figure 6.14. The magnetic triode. The electrons are submitted to combined electric and magnetic fields.

It was also proposed [46] that space charge neutralization could be achieved by the use of fission fragments in an argon plasma, but, because of the necessary high pressure, it is doubtful that a design of such a generator can be achieved in the near future.

6.8 Power Output, Losses and Efficiency

Power output. As seen in Section 6.6, when the space charge has been completely neutralized, the maximum power density output is given by Eq. (6.26). For an output voltage V larger than the contact potential V_{ac}, the current density is then given by Eq. (6.27), and the total output power can be written, by combining Eqs. (6.26) and (6.27), as

$$P_{out} = VJ_s \exp\left[-\frac{e(V - V_{ac})}{kT} \right],$$

or, letting $J_s' = J_s \exp (eV_{ac}/kT)$, it is found that

$$P_{out} = VJ_s' \exp\left(-\frac{eV}{kT} \right),$$

or, letting $V_0 = kT/e$ and $P_0 = V_0 J_s'$, then

$$\frac{P_{out}}{P_0} = \frac{V}{V_0} \exp\left(-\frac{V}{V_0} \right). \tag{6.36}$$

Equation (6.36) has been plotted in Fig. 6.15 where it is seen that the maximum power is obtained when $V = V_0$; this is the absolute maximum obtainable from a thermionic converter. In practice, this maximum power is not desirable, because of the corresponding very low output voltage and efficiency.

Losses. The losses in a high-vacuum thermionic converter can be listed as follows:

1. Losses in the plasma,
2. Losses at the collector,
3. Joulean losses due to the conducting materials,
4. Radiation loss from the cathode,
5. Heat flow in the conductor and the envelope,
6. Kinetic energy carried away from the cathode by the emitted electrons.

To these, the following losses should be added in the case of a gas-neutralized converter:

7. Heat conducted by the cesium vapor,
8. Energy to produce the cesium ions.

The losses in the plasma are due to the unavoidable collisions between the electrons traveling from the anode to the cathode. These collisions can be expressed by an internal electrical resistance in the plasma. These losses are important only in high-pressure converters and are equal to

$$P_1 = \frac{MS \, dv_R}{e^2 n_e} J_s^2,$$

where v_R is the index of refraction of the surface.

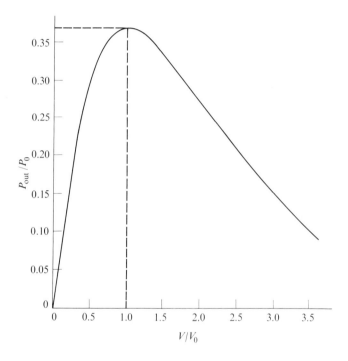

Figure 6.15. Power output of a thermionic converter. The maximum power is obtained when $eV = kT$.

The losses at the collector are due primarily to the electrical and thermal conductivity of the anode. The emission of the anode is generally negligible because of its lower temperatures.

Joulean losses are due mainly to the electrical connections between the electrodes and the load. The loss at the anode is usually extremely small, whereas the loss at the cathode can be studied with the heat lost by conduction.

The radiation loss from the cathode can become extremely large. It is given by Eq. (6.28). To diminish this loss, it is necessary to use materials having low emissivities for the cathode.

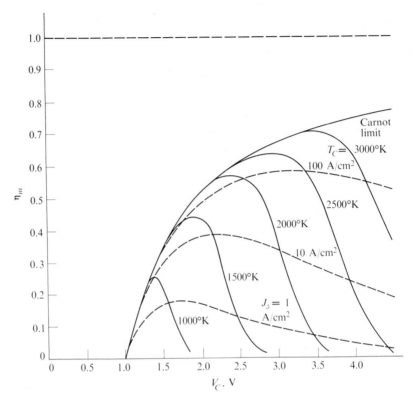

Figure 6.16. Maximum efficiency as a function of the cathode work function. The cathode temperature and saturated current densities are taken as parameters: $\Theta_c = 0.35$, $V_a = 1$ V. (From Cayless [16].)

Of all the heat lost by conduction, the most important is the heat lost through the lead wires joining the cathode to the anode. From the Wiedemann-Franz law,

$$P_5 = \tfrac{1}{2}J_s\left(\frac{\mathcal{L}T_c^2}{V_w} - V_w\right),$$

where V_w is the voltage drop along the wire and $\mathcal{L}$ is the Lorentz number, equal to

$$\mathcal{L} = \pi^2k^2/3e^2, \tag{6.37}$$

where k is Boltzmann's constant. This loss can be optimized by the choice of an adequate wire thermal conductivity.

The kinetic energy carried away by the emitted electrons is equal to

$$P_6 = 2kT_cJ_s/e.$$

This cannot be avoided, since it is directly proportional to the cathode temperature and current.

Efficiency. The efficiency is, by definition, equal to the ratio of the power output to the power input:

$$\eta = P_{\text{out}}/P. \tag{6.38}$$

The power output, when the space charge is neutralized, is equal to $P_{\text{out}} = J_s V_L$, whereas the voltage at the load is equal to the contact potential of the converter minus all the voltage drops:

$$V_L = V_{ac} - V_w - V_q - V_p,$$

where V_w, V_q, and V_p are, respectively, the voltage drops in the wires, the ion generation process, and the plasma.

The power input is equal to

$$P_{\text{in}} = J_s V_c + P_1 + P_4 + P_5 + P_6.$$

Replacing the different losses by their values and introducing them into Eq. (6.38) yields

$$\eta = \frac{1 - \left(\dfrac{V_a}{V_c} + \dfrac{V_w}{V_c} + \dfrac{V_q}{V_c} + \dfrac{V_p}{V_c}\right)}{1 + \dfrac{MS\,dv_R}{e^2 n_e V_c} + \dfrac{\mathscr{B}_1 \Theta_c T_c^4}{J_s V_c} + \dfrac{1}{2}\left(\dfrac{\mathscr{L} T_c^2}{V_w V_c} - \dfrac{V_w}{V_c}\right) + 2\,\dfrac{kT_c}{eV_c}}. \tag{6.39}$$

Equation (6.39) is plotted in Fig. 6.16; all the losses except losses by radiation have been neglected.

Many thermionic converters were experimented with in the U.S. [24]* and abroad [6, 14].† A typical high-vacuum converter [4] produces a power density of about 0.4 W/cm² at an output of about 0.5 V and a cathode temperature of 1500°K. Prototypes with lower emitter temperatures and higher power densities were also tested. Low-pressure gaseous converters yielded an output power of about 10 W/cm² at 10% efficiency, at cathode temperatures higher than 2000°K. The power outputs of such converters as adapted from Kihara [50] and his co-workers are plotted in Fig. 6.17. The reported power characteristic is given as a function of the output current and different values of the emitter temperature and the pressure.

---

* See also reports from General Atomic, San Diego, California; General Motors Research Labs., Warren, Michigan; Los Alamos Scientific Lab., Los Alamos, New Mexico; Martin-Marietta Corp. Aerospace Division, Baltimore, Maryland; General Electric Co., Special Purpose Nuclear Systems Operation, Pleasanton, California; Thomson Ramo Woolridge, Inc., Electromechanical Division, Cleveland, Ohio.

† See also "Direct Energy Conversion in the USSR, Thermionic Converters," Library of Congress, Aerospace Technology Division, Washington, D.C., ATD Rept. P-65-57, Sept. 22, 1965.

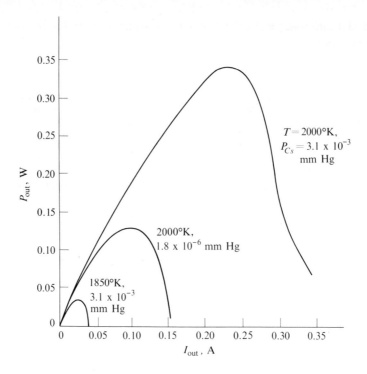

Figure 6.17. Experimental power output. The power output increases with increasing temperature and increasing pressure. (Adapted from Kihara, Oho, et al. [50, Figs. 6 and 7].)

6.9 Thermionic Systems

Thermionic systems can be classed according to the type of fuel used. Note the existence of three important types:

1. Fossil fueled systems [65],
2. Solar fueled systems [10, 12, 61],
3. Nuclear fueled systems [11, 87].

Most of the fossil fueled generators are flame heated. There are two major applications: in space and in terrestrial uses. In space, the system can be arranged in such a way as to use the exhaust flame of the missile to generate electricity during the powered flight time. This electricity can be used to run auxiliary equipment which is operative only during powered flight intervals, or it might be used to charge batteries for use later on a space mission. In the terrestrial converter described by Martini [60], a combustion chamber is arranged at the bottom of a molybdenum cylinder, the outside of which acts as the emitter. The anode is placed in front of the emitter. A screw-in diode is used to make electrical and thermal contact with a divided heater module. Gasoline is fed to individually controlled fuel injectors by gravity and a blower injects air as an oxidizer. This generator has been proposed for military applications because it is compact, has a relatively long life, and produces a power density of 5 W/cm², which leads to an overall power generation of 120 W for the four modules of 6 cm² each forming the system.

Solar energy can be used for thermionic power generation, and consideration has been given to the construction of a thermionic solar system powering a satellite.* Included in this system is a solar energy storage battery to avoid fluctuations in energy input. There are 18 modules, each made up of five thermionic cells and having its own solar concentrator. The power output density of this system is 8 W/cm², which leads to an overall power output of 3.12 kW, since the area of each cell is 4.33 cm². The overall efficiency is 11% for an emitter temperature of 2000°K. The overall system weighs 400 kg, leading to a relatively high specific power of 74 W/kg, which compares favorably with photovoltaic systems.

The fission energy of a nuclear reactor can be used directly or indirectly to heat the cathodes of a thermionic system. The heat generated within a uranium dioxide fuel element heats, through conduction, a molybdenum cylinder fitted around it. The outside of the cylinder acts as the cathode of a diode separated from the anodic tube by a small distance. This "in-pile" system has been experimented with, with some success, in the U.S. [71, 74]† and England [34],‡ reaching efficiencies of about 15%. This method promises much with advances in high-temperature nuclear reactor technology. It can yield power outputs in the megawatt range by connecting all the nuclear thermionic cells in series to form a thermionic fuel element. A thermionic-nuclear core can then be formed by assembling the different fuel elements. It is also clear that the in-pile method leads to higher efficiencies than the "out-of-pile" method considered [54] for the construction of a 10-kW nuclear-thermionic space power plant [8].§ In the latter, the converters are placed on the surface of the reactor, from which heat reaches the emitters by radiation.

6.10 Economic Aspects and Future Trends

Thermionic converters have many of the advantages of other forms of direct energy conversion. They are free from moving parts, making them quiet and potentially long lived. This makes thermionic conversion advantageous for space applications and use in remote, unattended generators. Their theoretical efficiencies of 30% or more also promises to increase overall efficiencies in conventional power plants.

---

*   Thomson Ramo Woolridge Inc., Advanced Solar Thermionic Power Systems, Report No. ASD-TDR-62-877 (AD 29517), 1962.

†   Los Alamos Labs.

‡   Harwell Labs.

§   Studied by General Electric Special Purpose Nuclear Systems Operation.

Space applications have received the most emphasis in the United States, and it is here that perhaps the greatest potential use of thermionic conversion lies. High power densities and power per weight ratios, together with structural rigidity and with the absence of moving parts, make thermionic power generation ideal for space use. All three types described above have been considered for space applications.

There are many terrestrial uses of thermionic power generation. These can be divided into three major classes:

1. Use in remote, unattended generators,
2. Use in motor vehicles,
3. Use in large power plant systems.

It has also been noted that a thermionic flame heated converter, due to its small size and simple construction, might be economically competitive with present methods of electrical generation, since such a system would not require any distribution network.

Like other direct conversion methods, both nuclear and solar fueled thermionic converters readily lend themselves to use in remote areas. Solar systems do require storage arrangements and, therefore, only nuclear systems which do not have the storage problem may find use where relatively large power outputs are needed in remote installations.

The second thermionic terrestrial application is use in motor vehicles. One line of development of cesium diodes was aimed at fossil fueled converters. Such a system requires a vacuum-tight envelope capable of being cycled up and down rapidly from room temperature to about 2000°K. If this and other problems are solved, the way to thermionically driven automobiles might be open, this being a major commercial application. The driving properties and the power per weight ratios are good, although much work needs to be done.

The third major earth bound application is in large power plants. In comparison with large fossil fueled plants, thermionic conversion can be disregarded as impractical. This is because the "natural" size of thermionic converters is about a few hundred watts and, even if this can be increased to a few kilowatts, they still are not feasible, because present power plants operate in the 500-MW range. Thus about 100,000 thermionic units would have to be added as toppers in order to be at all useful. Also, commercial fossil fueled power plants have efficiencies approaching 40 to 45% and any slight increase thermionic topping might provide would not justify the high capital cost of these units.

Although the practice of topping conventional steam turbine power plants does not seem practical in the case of fossil fueled plants, it is not yet clear whether this method might prove worthwhile in nuclear fueled power plants. The overall efficiency of nuclear plants now in operation is around 20 to 30%. Thermionic converter efficiencies are of the order of 2 to 12% in working models, with prospects of approaching 30%. It can be shown that 15% efficiency in a thermionic generator can boost the overall efficiency of a steam plant from 30% to over 40%. This increase, it is argued, would justify a fairly high capital cost for the additional equipment. Since, for most applications, low impedance levels are not tolerable, a way to step up the voltage delivered by thermionic conversion must be found. This can be done by connecting many units in series or by current modulation within the converter, followed by a transformer to match any load impedance desired.

In in-pile thermionic nuclear reactors, provisions must be made to remove the fission products accumulated in the interelectrode space. This poses a real problem in gas-filled diodes, as the diode gas must be replenished at a rate equal to that lost in the expulsion of the fission fragments. Since the efficiencies of nuclear power plants have risen greatly, the use of thermionic toppers in nuclear plants seems improbable. It would seem that the future of thermionic converters in nuclear plants is not at all certain and would appear to rely on the possibility of major breakthroughs in converter power capabilities and in production and installation costs.

Another alternative is using thermionic conversion in conjunction with thermoelectric conversion. In such an installation, the thermionic units would be toppers for the thermoelectric units, and such a plant would have the advantage of having no moving parts. Such devices have already been constructed.

The importance of the economic aspects of using thermionic converters depends on each individual case. In space applications, the cost is not as important a factor as are reliability and specific weight. Hence, thermionic conversion needs only compete with other direct energy conversion schemes for this application, and this it does well. The same general remarks apply for applications in remote areas. The motor vehicle application is far from a solution and no meaningful economic comparisons can be made. Finally, in large nuclear fueled terrestrial power plants, it appears that, due to the high efficiencies of the conventional plants, thermionic toppers will have to improve significantly so far as efficiency, unit cost, and unit output are concerned before this method will be economically competitive.

PROBLEMS

6.1 In the parallel-plane thermionic cell shown in Fig. 6.18, the cathode-anode spacing is d, and the anode is at a potential V_a with respect to the cathode. An electron leaves the cathode at time $t = 0$ with zero initial velocity in the x-direction. (a) What is the transit time of this electron? (b) How do the currents I_1 and I_2 in the external circuit depend on time?

6.2 Two parallel plates of emissive material, separated by a distance d, are placed in vacuum. At the steady state, a steady heat influx maintains the cathode at a temperature T_c and the anode at a lower temperature T_a. The potential distribution diagram is shown in Fig. 6.1(b). Calculate (a) the fraction of the cathode saturation current which reaches the anode, (b) the fraction of the anode saturation current which reaches the cathode, and (c) the net current flowing from the anode to the cathode.

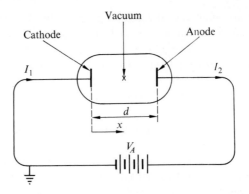

Figure 6.18. A parallel-plane thermionic cell.

6.3 In the cell of Problem 6.2 assume that $T_c = 1250°C$, $T_a = 500°C$, and $d = 10$ microns. Calculate the variation of the maximum power output per unit area with V_{ac}. In the light of your results, what kind of materials do you propose for the anode and for the cathode? Find the values of the output current density and the output voltage for $V_a = 0.3$ V.

6.4 By taking into account, in Eq. (6.29), the loss of heat by conduction, emission, and radiation of the anode as well as the loss by electrical mass transfer, calculate the efficiency of the thermionic cell. Try to find an equation analogous to Eq. (5.51). Here ZT is a factor of merit of the thermocouple. For a tungsten-impregnated thermionic converter $ZT = 15$ at about 1500°K.

6.5 Consider a thermionic close spaced high-vacuum cell having a barium impregnated tungsten emitter ($V_c = 1.7$ V) and a Cs/WO collector ($V_a = 0.71$ V). Assuming $d = 14$ microns, $T_c = 1200°C$, and $T_a = 600°C$, calculate (a) the saturation currents of the anode and the cathode and (b) the power output per unit area.

6.6 Using Eq. (6.30), what materials do you propose for the cathode and the anode to build a thermionic cell having a 15% efficiency at $T_c = 2000°K$?

6.7 Equation (6.34) yields the reduced potential versus the distance for Langmuir's distribution (curve A in Fig. 6.6). Using the equations of continuity, conservation of energy, conservation of momentum, and Poisson's law (see Chapter 4), calculate the reduced potential distribution in the case of numerous collisions (curve B in Fig. 6.6).

6.8 The cesium pressure vapor of a low-pressure thermionic converter is equal to 10^{-4} mm Hg. Find (a) the cesium pool temperature T_{cS}, and (b) the saturation current for $V_{ac} = 1.8$ V. [*Hint:* Use Fig. 6.9.]

6.9 Nottingham [70] found that the minimum cathode temperature for complete ionization of a liquid is given by

$$T_{min} = 3.6 T_{cS}.$$

What should be T_c for a cesium pressure vapor equal to 1 mm Hg? What type of thermionic converter is this? Is your result reasonable?

6.10 A magnetic field is established in the interelectrode space of a low-pressure cesium vapor thermionic cell. The electrons are assumed to be emitted perpendicularly to the cathode and the magnetic field. Find the trajectory of an emitted electron.

6.11 Treat Problem 6.10 by assuming an arbitrary direction of the field and the emitted electron. Discuss your results [82].

6.12 It is proposed to build a flame heated thermionic converter receiving a heat energy, $\mathcal{Q}_{in} = 2$ kW and yielding an optimum output voltage of $V_{out} = 1$ V. Choose the materials necessary for the electrodes, the type of converter, the distance between the electrodes, and the operating temperature.

6.13 Build a nuclear powered thermionic system with the cylindrical nuclear reactor having the following characteristics:

$$\text{Power produced (heat)} = 0.3 \text{ MW},$$
$$\text{Surface temperature} = 1200°C,$$
$$\text{Surface area} = 1000 \text{ cm}^2.$$

Choose the materials and the type of converter. Calculate the output voltage, current, and efficiency as well as the number of unit cells, their dimensions, and the mode of staging (if necessary).

6.14 Heat can be rejected in space only by radiation. Compare the volumes of two thermionic converters having the same materials and producing the same power output, one working on a spaceship and the other on earth. Calculate the ratio of the two volumes. [*Hint:* Consider the temperatures of the anodes.]

6.15 Can a thermionic converter produce ac power? One of the methods proposed is the modulation of the electron flow by varying the auxiliary ion supply. Criticize this method.

REFERENCES AND BIBLIOGRAPHY

1. Aamodt, R. L., L. J. Brown, and B. D. Nichols, "Thermionic Emission from Molybdenum in Vapors of Cesium and Cesium Fluoride," *J. Appl. Phys.*, Vol. 33, No. 6, p. 2080, 1962.

2. Angello, J. P., *A.C. Thermionic Converter Operation*, 18th Power Sources Conference, May 1964, PSC Publications Committee, Red Bank, N.J., 1964.

3. Auer, P. L., "Potential Distribution in a Low-Pressure Thermionic Converter," *J. Appl. Phys.*, Vol. 31, No. 12, p. 2096, 1960.

4. Beggs, J. E., "Vacuum Thermionic Energy Converter," *Adv. Energy Conversion*, Vol. 3, No. 2, p. 447, 1963.

5. Block, F. G., et al., *Thermionic Diodes for Direct Conversion Reactors*, United Nations 3rd International Conference on the Peaceful Uses of Atomic Energy, Geneva, 1964.

6. Bohdansky, J., C. A. Busse, and G. M. Grover, *The Use of a New Heat Removal System in Space Thermionic Power Supplies*, European Atomic Energy Community, Joint Nuclear Research Center, Ispra, Italy, EUR-2229, 1965.

7. Blue, E., and J. H. Ingold, *Investigation of a Thermionic Converter Operating with a High Temperature Collector*, International Conference on Thermionic Electrical Power Generation, London, 1965.

8. Blue, E., and J. H. Ingold, "Thermionic Energy Conversion," in *Direct Energy Conversion*, G. W. Sutton (ed.), McGraw-Hill, New York, 1966.

9. Boffi, V. C., V. G. Molinari, and D. E. Parks, "Thermoelectric Theory of a Plasma Diode," *Adv. Energy Conversion*, Vol. 7, No. 1, p. 1, 1967.

10. Bolan, P., R. Cohen, and D. Bentsen, *A Solar Space Power Plant with Liquid Metal Heated Thermionic Converters*, Paper (AIAA) 64-737, 3rd Biennial Aerospace Power Systems Conference, Philadelphia, Sept. 1964.

11. Bolan, P., R. Cohen, and G. Bordner, *An Engineering Evaluation of Advanced Nuclear Thermionic Space Powerplants*, 3rd Biennial Aerospace Power Systems Conference, Philadelphia, Sept. 1964.

12. Brosens, P. J., "Solar Thermionic Generators for Space Power," *Am. Soc. Mech. Eng. Ser. A. J. Eng. Power*, Vol. 86, No. 6, p. 281, 1965.

13. Bullis, R. H., and W. J. Wiegand, *Plasma Properties in a Thermionic Converter*, 24th Annual Conference on Physical Electronics, Cambridge, Mass., March 1964.

14. Busse, C. A., R. Caron, and C. M. Cappelletti, "Operating Experience with an Experimental Nuclear-Heated Thermionic Converter," *Adv. Energy Conversion*, Vol. 4, No. 11, p. 121, 1964.

15. Caulfield, H. J., "Estimation of Contact Potential Difference in Thermionic Energy Converters," *J. Appl. Phys.*, Vol. 35, No. 10, p. 2862, 1964.

16. Cayless, M. A., "Thermionic Generation of Electricity," *Brit. J. Appl. Phys.*, Vol. 12, No. 9, p. 433, 1961.

17. Champeix, M. R., "Transformation of Heat into Electrical Energy in Thermionic Phenomena," *Le Vide*, Vol. 6, p. 936, 1951 (French).

18. Chapman, R. A., H. J. Caufield, and H. W. Hemstreet, Jr., "Measurement of Electron Temperature in Low-Pressure Cesium Thermionic Energy Converters," *J. Appl. Phys.*, Vol. 35, No. 10, p. 2813, 1964.

19. Chapman, R. A., "Thermionic Work Function of Thin-Oxide-Coated Aluminium Electrodes in Vacuum and in Cesium Vapor," *J. Appl. Phys.*, Vol. 35, No. 10, p. 2832, 1964.

20. Chayka, G. E., "The Effect of High Fields on the Thermionic Emission of Semiconductors," *Radio Eng. Electr. Phys.*, Vol. 10, No. 6, p. 954, 1965.

21. Dirac, P. A., "Theory of Quantum Mechanics," *Roy. Soc. Proc. (London)*, Vol. 112, p. 661, 1926.

22. Dushman, S., "Thermionic Emission," *Rev. Modern Phys.*, Vol. 2, p. 381, 1930.

23. Dybua, B. C., "The Thermionic Emission of Some Refractory Metals in Barium Vapor," *Radio Eng. Electr. Phys.*, Vol. 10, p. 999, 1965.

24. Eastman, G. Y., and D. M. Ernst, *The Development of a Cesium-Vapor-Filled Thermionic Energy Converter*, RCA Electron Tube Division, Lancaster, Pa., NASA-CR-64614, Dec. 1962.

25. Edison, T. A., "A Phenomenon of the Edison Lamp," *Engineering*, p. 553, Dec. 12, 1884.

26. ElSaden, M. R., *Current-Voltage Relations for a High-Pressure Thermionic Converter*, American Society of Mechanical Engineers Winter Meeting, New York, Nov.–Dec. 1964.

27. Fermi, E., "Wave Mechanics of Collisions," *Zs. Phys.*, Vol. 40, No. 5, p. 399, 1926.

28. Fleming, J. A., "Rectification of Electric Oscillations by Means of a Vacuum Valve," *Proc. Roy. Soc. (London)*, Vol. 74, p. 476, 1905.

29. Fouad, A. A., and E. M. Welsh, "A Cyclic Analysis of the Gabor-Type Auxiliary Discharge Thermionic Converter," *Adv. Energy Conversion*, Vol. 15, No. 5, p. 71, 1965.

30. Gabor, D. A., "New Thermionic Generator," *Nature*, Vol. 189, No. 4768, p. 868, 1961.

31. Goldstein, C. M., "Limits on Collisionless Model of Thermionic Converter," *J. Appl. Phys.*, Vol. 35, No. 12, p. 3629, 1964.

32. Grodko, V. A., et al., "Rational Selection of Cathode Materials for a Thermionic Converter," *Poroshkovaya Metallurgiya*, Vol. 4, p. 79, 1963 (Russian).

33. Hansen, L. K., and C. Warner, *The Electron-Rich Unignited Mode of Thermionic Diodes*, Report of Thermionic Conversion Specialist Conference, Gatlinburg, Tenn., Oct. 1963, p. 44.

34. Harrowell, R. V., "The Thermionic Converter," in *Direct Generation of Electricity*, K. H. Spring (ed.), Academic Press, New York, 1965.

35. Hatsopoulos, G. N., and T. Kaye, "Analysis of Experimental Results of a Diode Configuration of a Novel Thermoelectron Engine," *Proc. IRE*, Vol. 46, No. 9, p. 1574, 1958.

36. Hatsopoulos, G. N., "Thermodynamics of Thermionic Engines," in *Direct Conversion of Heat to Electricity*, T. H. Welsh and I. Kaye (eds.), Wiley, New York, 1960.

37. Hatsopoulos, G. W., "Transport Effects in Cesium Thermionic Converters," *Proc. IEEE*, Vol. 51, No. 5, p. 725, 1963.

38. Hatsopoulos, G. W., in Report on 24th Annual Conference on Physical Electronics, MIT, Cambridge, Mass., March 1964, p. 307.

39. Hernqvist, K. G., M. Kanefsky, and F. H. Norman, "Thermionic Energy Converter," *RCA Rev.*, Vol. 19, No. 6, p. 244, 1958.

40. Hernqvist, K. G., "Thermionic Converters," *Nucleonics*, Vol. 17, No. 7, p. 49, 1959.

41. Hernqvist, K. G., "Analysis of the Arc Mode Operation of the Cesium Vapor Thermionic Energy Converters," *Proc. IEEE*, Vol. 51, No. 5, p. 748, 1963.

42. Herring, C., and M. H. Nichols, "Thermionic Emission," *Rev. Modern Phys.*, Vol. 21, p. 185, 1949.

43. Houston, J. M., "Theoretical Efficiency of the Thermionic Energy Converter," *J. Appl. Phys.*, Vol. 30, No. 4, p. 481, 1959.

44. Houston, J. M., and H. F. Webster, "Thermionic Energy Conversion," *Adv. Electronics*, Vol. 17, p. 125, Academic Press, New York, 1962.

45. Ingold, J. H., "Thermionic Properties of Some Refractory Metal Carbides," *J. Appl. Phys.*, Vol. 34, p. 2033, 1963.

46. Jablonski, F. E., et al., "Space Charge Neutralization by Fission Fragments in the Direct Conversion Plasma Diode," *J. Appl. Phys.*, Vol. 30, p. 2017, 1959.

47. Johnson, E. O., and W. M. Webster, "The Plasmatron, A Continuously Controllable Gas Discharge Development Tube," *Proc. IRE*, Vol. 40, No. 6, p. 645, 1952.

48. Keller, K., and H. L. McDonald, "The Possibility of Transfer of Electricity in a Thermionic Converter by Negative Ions," *Adv. Energy Conversion*, Vol. 7, No. 6, p. 113, 1967.

49. Kennedy, A. J., and P. K. Shefsiek, *Barium Vapor-Filled Thermionic Energy Converters*, Symposium on High Temperature Conversion—Heat to Electricity, University of Arizona, Tucson, Feb. 1964, p. 247.

50. Kihara, M., S. Oho, Y. Shibata, and Y. Koike, "Some Characteristics of a Cesium Plasma Cell," *Proc. IEEE*, Vol. 51, No. 5, p. 769, 1953.

51. Lafferty, J. M., "Boride Cathodes," *J. Appl. Phys.*, Vol. 22, p. 299, 1951.

52. Langmuir, I., "The Effect of Space Charge and Initial Velocities on the Potential Distribution and Thermionic Current Between Parallel Plane Electrodes," *Phys. Rev.*, Vol. 21, p. 426, 1923.

53. Leblanc, A. R., and W. W. Grannemann, "Thermionic Generator for Re-entry Vehicles," *Proc. IEEE*, Vol. 52, No. 11, p. 1302, 1964.

54. Leonard, A., *Out-of-Pile Thermionic Space Power Systems Using a Gaseous Heat-Transfer Fluid*, Rand Corp., Santa Monica, Cal., Report RM-4469-PR, AD-620-650, Sept. 1965.

55. Lewis, H. W., and J. R. Reitz, "Thermoelectric Properties of the Plasma Diode," *J. Appl. Phys.*, Vol. 30, No. 9, p. 1439, 1959.

56. Lewis, H. W., and J. R. Reitz, "Efficiency of the Plasma Thermocouple," *J. Appl. Phys.*, Vol. 31, No. 4, p. 723, 1960.

57. Long, J. D., J. Psarouthakis, and E Scicchitano, *Double-Diode Thermionic Energy Converter*, Symposium on High Temperature Conversion—Heat to Electricity, University of Arizona, Tucson, Feb. 1964, p. 74.

58. Lyczko, F. J., *Single Diode 100 Watt Ti-Converter Design*, 18th Power Sources Conference Proceedings, May 1964, PSC Publications Committee, Red Bank, N.J., 1964, p. 146.

59. McIntyre, R. G., "Effect of Anode Emission of Electrons on Space Charge Theory of the Plasma Thermionic Converter," *Proc. IEEE*, Vol. 51, No. 5, p. 760, 1963.

60. Martini, W. R., *Thermionic Back-Pack Power Supplies*, Symposium on High Temperatures Conversion—Heat to Electricity, University of Arizona, Tucson, Feb. 19–21, 1964, p. 344.

61. Menetrey, W. R., and A. Smith, "Solar Energy Thermionic Electrical Power Supplies," *J. Spacecraft and Rockets*, Vol. 1, p. 659, 1964.

62. Michaelson, H. B., "Work Function of the Elements," *J. Appl. Phys.*, Vol. 21, p. 536, 1950.

63. Miller, H., *In-Reactor Testing of Nuclear Thermionic Converters*, Symposium on High

Temperature Conversion—Heat to Electricity, University of Arizona, Tucson, Feb. 19–21, 1964, p. 384.

64. Morgan, K., "Thermionic Generation," *Inst. Elec. & Electr. Trans.*, Vol. E-7, No. 12, p. 175, 1964.

65. Monroe, J. L., Jr., and D. W. Stoffel, *Fueled Double Module Converters Fabrication, Testing and Evaluation*, Symposium on High Temperature Conversion—Heat to Electricity, University of Arizona, Tucson, Feb. 19–21, 1964, p. 360.

66. Morgulis, N. D., "Conversion of Thermal into Electrical Energy by Thermionic Emission," *Adv. Phys. Sci. (Moscow, URSS)*, Vol. 70, p. 679, 1960 (Russian).

67. Moss, H., "Thermionic Diodes as Energy Converters," *J. Electronics and Control*, Vol. 2, p. 305, 1957.

68. Nottingham, W. B., "Thermionic Emission," *Handbuch Der Physik*, Vol. 21, Springer-Verlag, Berlin, 1956.

69. Nottingham, W. B., *Cesium Plasma Diode as a Heat-to-Electrical Power Transducer*, 4th International Conference on Ionization Phenomena in Gases, Uppsala, Sweden, Aug. 1959.

70. Nottingham, W. B., *Emitter Materials for High Temperature Energy Conversion*, International Symposium on High Temperature Technology, Butterworths, Washington, D.C., 1963, p. 389.

71. Ogle, H. M., "In-reaction Tests of Plasma Diodes," Paper 5 of *Direct Conversion*, American Institute of Electrical Engineers, New York, Sept. 1962.

72. Picquendar, J. E., "The Thermionic Conversion of Energy," *Rev. Generale Electronique*, No. 216, p. 19, 1964 (French).

73. Pidd, R. W., et al., "Characteristics of UC, ZrC and (ZrC) (UC) as Thermionic Emitters," *J. Appl. Phys.*, Vol. 30, No. 10, p. 1861, 1959.

74. Ranken, W. A., "In-Core Thermionic Converter Testing at Los Alamos," *Nuclear News*, Vol. 7, No. 7, p. 3, 1964.

75. Rasor, N. S., "Emission Physics of the Thermionic Energy Converter," *Proc. IEEE*, Vol. 51, No. 5, p. 733, 1963.

76. Reinchelt, W. H., "Cesium Plasma Diodes Utilizing Fission Energy as An Emitter Heat Source," Paper 6 of *Direct Conversion*, American Institute of Electrical Engineers, New York, Sept. 1962.

77. Reinchelt, W. H., *Experimental Observations with a Cesium Plasma Cell Which has a Uranium Carbide Emitter*, Symposium on High Temperature Conversion—Heat to Electricity, University of Arizona, Tucson, Feb. 19–21, 1964, p. 132.

78. Richards, H. K., *DC and High Frequency Voltage and Power Output and Interaction in Cesium Potassium and Sodium Thermionic Converters*, Symposium on High Temperature Conversion—Heat to Electricity, University of Arizona, Tucson, Feb. 19–21, 1964, p. 161.

79. Richardson, O. W., *Emission of Electricity from Hot Bodies*, Longmans Green, London, 1921.

80. Schlichter, W., *Die Spontane Electronemission von Gluhender Metalle und das Gluhelektrische Element*, Doctorate Dissertation, Gottingen University, Germany, 1915.

81. Schottky, W., H. Rothe, and H. Simon, "Physik der Gluhelektroden," in *Handbuch der Experimentalphysik*, Vol. 13, No. 2, Akademische Verlagsgesellschaft, Leipzig, 1928.

82. Schock, A., "Effect of Magnetic Fields on Thermionic Power Generators," *J. Appl. Phys.*, Vol. 31, No. 11, p. 1978, 1960.

83. Shaw, D. T., "On the Diffusion Theory of an Ignited Mode Thermionic Converter," *Adv. Energy Conversion*, Vol. 7, No. 1, p. 23, 1967.

84. Steele, H., "Energy Converters Using Low Pressure Cesium," in *Direct Conversion of Heat to Electricity*, J. H. Welsh and J. Kaye (eds.), Wiley, New York, 1960.

85. Taylor, J. B., and I. Langmuir, "The Evaporation of Atoms, Ions and Electrons from Cesium Films on Tungsten," *Phys. Rev.*, Vol. 44, p. 423, 1933.

86. Taylor, J. B., and I. Langmuir, "Vapor Pressure of Cesium by the Positive Ion Method," *Phys. Rev.*, Vol. 51, p. 753, 1937.

87. Van Hoomissen, J. E., and J. M. Case, *Application of Nuclear Thermionics to Undersea Technology*, American Society of Mechanical Engineers Underwater Technology Conference, New London, Conn., May 5–7, 1965.

88. Warner, C., and L. K. Hansen, 23rd Annual Conference on Physical Electronics, MIT, Cambridge, Mass., March 1963, p. 400.

89. Warner, C., *The Ion-Rich, Unignited Mode of Thermionic Converters*, Report of the Thermionic Conversion Specialist Conference, Gatlinburg, Tenn., Oct. 1963, p. 51.

90. Webster, H. F., "Calculation of a High-Vacuum Thermionic Energy Converter," *J. Appl. Phys.*, Vol. 30, No. 4, p. 488, 1959.

91. Webster, H. F., "Thermionic Converter Research," Paper 4 of *Direct Conversion*, American Institute of Electrical Engineers, New York, Sept. 1962.

92. Welsh, J. A., G. N. Hatsopoulos, and J. Kaye, "Theoretical Analysis of a Magnetic Triode as a Thermionic Engine," in *Direct Conversion of Heat to Electricity*, J. A. Welsh and J. Kaye (eds.), Wiley, New York, 1960.

93. Wilson, R. G., and J. Lawrence, "Operating Characteristics of Two Thermionic Converters having Rhenium-Nickel and Tungsten-Nickel Electrodes," *Adv. Energy Conversion*, Vol. 4, No. 12, p. 195, 1964.

94. Wilson, R. G., "Electron and Ion Emission from Polycristalline Surface of Nb, Mo, Ta, W, Re, Os and Ir in Cs Vapor," *J. Appl. Phys.*, Vol. 37, No. 11, p. 4125, 1966.

95. Wilson, V. C., "Conversion of Heat to Electricity by Thermionic Emission," *J. Appl. Phys.*, Vol. 30, No. 4, p. 475, 1959.

96. Wilson, V. C., "The Gas-Filled Thermionic Converter," in *Direct Conversion of Heat to Electricity*, J. H. Welsh and J. Kaye (eds.), Wiley, New York, 1960.

97. Wright, D. A., "A Survey of Present Knowledge of Thermionic Emitters," *Proc. Inst. Electr. Eng. (London)*, Vol. 100, Part 3, p. 125, 1953.

98. Wright, J. K., and R. V. Harrowell, "Thermionic Power Generation," *Electrical Rev. (London)*, Vol. 172, p. 217, 1963.

99. Yang, L., et al., "Some Critical Materials Problems of Thermionic Cathode Systems for Fission-Heat Conversion," Symposium on Thermionic Power Conversion, Colorado Springs, *Adv. Energy Conversion*, Vol. 3, No. 1, p. 93, 1962.

100. Zgorzelski, M., "Experimental Investigations of the Magnetic-Triode Type Thermionic Energy Converter," *Acad. Polon. Sci. Bull. Ser. Sci. Techniques*, Vol. 12, p. 449, 1964 (Polish).

101. Zollweg, R. J., and M. Gottlieb, "Radio-Frequency Oscillations in Thermionic Diodes," *Proc. IEEE*, Vol. 51, No. 5, p. 754, 1963.

7 FUEL CELLS

7.1 Introduction

A fuel cell is a device which converts chemical energy directly into electrical energy. This converted chemical energy is due to oxidation of the fuel. In a heat engine, this same energy is usually transformed into heat which is transformed into electrical energy through a mechanical energy path. For this reason, electromechanical devices are subject to the Carnot efficiency limitation. Since in a fuel cell system the conversion of energy can be carried out isothermally, the Carnot limitation does not apply, and it is safe to seek, at least theoretically, efficiencies much higher than those furnished by conventional thermoelectric plants.

This same concept is widely used for the storage of electrical energy in electrical batteries. An electric current flows through an electrode, an electrolyte, and then another electrode. If the electrolyte is water, hydrogen is produced at the anode and oxygen at the cathode. By reversing this process, oxygen will react with hydrogen to form water, and electricity is produced. An electric current will continue to flow as long as there is a continuous flow of oxygen and hydrogen, and the battery is now called a fuel cell.

The first fuel cell was built in 1801 by Sir Humphrey Davy [27] of England. In his experiment, he built a cell using zinc and oxygen in an electrolyte to generate electricity directly from chemical energy; the product of oxidation was a sodium zincate. It was in 1839 that W. R. Grove [41] built a fuel cell using hydrogen as a fuel. He was the first worker to point out some of the problems facing this method of generating power. Work on fuel cells and batteries continued for some years, but due to the overwhelming success of electromechanical devices, progress in the fuel cell field stopped for almost a century, until after World War II. Since then, the work revived by Justi [56–58] in Germany, and Bacon [9, 10] in Great Britain, among others [6, 19], was continued at an accelerated pace.

7.2 Thermodynamics of a Fuel Cell

Energy. The oxidation of fuel is a chemical reaction which, in the case of a fuel cell device, is carried out at constant pressure and temperature. The maximum

222

heat available can be described either by the change of enthalpy or by the change of entropy ($\Delta \mathscr{H}$ or $\Delta \mathscr{S}$). It is necessary, therefore, to define a quantity which will express the maximum energy available to a system. This will be obtained only from a reversible system and can be described by Gibbs free energy $\mathscr{G}$:

$$\mathscr{G} = \mathscr{H} - T\mathscr{S}.$$

The change of Gibbs free energy for a reaction at constant temperature is

$$\Delta \mathscr{G} = \Delta \mathscr{H} - T \Delta \mathscr{S}. \tag{7.1}$$

The available energy is equal to the change in the enthalpy minus the heat lost to the system during the reaction. The entropy of a reaction at constant pressure can be defined as $\mathscr{S} = -\partial \mathscr{G}/\partial T|_p$, and replacing $\mathscr{S}$ by this value in Eq. (7.1), one obtains

$$\Delta \mathscr{G} = \Delta \mathscr{H} + T \frac{\partial(\Delta \mathscr{G})}{\partial T}\bigg|_p. \tag{7.2}$$

This equation is known as the *Gibbs-Helmholtz equation*. Here $\Delta \mathscr{G}$ is a measurable quantity; it is a function of temperature and represents the maximum useful work W of the fuel cell. Thus, $W = -\Delta \mathscr{G}$. But $W = n_1 eV$, where n_1 is the number of charges, e is the electronic charge and V is the electromechanical potential in volts. If the free energy is expressed as a function of the number n of electrons transferred per mole of oxidized fuel,

$$\Delta \mathscr{G} = -nF_n V, \tag{7.3}$$

where F_n is the number of coulombs per mole of reaction, it is called the Faraday, and its value is $F_n = N_0 e = 0.965 \times 10^5$ Cb/mole, where $N_0 = 6.023 \times 10^{23}$ is Avogadro's number and $e = 1.6021 \times 10^{-19}$ is the electronic charge. Introducing Eq. (7.3) into Eq. (7.2), one obtains for the increase of enthalpy

$$\Delta \mathscr{H} = -nF_n \left[V - T\left(\frac{\partial V}{\partial T}\right)_p \right].$$

Since the potential can be measured easily through a certain temperature range, both $\Delta \mathscr{H}$ and $\Delta \mathscr{G}$ may be obtained.

Potential. Combining the first and second laws of thermodynamics, stated in Chapter 2, one obtains

$$dU = T\,d\mathscr{S} - p_r\,d\mathscr{V}, \tag{7.4}$$

where $\mathscr{V}$ is the volume. On the other hand, by replacing the enthalpy by its value, Eq. (7.4) yields

$$\mathscr{G} = U + p_r\mathscr{V} - T\mathscr{S}. \tag{7.5}$$

The derivative of Eq. (7.5) yields, for constant temperature,

$$d\mathscr{G} = dU + p_r\,d\mathscr{V} + \mathscr{V}\,dp_r - T\,d\mathscr{S}.$$

Combining Eqs. (7.5) and (7.4) leads to $d\mathscr{G} = \mathscr{V} \, dp_r$ and, after integration,

$$\Delta\mathscr{G} = \mathscr{G} - \mathscr{G}_0 = \int_{p_0}^{p} \mathscr{V} \, dp_r.$$

For a perfect gas, from Chapter 2, $p_r\mathscr{V} = \dot{\mathscr{R}}T$, where $\dot{\mathscr{R}} = n'\mathscr{R}$ with n' being the number of moles and $\mathscr{R}$ the gas constant per mole. Thus,

$$\Delta\mathscr{G} = \int_{p_0}^{p} \dot{\mathscr{R}}T \frac{dp_r}{p_r} = \dot{\mathscr{R}}T \ln \frac{p_r}{p_0}.$$

Consider the following electrochemical reaction:

$$X + Y^{n+} \rightleftharpoons X^{n+} + Y. \tag{7.6}$$

One may write $\Delta\mathscr{G} = \Delta\mathscr{G}_1 - \Delta\mathscr{G}_2$, where $\Delta\mathscr{G}_1$ and $\Delta\mathscr{G}_2$ are the changes of the free energy for the products of the reaction and the reactants, respectively. For Reaction (7.6),

$$\Delta\mathscr{G}_1 = \dot{\mathscr{R}}T \ln (p_{X^{n+}}) - \dot{\mathscr{R}}T \ln (p_{0X^{n+}}) + \dot{\mathscr{R}}T \ln (p_Y) - \dot{\mathscr{R}}T \ln (p_{0Y})$$

and

$$\Delta\mathscr{G}_2 = \dot{\mathscr{R}}T \ln (p_X) - \dot{\mathscr{R}}T \ln (p_{0X}) + \dot{\mathscr{R}}T \ln (p_{Y^{n+}}) - \dot{\mathscr{R}}T \ln (p_{0Y^{n+}}).$$

Thus, after some rearrangement,

$$\Delta\mathscr{G} = -\dot{\mathscr{R}}T \ln \left(\frac{p_{0X^{n+}} p_{0Y}}{p_{0X} p_{0Y^{n+}}} \right) + \dot{\mathscr{R}}T \ln \left(\frac{p_{X^{n+}} p_Y}{p_X p_{Y^{n+}}} \right). \tag{7.7}$$

Assuming now that the gas is not ideal, Eq. (7.7) is valid only if the pressure is multiplied by a certain coefficient called the *activity coefficient* Λ, such that $f = \Lambda p$, where f is the *fugacity*. The activity a of a product is defined as the ratio of the fugacity of the product to some standard fugacity f_0. Thus $a = f/f_0$. Taking this into consideration, Eq. (7.7) becomes

$$\Delta\mathscr{G} = \Delta\mathscr{G}_0 - \dot{\mathscr{R}}T \ln \left(\frac{a_X a_{Y^+}}{a_{X^+} a_Y} \right), \tag{7.8}$$

where

$$\Delta\mathscr{G} = -\dot{\mathscr{R}}T \ln \mathscr{C}_E, \tag{7.9}$$

where $\mathscr{C}_E$ is the equilibrium constant of the reaction. Now, replacing $\Delta\mathscr{G}$ by its value in Eq. (7.3), one obtains the well-known Nernst equation,

$$V = V_0 + \frac{\dot{\mathscr{R}}T}{nF_n} \ln \left(\frac{a_X a_{Y^+}}{a_{X^+} a_Y} \right), \tag{7.10}$$

where V_0 is called the *standard potential* of the cell. It is the measured potential when all losses are neglected. Its value is given by combining Eqs. (7.3) and (7.9); then

$$V_0 = \frac{\dot{\mathscr{R}}T}{nF_n} \ln (\mathscr{C}_E). \tag{7.11}$$

This is the maximum theoretical potential which can be given by a fuel cell. As will be seen later, V_0 is much larger than the actual output voltage; this is due to the losses in the cell. To have a basis of comparison between the different materials in a fuel cell, it is wise to define a standard potential for a single electrode rather than for the entire cell. Note that Reaction (7.6) can also be written

$$X + Y^{n+} + ne \rightleftharpoons X^{n+} + ne + Y. \qquad (7.12)$$

This can be considered as the sum of the following two reactions:

$$Y^{n+} + ne \rightleftharpoons Y$$

and

$$X \rightleftharpoons X^{n+} + ne.$$

Define V_{OX} to be the standard potential of element X and V_{OY} to be the standard potential of element Y; then for the total standard potential of the cell, $V_0 = V_{OY} - V_{OX}$ and, similarly, $V = V_Y - V_X$.

7.3 Normal Fuel Cells and Regenerative Fuel Cells [59]

It is important to know how the electric current is generated by a fuel cell. As seen above, the phenomenon is exactly the reverse of the one happening in a battery.

Figure 7.1 represents a schematic fuel cell. The fuel gas is introduced on the anode side; it then diffuses through the anode. At the anode it is oxidized and the electrons are transferred to the oxidizer through an external circuit. The oxidized fuel and the reduced oxidizer are thus rejected as products of oxidation. To permit the diffusion of the different elements, both electrodes should be porous. It is important to note that the components of the cell, i.e., the electrodes and the

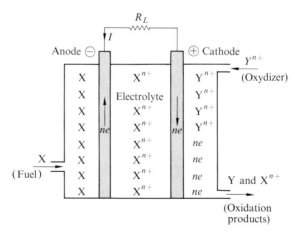

Figure 7.1. Schematic fuel cell. X = fuel, ne = a certain number of electrons, X^{n+} = fuel ions, and Y^{n+} = oxidizer ions.

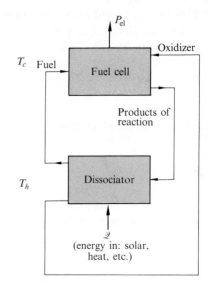

Figure 7.2. A regenerative fuel cell system. The overall system is subjected to Carnot's limitation since the dissociator is at a higher temperature than the fuel cell.

electrolyte, are not consumed in the process of power conversion. The ions passing through the electrolyte may be either positively or negatively charged, depending on the type of fuel cell.

Theoretically, a fuel cell can be used as a rechargeable cell in the same manner as a secondary battery. However, this is possible only for certain types of fuel cells. It is necessary that the electrochemical couple at the electrodes obtain a certain degree of reversibility for the fuel cell to store electrical energy. This type of fuel cell is known as a closed cycle regenerative fuel cell. It has attracted the greatest interest, since the energy necessary to recharge the cell can come from different sources: the sun, radioisotopes, a nuclear reactor, or any other source of heat. A regenerative fuel cell system is shown in Fig. 7.2. The energy to be transformed is here used to dissociate the products of the reaction into fuel and oxidizer. In this case, it is obvious that the dissociator will be a heat source at a high temperature T_h and the fuel cell will be a heat sink at a temperature T_c. Therefore the products of the reaction act as a working fluid in a thermal cycle. In this cycle, heat is changed into electrical power. Its efficiency is subject to the Carnot limitation. This does not contradict the statement that the fuel cell is not subject to the Carnot limitation since the *cell itself* is at the same temperature whereas the *entire system* is not.

7.4 Performance Characteristic $V = f(J)$

The open circuit voltage of a fuel cell has been calculated, and is given by the Nernst equation (Eq. 7.10). However, the overall reaction of a fuel cell can be affected by the presence of a load. The output voltage will decrease with increasing output current due to polarizations due to losses in the anode, in the cathode, and in the electrolyte. Follow the path of an electron from the time it is with the fuel to the time it is rejected with the products of oxidation. Note that there occur several effects, all of which consume a certain amount of energy. The fuel is transported to the porous anode where it is adsorbed at the anode surface, and then it is dissociated into electrons (which flow through the load) and ions (which flow through the electrolyte). This requires the transfer of the electrons to the anode, the desorption* of the ions, and their transport through the electrolyte to the cathode. The cathode is the meeting place of the electrons which have done useful work on their way through the load, of the ions, and of the injected oxidizer; then the same phenomena happen again in reverse at the cathode surface to form the products of oxidation which will be rejected.

Every step described above leads to a loss of voltage due to the phenomena of mass transport, charge transport, and adsorption which have occurred. These polarizations are usually classified as:

1. Activation polarization,
2. Concentration polarization,
3. Resistive polarization,
4. Other polarizations.

Activation polarization. From Reaction (7.12), note that for the oxidation of X to take place, X should be transported and adsorbed at the surface of the anode. This will require a shift in free energy equal to

$$\Upsilon \, \Delta \mathcal{G} = - \Upsilon n \, \Delta V_a F_n,$$

where Υ is called the *transfer coefficient*, or the fraction of the free energy necessary for the oxidation of X. For X^{n+}, obviously,

$$(\Upsilon - 1) \, \Delta \mathcal{G} = (1 - \Upsilon) n \, \Delta V_a F_n.$$

Therefore, the reaction rates and, consequently, the current densities J_1 and J_2 for the reactions $X \rightarrow X^{n+}$ and $X^{n+} \rightarrow X$ are, respectively,

$$J_1 = J_0 \exp \left(\frac{- \Upsilon \, \Delta \mathcal{G}}{\mathcal{R} T} \right)$$

or

$$J_1 = J_0 \exp \left(\frac{\Upsilon n \, \Delta V_a F_n}{\mathcal{R} T} \right) \tag{7.13}$$

and

$$J_2 = J_0 \exp \left[\frac{-(1 - \Upsilon) n \, \Delta V_a F_n}{\mathcal{R} T} \right], \tag{7.14}$$

where J_0 is the current density flowing in both directions at the equilibrium

---

* The freezing of gas from the electrode surface, and a reversal of gas adsorption.

state. The net current at the anode will be

$$J = J_1 - J_2 - J_0,$$ (7.15)

or, by combining Eqs. (7.13), (7.14), and (7.15) and solving for ΔV_a,

$$\Delta V_a = -\left(\frac{\dot{\mathscr{R}} T}{\Upsilon n F_n}\right) \ln (J_0) + \left(\frac{\dot{\mathscr{R}} T}{\Upsilon n F_n}\right) \ln (J + J_0)$$
$$-\left(\frac{\dot{\mathscr{R}} T}{\Upsilon n F_n}\right) \ln \left[1 + J_0 \exp \left(\frac{n \Delta V_a F_n}{\dot{\mathscr{R}} T}\right) \right].$$ (7.16)

The third term of this equation can be neglected for relatively large values of J. Therefore, Eq. (7.16) can be written in a simplified manner, giving the Tafel equation,

$$\Delta V_a = V_1 + V_2 \ln (J + J_0),$$ (7.17)

where

$$V_1 = -\left(\frac{\dot{\mathscr{R}} T}{\Upsilon n F_n}\right) \ln (J_0)$$

and

$$V_2 = \frac{\dot{\mathscr{R}} T}{\Upsilon n F_n}.$$

Large values of J_0 lead to low polarization and vice versa.

Tafel's equation states that the activation polarization is logarithmic. The current density J_0 is a very small quantity relative to J, and usually is omitted in Eq. (7.17). However, this omission will be unacceptable for values of J of the same order of magnitude as J_0.

To eliminate the effect of activation polarization, it is necessary to maximize the area of the fuel electrode-electrolyte interface. For this purpose, catalysts are used along with gas-diffusing electrodes.

Concentration polarization. When the fuel is oxidized at the anode and the oxidizer reduced at the cathode, the concentration of the adjacent ions tends to decrease. This creates an overvoltage known as the concentration polarization ΔV_c.

Assuming that d is the equivalent thickness of a diffusion layer at the surface of the electrode, Fick's law of diffusion for that layer is

$$\dot{n}_x = -\mathscr{D}_x \nabla n_x,$$

where $\dot{n}_x$ is the mass transfer rate, $\mathscr{D}_x$ is the diffusion coefficient, and n_x is the electrolyte concentration. After integration,

$$\dot{n}_x = \mathscr{D}_x (n_{0x} - n_{lx})/d,$$ (7.18)

where n_{0x} and n_{lx} are the ionic concentration out of and in the layer, respectively. The mass transfer rate can be expressed as a function of the current density by the relation

$$\dot{n}_x = J/n F_n.$$ (7.19)

Introducing Eq. (7.19) into Eq. (7.18) yields

$$J = nF_n \mathcal{D}_x (n_{0x} - n_{lx})/d.$$

The limiting current density is given for $n_{lx} = 0$; thus

$$J_l = nF_n \mathcal{D}_x n_{0x}/d,$$

and, consequently,

$$(J/J_l) = 1 - (n_{lx}/n_{0x}). \qquad (7.20)$$

Thus, it can be proven [8] that the concentration polarization is

$$\Delta V_c = \frac{\dot{\mathcal{R}}T}{nF_n} \ln \left(\frac{n_{lx}}{n_{0x}} \right),$$

or, taking account of Relation (7.20),

$$\Delta V_c = -\frac{\dot{\mathcal{R}}T}{nF_n} \ln \left(\frac{J_l}{J_l - J} \right). \qquad (7.21)$$

It is clear that an ion deficiency (at the cathode) may occur, in which case J_l and ΔV_c are positive, or an ion excess (at the anode), in which case both J_l and ΔV_c are negative.

Concentration polarization can be reduced greatly by increasing the limiting current density. This can be obtained by one or both of the two following methods:

1. Stirring the electrolyte,
2. Increasing its temperature.

These tend to accelerate the rate of diffusion and transport of the ions.

Resistive polarization. The materials used in a fuel cell are not perfect conductors. Therefore, due to the nonzero electrical resistivities of the electrodes and the electrolyte, a voltage drop will occur. For these resistance polarizations, in general,

$$\Delta V_R = \ell_{eq} J/\sigma_{eq}, \qquad (7.22)$$

where ℓ_{eq} and σ_{eq} are the equivalent length and conductivity, respectively, of the fuel cell.

To reduce these polarizations it is necessary to reduce the resistances of the cell elements. This can be done by the choice of a good cell design, i.e., minimum volume for maximum power output of the cell.

Other polarizations. There are many less important sources of polarizations, but these are usually associated with the three described above.

If the fuel transport rate to the surface of the electrode is smaller than the rate of reaction there will be a difference between the partial pressure of the bulk of the gas and the pressure of the gas in the pores of the electrodes where the reaction

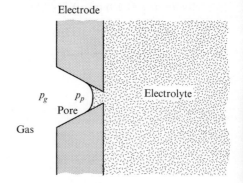

Figure 7.3. The gas transport polarization due to the difference between p_g and p_p.

takes place. This will lead to a gas transport polarization equal to

$$\Delta V_g = \frac{\mathscr{R}T}{nF_n} \ln\left(\frac{p_p}{p_g}\right), \tag{7.23}$$

where p_p and p_g are the partial pressures as shown in Fig. 7.3.

The utilization of diaphragms in the electrolyte is also a source of polarizations due to the increase of the equivalent resistance of the fuel cell. This may also be increased by electrode processes which may grow with time due to the corrosion of the electrodes, thus reducing the lifetime of the cell.

Total voltage output. The terminal voltage of a cell is equal to the difference between the reversible voltages as given by Eq. (7.10) and the sum of all the different polarizations. The activation and concentration polarizations exist directly at the electrodes, whereas the resistance polarization may be significant at both the electrodes and the electrolyte. Therefore,

$$V_{\text{out}} = \Delta V - \sum \text{polarizations.} \tag{7.24}$$

The equation of the performance characteristic (V versus J) will be complicated. The calculations presented above should be taken only on a qualitative basis, and much of the data still rely on experimentation. However, by combining Eqs. (7.17), (7.21), and (7.22) for the electrodes and the electrolyte, and neglecting all the other polarizations,

$$V_{\text{out}} = \Delta V - \left[V_1 + V_2 \ln(J + J_0) - \frac{\mathscr{R}T}{nF_n} \ln\left(\frac{J_l}{J_l - J}\right) \right.$$
$$\left. + \frac{\mathscr{R}T}{nF_n} \ln\left(\frac{J_l + J}{J_l}\right) + \frac{\ell_{eq}J}{\sigma_{eq}} \right]. \tag{7.25}$$

This equation is illustrated qualitatively in Fig. 7.4. Note that the activation polarization is important at low current densities whereas the concentration polarization becomes important at higher densities. On the other hand, Eq. (7.25) shows that the output voltage is affected by the temperature, the pressure, and the

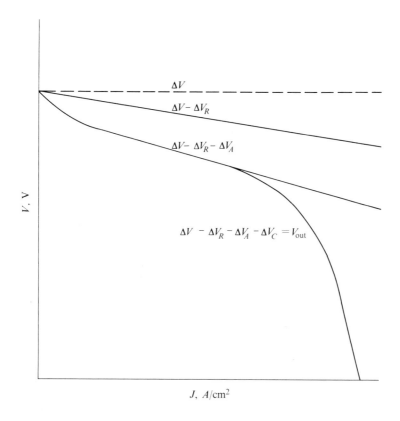

Figure 7.4. Typical performance characteristic of a fuel cell.

electrolyte concentration. Figures 7.5, 7.6, and 7.7 show that V_{out} decreases with those quantities decreasing; however, this result will depend on the individual coefficients appearing in Eq. (7.25), and will then change from one type of fuel cell to another. The results shown here are for a hydrogen-oxygen fuel cell having KOH as an electrolyte.

7.5 Different Types of Fuel Cells

The number of types of fuel cell systems is rapidly increasing as engineers and scientists are trying to solve the many problems encountered in the search for higher efficiency. Many types are developed from a particular need for a particular use. Thus, there are many ways of classifying existing fuel cells. One of the most useful classifications is based on the type of fuel utilized by the cell. However, many workers use a classification based on the type of electrolyte; others use temperature as a criterion.

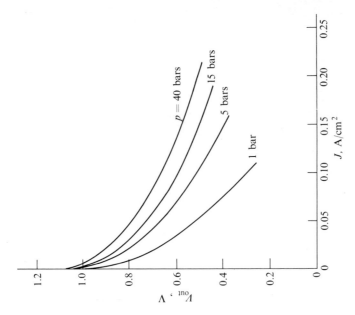

Figure 7.6. Influence of pressure on the performance characteristic of a hydrogen-oxygen fuel cell at $T = 110°C$ and an electrolyte concentration of 30% KOH. The output voltage increases with increasing pressure. Experimental results adapted from Laroche [63].

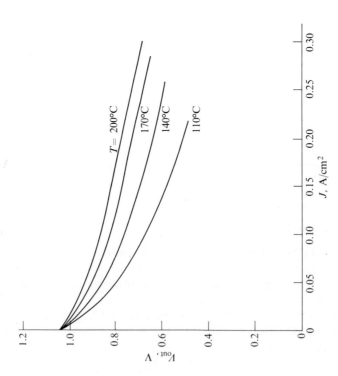

Figure 7.5. Influence of temperature on the performance characteristic of a hydrogen-oxygen fuel cell at $p_r = 40$ bars and an electrolyte concentration of 30% KOH. The output voltage increases with increasing temperature. Experimental results adapted from Laroche [63].

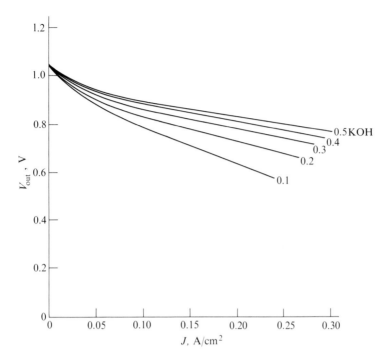

Figure 7.7. Influence of electrolyte concentration on the performance characteristic of a hydrogen-oxygen fuel cell at $T = 200°C$ and $p_r = 40$ bars. The output voltage increases with increasing KOH concentration. Experimental results adapted from Laroche [63].

The most important types can be listed as follows:

a) Fuel classification
    1. hydrogen fuel cells
    2. hydrazine fuel cells
    3. hydrocarbon fuel cells
    4. alcohol fuel cells
    5. ammonia fuel cells
    6. coal and natural gas fuel cells
    7. biochemical fuel cells
    8. redox fuel cells

b) Electrolyte and temperature classifications
    1. aqueous electrolytes: $-20°C < T < 250°C$
    2. molten salt electrolytes: $500°C < T < 700°C$
    3. solid electrolytes: $700°C < T < 1100°C$.

Table 7.1
Characteristics of several fuels

Fuel	$\Delta \mathcal{H}$, kcal/mole	$\Delta \mathcal{G}$, kcal/mole	V_0, V	Products	Reactivity
Hydrogen H_2 (g)	0		1.23	H_2O	Very good
Hydrazine N_2H_4 (l)	+12.05		1.56	$N_2 + H_2O$	Very good
Ethylene	+12.50		1.20	$H_2O^{++} + CO_2$	Fair
Methane CH_4 (g)	−17.89	−12.14	1.01	$H_2O + CO_2$	Bad
Propane C_3H_8 (g)	−24.75		1.10	$H_2O + CO_2$	Fair
Natural gas (g)	−50.00		1.20	$H_2O + CO_2$	Bad
Diesel fuel (l)	−55.00		1.20	$H_2O + CO_2$	Bad
Methanol CH_3OH (l)	−57.04		1.21	$H_2O + CO_2$	Good
Ammonia NH_3 (g)	−11.04		1.13	$N_2 + H_2O$	Good
Coal C (s)	0		1.02	CO_2	Almost none
Carbon monoxide CO (g)	−26.42	−32.81	1.33	CO_2	Bad

The choice of fuel. There are three important factors which influence the choice of a fuel: the cost, the amount of energy stored, and the reactivity of the fuel. It is obvious that in some applications some factors will be more important than others. For instance, if the overall weight of the fuel cell system should be as small as possible, it is necessary to use the fuel which stores the highest amount of energy, and the cost factor becomes secondary.

The amount of energy stored in a fuel can be represented by the maximum increase in free energy ($\Delta \mathcal{G}$) that can be obtained from the oxidation of the fuel. However, the total energy available in the fuel is given by ($\Delta \mathcal{H}$). The maximum electrochemical efficiency of a fuel cell can be expressed as

$$\eta = \Delta \mathcal{G} / \Delta \mathcal{H}.$$

The values of $\Delta\mathscr{G}$ and $\Delta\mathscr{H}$ for the different fuels are listed in Table 7.1. In the same table are expressed the standard potential resulting from the oxidation of the fuel and the products of oxidation.

Hydrogen fuel cells. Hydrogen is a very active fuel which can be utilized for the conversion of energy at relatively low temperatures. Fuel cell systems using hydrogen have the advantage of being simple and practical. Air can be used directly as an oxidizer and the oxidation reaction can take place in an alkaline or acidic aqueous electrolyte. For instance, in an alkaline potash (KOH) solution, the basic reaction at the cathode is

$$O_2 + 2H_2O + 4e \to 4OH^-,$$

and, at the anode,

$$2H_2 + 4OH^- \to 4H_2O + 4e.$$

The overall reaction is

$$O_2 + 2H_2 \to 2H_2O.$$

Water is produced, and this can be a very important advantage in some applications. The potash medium works as a conductor for the hydroxyl ions.

Such fuel cell systems were used in many space projects, such as Apollo and Gemini. The Bacon fuel cell is the most efficient electrochemical converter, as is clearly shown by its performance characteristic (Fig. 7.8, curve 1). The working temperature of this cell is around $200°C$ at 600 psi pressure and about 44% KOH. The Kordesh fuel cell is similar to Bacon's with a concentration of 30% KOH. Other fuel cells of this type were built by Union Carbide (0.25 KOH), Allis Chalmers, and Pratt and Whitney (0.85 KOH).

In Europe, apart from Bacon's [10] work in England,* progress is continuing in Germany, where Justi [56, 57] developed his double skeleton electrode (DSK) fuel cell,† in the Soviet Union [11], and in France [63].‡ Curve 2 of Fig. 7.8 represents another hydrogen fuel cell in which sulfuric acid ($5NH_2SO_4$) is the aqueous electrolyte [42].

At temperatures below the evaporation point of water ($100°C$), the problem of removing this product of oxidation becomes very important. Two solutions are proposed: adjusting the fuel recirculating rate, and using an ion-exchange membrane electrolyte. This membrane acts as a sponge and the water can be drained away without much difficulty. There are many types of ion-exchange membranes; they are all solid and have very good mechanical characteristics. An example is a sulphonated polystyrene. Fuel cells of this type were developed by• Grubb and Niedrach [42, 71] for the Gemini project. One of the disadvantages

---

*   Shell, Ltd.

†   Varta, Frankfurt am Main.

‡   Industrial National Office of Nitrogen (ONIA).

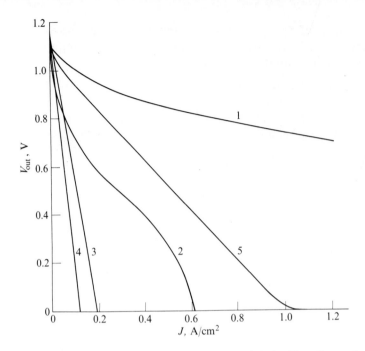

Figure 7.8. Performance characteristics of several types of fuel cells. Curve 1: Bacon hydrogen-oxygen cell, $T = 200°C$, electrolyte concentration 45% KOH [1]; curve 2: hydrogen-air cell, electrolyte $5NH_2SO_4$ [42]; curve 3: hydrogen-oxygen cell, $T = 1094°C$, electrolyte $(ZrO_2)(CaO)$ [15]; curve 4: Battelle hydrogen-oxygen cell, solid electrolyte $Zr_{0.85}Ca_{0.15}O$ [15]; curve 5: hydrazine-oxygen cell, $T = 70°C$, electrolyte KOH [86].

of the ion-exchange membrane electrolyte is the higher resistive polarization and the higher price of the electrolyte.

At higher temperatures ($500°C < T < 700°C$) molten salt electrolytes are used. These molten salts are usually mixtures of molten carbonates, e.g., those of Li, Na, K, and Ca, or those of Li and K for higher temperatures.

The reactions taking place in such cells can be expressed as follows: At the cathode,

$$\tfrac{1}{2}O_2 + CO_2 + 2e \rightarrow CO_3^{--},$$

and, at the anode,

$$H_2 + CO_3^{--} \rightarrow H_2O + CO_2 + 2e.$$

The overall reaction is

$$H_2 + \tfrac{1}{2}O_2 \rightarrow H_2O.$$

Fuel cells of this type were built [81, 83], using Li, Na, and K mixture carbonates as the electrolyte. Many studies were also performed in Europe by Chambers [22] in Great Britain (Sondes place cells) and Broers and Ketelaar [20] in the Netherlands among others.

At much higher temperatures it becomes more practical to use solid electrolytes. These are good conductors at temperatures around 1000°C due to the mobility of some ionic components. The electrolytes usually used are mixtures of zirconia (ZrO_2), yttrium oxide (yttria, Y_2O_3), and calcium oxide (CaO). The latter two materials serve as stabilizers for the zirconia. Two types are represented in Fig. 7.8, curves 3 and 4.

The reactions in this case are

$$\tfrac{1}{2}O_2 + 2e \rightarrow O^{--},$$

at the cathode and, at the anode,

$$H_2O + O^{--} \rightarrow H_2O + 2e.$$

Overall, then,

$$\tfrac{1}{2}O_2 + H_2 \rightarrow H_2O.$$

Fuel cells of this type were built by Westinghouse [5] (Bell and Spigot design), General Electric [75] (1100°C), and the Battelle Institute [14] in Germany.

Hydrazine fuel cells. One of the most important disadvantages of hydrogen is the fact that it is a gas. Numerous difficulties of storage are created, requiring either use of cryogenic techniques or operation at high pressures. However, these difficulties can be avoided by using a liquid fuel such as hydrazine (NH_2NH_2). Hydrazine is highly reactive and is soluble in aqueous electrolytes such as KOH. It has, however, the disadvantage of being costly and dangerously toxic, and if fuel is not prevented from reaching the cathode, direct oxidation may occur, leading to a loss in efficiency.

The basic reactions of a hydrazine fuel cell are

$$O_2 + 2H_2O + 4e \rightarrow 4OH^-,$$

at the cathode, and, at the anode,

$$NH_2NH_2 + 4OH^- \rightarrow N_2 + 4H_2O + 4e.$$

The overall reaction is

$$O_2 + NH_2NH_2 \rightarrow N_2 + 2H_2O,$$

with a standard potential equal to 1.56 V.

The most important hydrazine fuel cell systems built in this country are the work of Allis Chalmers [57, 86], using oxygen as the oxidizer and potash as the electrolyte. The power output of their batteries ranges from 100 W to several kilowatts. A performance characteristic of such a hydrazine-oxygen cell is shown in Fig. 7.8, curve 5, at a working temperature of 70°C. Some European systems [37]* work at much lower temperatures ranging from -20 to $+50°C$.

---

*   Built by Chloride Technical Services Ltd. and Shell, Ltd., England.

Hydrocarbon fuel cells. There are two kinds of hydrocarbons: unsaturated hydrocarbons such as ethylene, and saturated hydrocarbons whose chemical equations are given by the general expression C_nH_{2n+2}, where n equals 1 for methane (CH_4), 2 for ethane (CH_3CH_3), 3 for propane ($CH_3CH_2CH_3$), and higher numbers for the higher hydrocarbons that constitute Diesel oil and gasoline. The difference between the unsaturated and the saturated hydrocarbons is that the former are easily broken down into carbon-carbon double bonds. They are artificial products obtained in gas refineries. The latter are directly obtainable from natural products and are much more interesting, due to their low cost. However, the reactivity of a hydrocarbon fuel decreases with increasing n. This reactivity is influenced greatly by many other factors, such as the temperature, the electrolyte, and the electrocatalyst.

At temperatures below 150°C, hydrocarbon fuel cells perform poorly; however, by using phosphoric or sulfuric electrolytes and black platinum electrocatalysts, the performance of methane and propane fueled cells has been greatly improved.

Good performance has been obtained for propane at temperatures slightly higher than 150°C. At atmospheric pressure and in the presence of water,

$$CH_3CH_2CH_3 + 6H_2O \rightarrow 3CO_2 + 20H^+ + 20e.$$

Another attractive possibility is to break the hydrocarbon fuel into hydrogen and carbon oxides. The hydrogen can then be the primary fuel for the cell. This can be obtained through a steam reforming process by the following reactions:

$$C_nH_{2n+2} + H_2O \rightarrow nCO + (2n + 1)H_2, \tag{7.26}$$

and

$$nCO + nH_2O \rightarrow nCO_2 + nH_2, \tag{7.27}$$

leading to the overall reaction,

$$C_nH_{2n+2} + (n + 1)H_2O \rightarrow nCO_2 + 2(n + 1)H_2. \tag{7.28}$$

Hydrogen can be obtained in a pure form by passing the products of Reaction (7.27) in a palladium alloy diffusion cell.

The performance is much better at higher temperatures (between 500 and 700°C) by using molten salt electrolytes. This creates new problems such as electrode deterioration, gasketing of the cells, and destruction of the insulation. If methane is used, the following reactions occur: at the cathode,

$$2O_2 + 4CO_2 + 8e \rightarrow 4CO_3^{--},$$

at the anode,

$$CH_4 + 4CO_3^{--} \rightarrow 2H_2O + 5CO_2 + 8e,$$

and the overall reaction is

$$2O_2 + CH_4 \rightarrow 2H_2O + CO_2.$$

The high temperatures favor a rapid electrochemical oxidation of the fuel, and the fused carbonate is a good CO_2-rejecting electrolyte.

Fuel cells of this type have been studied by Broers and Ketelaar [20], Chambers [22], Peattie [72–75], Sandler [81], among others [12] in the U.S. and abroad.

For temperatures around 1000°C, solid electrolytes such as K_2CO_3 or ZrO_2 are used. The cell of Weissbart and Ruka [92, 93], which utilizes methane as a fuel and works at a temperature of 1020°C, is an example of such a cell. Others use a propane fuel diluted in water, nitrogen, and carbon dioxide compositions [14].

Alcohol fuel cells. Alcohol fuels such as methanol (CH_3OH) have the attractive characteristic of being cheap and storable in liquid form. Their reactivity is acceptable and they can fuel low-temperature (around 60°C) cells.

The electrolytes employed are either potassium hydroxide compounds or acid electrolytes. The electrolyte reacts at the anode with the carbon dioxide to produce carbonates. Because of their low solubility, their life is relatively short and they are considered only for special applications such as buoys and signal lights. Work on such fuel cell systems has been performed by Heath [47, 50], Tarmy [85], and Vielstich [89].

In such systems, direct oxidation of the fuel at the cathode creates a serious source of loss of generated electric power. The use of concentrated cesium carbonate and a good design of the fuel-electrolyte flow system can limit these losses.

Ammonia fuel cells. Ammonia (NH_3) is also a cheap fuel with good reactivity and it is easily storable in liquid form by using pressure storage tanks. Fuel cells using ammonia are very similar to hydrazine fuel cells using caustic electrolytes and working at low temperatures. However, their development is not very encouraging due to their high vapor pressure and their toxicity. Work in this field has been performed by Eisenberg [32] and Wynveen [98].

Coal and natural gas fueled cells [53]. For obvious economic reasons it is extremely attractive to build fuel cells using natural gas or coal as the primary fuel. The natural gas is usually a mixture of carbon monoxide, hydrogen, and methane. Coal cannot be used directly, but can be partially oxidized to yield a carbon monoxide gas.

At low temperatures, alkaline electrolytes are used, but difficulties arise due to the solubility of the carbon dioxide in the electrolyte. This can be avoided by using integrating fuel cells with gasifiers* as proposed by Bauer [13], and Bréelle and Degobert [18].

At much higher temperatures a solid electrolyte (zirconia) fuel cell using natural gas has been built by General Electric [21]. The gas is pyrolized at a temperature of 1100°C to form hydrogen and carbon. This deposited carbon forms the anode of the cell whereas the cathode is made up of molten silver. Power densities of 0.2 W/cm^2 have been achieved for these cells.

---

* A gasifier is a product or a device that can produce gas by synthesis from coal.

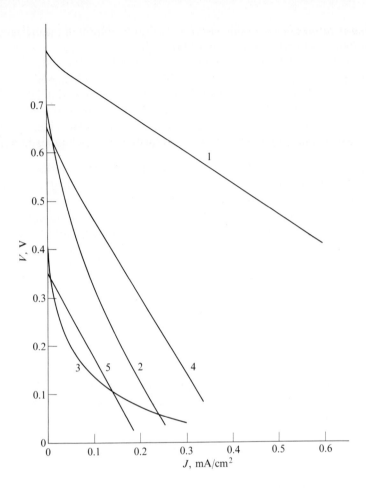

Figure 7.9. Performance characteristics of biochemical fuel cells. Curve 1: fresh mushroom in modified Hoagland's nutrient; curve 2: fresh mushroom in KOH; curve 3: fresh mushroom in Sisler's medium; curve 4: lyophyllized algae in modified Hoagland's nutrient; curve 5: sucrose in modified Hoagland's nutrient. (Adapted from Colichman [23].)

Biochemical fuel cells [16, 23].   A biochemical fuel cell is a device which converts electrochemical energy into electrical energy by using bio-organic matter as the source of fuel. Air is injected at the cathode as an oxidizer and bio-organic matter at the anode as the fuel. Microorganisms are then employed to catalyze the oxidation of the fuel. The electrolyte is usually an organic solution or an aqueous medium such as potassium hydroxide. The bacteria create new losses by consuming about half of the fuel. This consumption is necessary for their nutritional requirements.

The electrical power output of a biochemical fuel cell is proportional to the bacterial metabolism rate. Theoretically, this rate may be extremely high, leading to biochemical performances similar to those of regular fuel cells. However, the biochemical fuel cell still remains a low-power generator, mainly because of a lack of knowledge of biochemical kinetics.

The field is open to research and is encouraged by the prospect of utilizing extremely cheap or even worthless fuels, such as sewage, garbage, grasses, sawdust, leaves, etc. The production of power in this case may be a solution to the garbage collection problem in our cities. Space researchers are also extremely interested in these generators in connection with the many closed ecological schemes for sustaining human life in space.

Performance characteristics of several biochemical fuel cells have been plotted in Fig. 7.9. Curves 1, 2, and 3 represent fresh mushrooms in different media; curves 4 and 5 represent algae and sucrose ($C_{12}H_{22}O_{11}$) fuel cells, respectively. It is seen that the mushroom biofuel with a modified Hoagland's nutrient* electrolyte constitutes the best biochemical fuel cell, with a maximum power density of about 0.2 mW/cm^2.† This field is progressing rapidly, and, for more detail, references should be consulted.

Redox fuel cells (reduction and oxidation) [91]. A redox fuel cell is a regenerative fuel cell of the same type as the one described in Section 7.3. The difference is that the redox cell is an indirect regenerative cell.

By definition, a redox fuel cell is an electrochemical generator on the electrodes of which react two oxydo-reductor systems.‡ This is shown in Fig. 7.10.

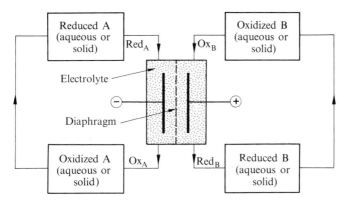

Figure 7.10. Example of a redox fuel cell. On each electrode of the cell reacts an oxydo-reductor system.

---

* Hoagland's nutrient is a medium consisting of three solutions: solution I: EDTA diNa$^+$ salt; solution II: microelements; solution III: microelements freshly prepared [23].

† It should be noted that the bio-organic matter (mushroom, sucrose, etc.) serves as fuel for the bacteria, the result of the combustion (by the microorganisms) being here electricity instead of heat.

‡ An oxydo-reductor system is one that brings the cell to its initial degree of oxidation either by reduction or by oxidation [91].

The fuels most used in these systems are metal hydrides such as lithium and calcium hydrides, while the electrolytes are mainly highly concentrated aqueous solutions. The latter created several problems of polarizations and regeneration, and electrolytes other than the strongly acidic were considered. In several experiments, electrodes were made of precious metals but nonprecious compounds such as bromine-bromide were also considered.

The source of high temperatures is either nuclear or solar, leading to the possible application of redox fuel cell systems as power stations in space.

7.6 Electrodes and Electrocatalysts

Electrodes are important components in a fuel battery. They collect the electrons and carry the electric current. They are the seat of the electrochemical reactions and provide the fuel-electrolyte interface. Therefore, many requirements are involved in the choice of an electrode, such as good electrical conductivity, good catalytic properties, acceptable mechanical characteristics, and chemical stability in a corrosive electrolyte.

The catalytic properties of several materials were studied [38, 48, 102], and it was found that the rate of reaction is inversely proportional to the free energy of adsorption of the hydrogen atoms. Electrodes with high catalytic activity are required at low temperatures. At higher temperatures, a lower catalytic activity is acceptable, since the activation energy barriers for the electrochemical reactions are overcome by the higher thermal energy.

The catalytic properties of a metal can also be explained in terms of the number of vacant orbitals* that react with an electron donor. It was found that the catalytic activity increases with the number of reacting vacant orbitals decreasing. This is shown in Fig. 7.11, where hydrogen oxidation is assumed to be the same for each metal.

The theory of electrocatalytic processes is still in its infancy, and the selection of a suitable catalytic surface is based mainly on experimental evidence. Data have been obtained by treating surfaces of different metals. When the best metal for the given electrolyte had been chosen, the surface configuration was investigated.

These investigations led to many types of porous electrodes. Two important methods have been devised to obtain a stable interface configuration. The first method is to use a larger pore radius at the fuel side of the electrode than the pore radius at the electrolyte side. The second method is to utilize controlled wetting by partially treating the electrode with a hydrophobic substance. This will enable the contact angle to increase in the vicinity of the reaction site.

When the cell is loaded, the equilibrium established at the interfaces may be affected, but this problem is partially overcome since stable interfaces can be maintained for a thousand-hour operation.

---

*   An orbital is an electron that revolves in an orbit about the nucleus of an atom (Rutherford and Bohr theories).

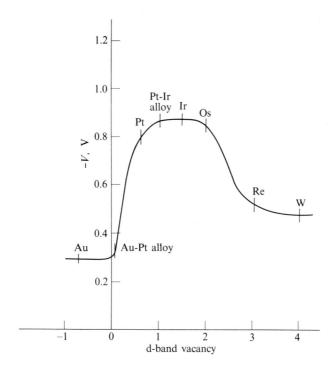

Figure 7.11. The hydrogen standard potential as a function of d-band vacancy of several electrode materials. (Adapted from Young and Rozelle [100].)

If the catalytic properties of an electrode are deficient, it becomes necessary to use a catalyst to activate the electrode surface and thus allow the electrochemical reactions to occur with the highest efficiency possible. The choice of a catalyst depends on the nature of the fuel at the anode and of the oxidizer at the cathode. The catalyst should be chemically stable and should not react with the electrolyte or change its state. Its catalytic properties should remain fairly constant with time.

Most of the materials providing good catalytic surfaces are in group VIII of the periodic table. These are iron, cobalt, nickel, platinum, rhodium, palladium, and iridium. Others, gold and silver, are in group Ib of the table.

The techniques of electrode construction are numerous. Many of them were reported in the literature [53]. Some special electrode structures are among the reasons for success of many fuel cell systems such as the monoskeleton (MSK) electrode [58],* the double-skeleton electrode (DSK) of Justi [56–58], and the Porvic base electrode of Shell, Ltd. [95].

---

* At Brown Boveri, Baden, Switzerland.

7.7 Power Output and Efficiency of a Fuel Cell

The output voltage was calculated in Section 7.4 (Eq. 7.25). It is clear that the density of power generated by a fuel cell will be given by

$$P = V_{\text{out}}J. \tag{7.29}$$

Combining Eqs. (7.25) and (7.29) yields

$$P = \Delta V J - \left[V_1 J + V_2 J \ln(J + J_0) - \frac{\mathcal{R}TJ}{nF_n} \ln\left(\frac{J_l}{J_l - J}\right) \right.$$
$$\left. + \frac{\mathcal{R}TJ}{nF_n} \ln\left(\frac{J_l + J}{J_l}\right) + \frac{\ell_{\text{eq}}J^2}{\sigma_{\text{eq}}} \right]. \tag{7.30}$$

Note in Fig. 7.4 that the power density will go to zero for $J = 0$ and for $V = 0$. Therefore, there will be a certain value of current density for which the power

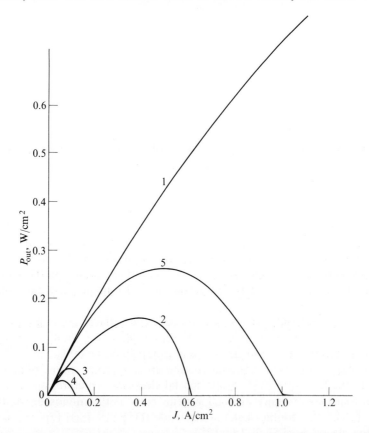

Figure 7.12. Power output versus current density for several fuel cells. Curve 1: Bacon fuel cell; curve 2: hydrogen-oxygen fuel cell, electrolyte $5NH_2SO_4$; curve 3: hydrogen-oxygen fuel cell, solid electrolyte; curve 4: Battelle hydrogen-oxygen fuel cell; curve 5: hydrazine-oxygen fuel cell. (Adapted from Fig. 7.8.)

generated is maximum. Figure 7.12 illustrates the power density versus current density curves corresponding to the performance characteristics of Fig. 7.8. Here note that for a given current density, the power generated by a fuel cell increases with increase of temperature, pressure, and electrolyte concentration as illustrated by Fig. 7.13.

This power output is related to Gibbs free energy, which represents the maximum energy obtainable from a chemical reaction in a fuel cell. If t is the time during which the energy flows,

$$\Delta \mathcal{G} = -nF_nV = -IVt|_{t=0},$$

or $\Delta \mathcal{G} = -qV$, where q is the total electric charge at time $t = 0$. Since the total energy available to the cell is equal to the heat of the reaction, as expressed by the enthalpy, the total efficiency of the reaction can be written as $\eta_i = \Delta \mathcal{G}/\Delta \mathcal{H}$ or, taking into consideration Eq. (7.1), as

$$\eta_i = 1 - T\Delta \mathcal{S}/\Delta \mathcal{H}. \tag{7.31}$$

Here $\Delta \mathcal{S}$ is indeed a measure of the unavailable energy in the reversible system.

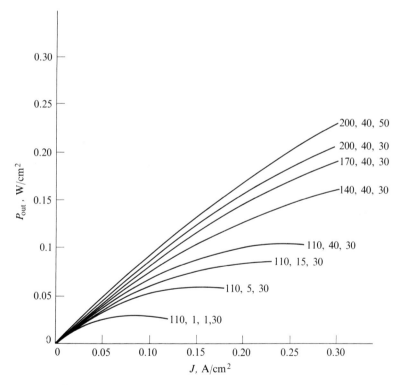

Figure 7.13. Effect of temperature, pressure, and electrolyte concentration on the power output of hydrogen-oxygen fuel cell with KOH electrolyte. The parameters are temperature in degrees centigrade, pressure in bars, and concentration in percent KOH, respectively. (Adapted from Figs. 7.5 and 7.6.)

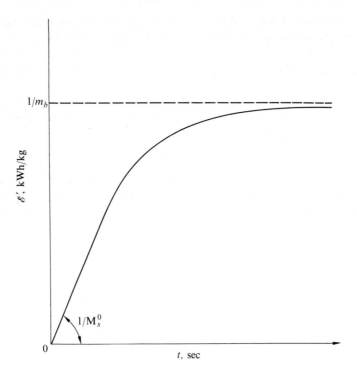

Figure 7.14. Energy density versus time of discharge for a fuel cell battery.

Carnot's efficiency is $\eta_c = 1 - T_c/T$. A comparison of Carnot's efficiency of a heat engine to the electrochemical efficiency of Eq. (7.31) leads to an interesting result: an increase in temperature will lead to higher efficiency in a heat engine but to a lower efficiency in a fuel cell.

The overall efficiency of the cell under charge is equal to the ratio of the output energy to the input energy, i.e.,

$$\eta_{out} = -\frac{P_{out}t}{\Delta \mathcal{H}} = -\frac{\mathcal{E}}{\Delta \mathcal{H}}. \tag{7.32}$$

Or, using the first law of thermodynamics,

$$\eta_{out} = 1 - \frac{\delta \mathcal{Q}}{\Delta \mathcal{H}}. \tag{7.33}$$

Therefore the losses due to the different polarizations can be expressed as irreversible heat transfer from the fuel cell.

As seen from Eq. (7.32), the total energy produced by a fuel cell is given by $\mathcal{E} = P_{out}t$ (watt-hours). If M_s^0 is the specific weight of the fuel system excluding the fuel battery itself expressed in grams per watt-hours, m_b is the specific weight of the fuel cell battery expressed in grams per watt, and M_t is the total weight of the system expressed in grams,

$$M_t = \mathscr{E} M_s^0 + \mathscr{E} m_b/t.$$

If one defines the energy density $\mathscr{E}'$ of the battery to be $\mathscr{E}' = \mathscr{E}/M_t$ (kWh/kg), then

$$\mathscr{E}' = \frac{1}{M_s^0/t + m_b}. \tag{7.34}$$

This is an important characteristic of fuel cell systems. It is plotted in Fig. 7.14 as a function of time. Note that for missions of long duration, the power density tends toward $1/m_b$.

7.8 Fuel Cell Batteries and Systems

To obtain a high power output it is necessary to connect several cells into a battery in the same manner as in thermoelectric power generators. This creates new technological problems, such as electrolyte circulation, removal of the products of reaction, and dual pumping.

There exist two basic types of fuel cell batteries, as far as the arrangement of electrodes is concerned:

1. The storage battery,
2. The filter battery.

In the storage batteries the cells can be connected either in series or in parallel, as shown in Fig. 7.15. The filter batteries are usually in series, as shown in Fig. 7.16. These batteries make possible the attainment of high voltages in the minimum volume available, since the entire battery acts as it were a larger cell with many electrodes. This method avoids the larger cell body of a storage battery and leads to better mechanical characteristics. The two types shown in Fig. 7.16 are the homopolar series type and the bipolar series type. In the latter, a diaphragm capable of passing the electric current from one cell to the other is necessary. It should of course remain chemically stable in the electrolyte medium.

A complete fuel cell system consists mostly of the fuel cell battery itself, a source of fuel with the fuel circulation, a source of oxidizer with the oxidizer circulation, and a system for removing the products of the reaction.

As an example, the fuel battery system for Project Gemini [82] is shown in Fig. 7.17. In this system there are two fuel cell sections, each one consisting of three stacks of 32 individual cells. A cryogenic system supplies the fuel and the oxidizer. The water product, once it is removed and stored in an accumulator, is potable and can be used for drinking purposes. The peak load rating of the Gemini battery is 2 kW.

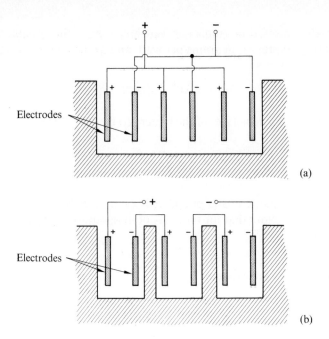

Figure 7.15. Storage type batteries: (a) batteries in parallel, (b) batteries in series.

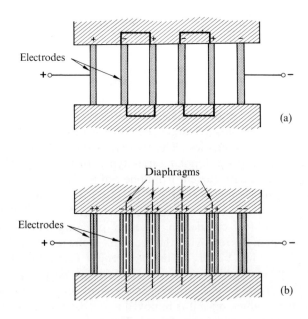

Figure 7.16. Filter type batteries: (a) homopolar series batteries, (b) bipolar series batteries.

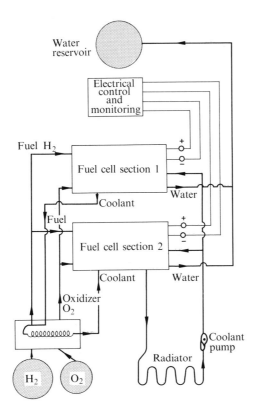

Figure 7.17. Fuel battery system for Project Gemini.

7.9 Applications and Future Trends

Fuel cells are now used in many applications where they are the only or the most competitive converters of energy. For fuel battery systems to be used commercially, the price of the power produced should be lowered to an acceptable level and many other technical improvements should be introduced. To understand these problems, let us review the different areas of application.

Fuel cells for cars [26, 36, 69]. Fuel cell batteries producing powers ranging from several hundred watts to several dozen kilowatts have been considered for car propulsion. The advantages of using fuel cells for this application are numerous, and include reduction of atmospheric pollution due to the absence of exhaust gases and reduction of the noise level which may make life in big cities nearly intolerable. These alone are good reasons for encouraging extended research work in this field.

Two problems remain to be solved in a satisfactory manner. The weight per unit power produced by a fuel battery is still very large—around 60 kg/kW for hydrogen-oxygen fuel cells at the ambient temperature. This seems unacceptable as compared to the 5 kg/kW of conventional automobile engines. However, if the temperature is raised, the weight to power ratio can be lowered to about 20 kg/kW. With many possible technological improvements, this figure could be lowered further.

The second serious problem is the price of the power produced. The price could be lowered from the present value of $80/kW to around $10/kW. Taking into account the fact that the efficiency of a fuel battery motor could be made at least three times higher than the efficiency of conventional cars, the day when electric cars are on the market may not be far away.

Fuel cells for power stations. It does not seem that power stations based solely on fuel cells will be feasible in the near future. However, fuel cell systems coupled with conventional power stations present a very attractive possibility. They can be utilized for storing energy, thus making the consumption of electric power much more uniform throughout the day. Another possibility is their use in combination with thermoelectric stations. New sources of energy, solar or nuclear, can thus be utilized in conjunction with the conventional fuels of the thermoelectric plants.

Fuel cells for rail traction. This concerns generators of the order of a megawatt. There is, of course, no question of replacing the electrified lines by independent fuel cell locomotives. Fuel cells could be applied only to the nonelectrified lines to eventually replace the Diesel engines. The efficiencies of the latter are low, of the order of 25%, and the use of fuel cells is considered by many workers. It seems that low-temperature fuel batteries are most seriously considered for this application.

Fuel cells for space [66]. Fuel cell systems producing electrical power of the order of a kilowatt were employed in space applications, the most important being in the Gemini [82] and the Apollo [84] projects, where hydrogen-oxygen batteries were the generators of electric power.

Two important advantages should be noted here in conjunction with space applications: the production of potable water as a result of the chemical reaction, and the storage of energy in case the spaceship cannot immediately use the solar energy.

Fuel cells for defense needs [34]. Air [52], naval [51], and surface [44] forces are all potential heavy users of fuel cells. A primary use for the Air Force is in space elements, and this has achieved significant progress. In addition to uses on the Apollo project, other important contributions to this field have been made, especially by Shell Research Ltd. [95];* they demonstrated a cell which operated on hydrogen produced from methanol. The overall power unit was reasonably small and will ultimately be used by the British Ministry of Aviation. An extensive

---

* See Chapter 12.

research effort is under way in the U.S. and in Europe to utilize electrochemical power in submarines. One of the most interesting developments is the fuel cell system developed at ASEA Laboratories [95]* (Sweden) to power a submarine which operates with liquid ammonia. Along the same line, Allis-Chalmers† has a one-man submarine operating on hydrogen. Because of low-noise-level considerations, fuel cell powered surface vehicles appear very attractive. The engine and gearbox, as well as the transmission shaft and differential unit of the vehicle, may be replaced by a fuel cell system providing power. The use of a methanol-air fuel cell with an acid electrolyte is a likely possibility.

Independently of the capital and operating costs, two special factors are of the greatest importance in military applications: low noise levels and compactness. Other factors may be of crucial importance in special applications; some of them are weight, reliability, and integration of the fuel into the system.

Conclusions. Taking into account all the needs involved and the characteristics desired, the efficiency of a fuel cell appears to be directly related to and influenced by a host of factors, mainly related to the different types of polarization, such as the factors affecting mass and ion transport as well as electron transfer phenomena. Other factors are no less important: these are the reactivity of the fuel, the invariance of the cell characteristics, the catalysis of electrode processes, and electrode poisoning. These phenomena are often interrelated, and definite progress has been made toward their understanding.

PROBLEMS

7.1 Consider a hydrogen-oxygen fuel cell in which the following reactions occur: at the anode,

$$H_2 \rightarrow 2H^+ + 2e^-;$$

at the cathode,

$$2e^- + \tfrac{1}{2}O_2 + 2H^+ \rightarrow H_2O(l).$$

In these reactions $\mathscr{G} = -56.7$ kcal at 25°C and 1 atm, and $\mathscr{H} = -68.3$ kcal. Find (a) the overall reaction, (b) the ideal emf, and (c) the change in entropy.

7.2 Answer the same questions as in Problem 7.1 for a fuel cell having the following reactions: at the anode,

$$C + 2O^{--} \rightarrow CO_2 + 4e^-;$$

at the cathode

$$4e^- + O_2 \rightarrow 2O^{--}.$$

In these reactions $\mathscr{G} = -94.26$ kcal at 25°C and 1 atm, and $\mathscr{H} = -94.05$ kcal.

---

* See Chapter 12.

† "A Look at Allis-Chalmers Fuel Cells Today," Advertising paper No. 48B2663. Allis-Chalmers, Marketing, Space and Defense Sciences Dept. Milwaukee, Wis., 1966.

7.3 Daniell's cell is characterized by the following reactions: at the anode,

$$Zn(s) \rightarrow Zn^{++} + 2e^-;$$

at the cathode,

$$Cu^{++} + 2e^- \rightarrow Cu(s).$$

The total standard potential of the cell is 1.1 V at 25°C. (a) Find the potential of each half reaction mentioned above. (b) Calculate the change in Gibbs free energy. (c) What is the equilibrium constant of the overall reaction?

7.4 A regenerative fuel cell system operates between a regeneration temperature T_h and a conversion temperature T_c. Find the maximum efficiency possible of such a system as a function of the characteristics of the cell.

7.5 Prove Eq. (7.23) in a manner similar to that followed for Eq. (7.17).

7.6 For some fuel cells the heat capacitance at constant pressure can be given as a function of temperature by

$$C_p = \mathscr{A}_1 + \mathscr{A}_2 T + \mathscr{A}_3/T^2,$$

where $\mathscr{A}_1, \mathscr{A}_2$, and $\mathscr{A}_3$ are known constants. Find the output voltage as a function of temperature for such cells.

7.7 From the result of Eq. (7.25) find an expression for the maximum power output of a fuel cell.

7.8 The results reported in Figs. 7.5 and 7.6 are experimental. Can you predict these results theoretically for a hydrogen-oxygen fuel cell using a 30% KOH electrolyte?

7.9 Experimentally, the standard potential of a hydrazine fuel cell has been found to be about one-third smaller than the value mentioned in the text. It has been suggested [30] that the following secondary reaction occurs at the anode and is the cause of the discrepancy:

$$N_2H_4 \rightarrow N_2 + 2H_2,$$

$$2H_2 + 4OH^- \rightarrow 4H_2O + 4e^-.$$

Can you prove the validity, or the invalidity, of this suggestion?

7.10 Hydrocarbons usually become unstable at high temperatures. To prevent the deposition of carbon in high temperature solid electrolyte cells, parts of the products of the reaction have been mixed with the incoming fuel (Westinghouse). These products (CO_2 and H_2O) reform with the fuel, thus producing CO and H_2, which are then oxidized as fuel to produce power. Study the feasibility of such a cell and calculate its output voltage.

7.11 The reactions involved in a biochemical fuel cell can be represented as follows: bacterial digestion,

$$2CH_2O + H_2SO_4 + DSV \rightarrow 2CO_2 + 2H_2O + H_2S + DSV,$$

anode reaction,

$$H_2S \leftrightharpoons 2H^+ + S^{--} \rightarrow 2H^+ + S + 2e^-;$$

cathode reaction,

$$\tfrac{1}{2}O_2 + H_2O \rightarrow 2OH^- - 2e^-.$$

Find the total electrochemical reaction and the overall net reaction of this biochemical cell. $2CH_2O + H_2SO_4$ is the biofuel and DSV (desulfovibrio) is the anaerobe.

7.12 Justify the experimental curves given in Fig. 7.9. Follow a calculation similar to that of Section 7.4.

7.13 (a) Under which assumptions can the power output of a fuel cell be considered as given by the relation

$$P_{out} = J(\mathscr{A}_1 - \mathscr{A}_2 J),$$

where $\mathscr{A}_1$ and $\mathscr{A}_2$ are constants? (b) Are those assumptions acceptable for an oxygen-hydrogen fuel cell? (c) Find the values of $\mathscr{A}_1$ and $\mathscr{A}_2$ in terms of the parameters given in Eq. (7.30).

7.14 Design an oxygen-hydrogen fuel cell producing 30 W of power and using a 30% solution of KOH at 25°C and 1 atm. Find (a) the different dimensions of the cell, (b) the distance between the electrodes, (c) the efficiency of the cell, and (d) the amounts of oxygen and hydrogen necessary for the reactions.

7.15 (a) Using the cells of Problem 7.14, design a fuel cell battery for car propulsion producing 30 kW of electric power. (b) Estimate the overall volume of such a battery.

REFERENCES AND BIBLIOGRAPHY

1. Adams, A. M., F. T. Bacon, and R. G. Watson, "The High Pressure H-O Cell," in *Fuel Cells*, W. Mitchell, Jr. (ed.), Academic Press, New York, 1963, p. 129.

2. Adams, A. M., "Recent Developments in Fuel Cells," *J. Inst. Fuel*, Vol. 27, No. 7, p. 366, 1964.

3. Allen, J. J., *Organic Electrode Processes*, Chapman and Hall, London, 1958.

4. Andrieth, L. F., and J. Kleinberg, *Non-Aqueous Solvents*, Wiley, New York, 1953.

5. Archer, D. H., et al., *An Investigation of Solid Electrolyte Fuel Cells*, Westinghouse Electric Co. 2nd Quarterly Progress Report, U.S. Air Force Contract AF33(657)-8251, 1962.

6. Austin, L. G., "Fuel Cells," *Scientific American*, Vol. 201, No. 4, 1959.

7. Austin, L. G., "Electrode Kinetics of Low Temperature H-O Fuel Cells," in *Fuel Cells*, G. J. Young (ed.), Reinhold, New York, 1960.

8. Austin, L. G., "Electrode Kinetics and Fuel Cells," *Proc. IEEE*, Vol. 51, No. 5, p. 820, 1963.

9. Bacon, F. T., and J. S. Forrest, "Recent Research in Great Britain on Fuel Cells," *Transactions 5th World Power Conference*, Vienna, 1956.

10. Bacon, F. T., "The High Pressure Hydrogen-Oxygen Fuel Cell," in *Fuel Cells*, G. J. Young (ed.), Vol. 1, Reinhold, New York, 1960, p. 78.

11. Bagotskii, V. S., and Y. B. Vacilev, *Fuel Cells, Their Electrochemical Kinetics*, Nauka Press, Moscow, 1964; translated by Consultants Bureau, New York, 1966.

12. Baker, B. S. (ed.), *Hydrocarbon Fuel Cell Technology*, Academic Press, New York, 1965.

13. Bauer, E., W. D. Treadwell, and G. Trumpler, "Ausfurhrungsformen von Brennstoff-kelten bei Hoher Temperatur," *Zs. Elektrochem.*, Vol. 27, p. 199, 1921.

14. Binder, H., et al., "Electrochemische Oxydation von Kohlenwasserstoffen in Einer Festelectrolyt-Brennstoffzele bei Temperaturen von 900–1000°C," *Electrochim. Acta*, Vol. 8, No. 10, p. 781, 1963.

15. Binder, H., A. Kohling, and G. Sandstede, "Elektrochemische Oxydation von Ameisensaure am Platinkatalysator mit und ohne Schwefalsorbat in Alkalischem und Saurem Electrolyter," *Adv. Energy Conversion*, Vol. 7, No. 3, p. 121, 1967.

16. Blasco, R. J., and E. Gileadi, "An Electrochemical and Microbiological Study of the Formic Acid-Formic Dehydrogenlyase System," *Adv. Energy Conversion*, Vol. 4, No. 3, p. 179, 1964.

17. Bloch, O., J. Jacq, and M. Prigent, "Mecanismes du Fonctionnement des Electrodes à Reactifs Dissous," in *Les Piles à Combustibles*, Editions Technip, Paris, 1965, p. 127.

18. Bréelle, Y., and P. Degobert, "Application Perspectives for Dissolved-Fuel Cells," *Adv. Energy Conversion*, Vol. 5, No. 4, p. 270, 1965.

19. Broers, G. H., *High Temperature Galvanic Fuel Cells*, Ph.D. Dissertation, Municipal University of Amsterdam, The Netherlands, 1958.

20. Broers, G. H., and J. A. Ketelaar, "High Temperature Fuel Cells," in *Fuel Cells*, G. J. Young (ed.), Vol. 1, Reinhold, New York, 1960, p. 78.

21. Carter, "Fuel Cell 'Burns' Natural Gas," *Machine Design*, Vol. 35, p. 10, Jan. 17, 1963.

22. Chambers, H. H., and A. D. Tantram, "Carbonaceous Fuel Cells," in *Fuel Cells*, G. J. Young (ed.), Vol. 1, Reinhold, New York, 1960, p. 94.

23. Colichman, E. L., "Preliminary Biochemical Cell Investigations," *Proc. IEEE*, Vol. 51, No. 5, p. 812, 1963.

24. Conway, B. E., *Electrochemical Data*, Elsevier, Amsterdam, 1952.

25. Cotton, J. B., and I. Dugdale, "A Survey of Possible Uses of Titanium in Batteries," in *Batteries*, D. H. Collins (ed.), Pergamon Press, New York, 1963, p. 297.

26. Dantrowitz, P., and L. Gaddy, *A State-of-the-Art Automotive Fuel Cells*, Symposium on Power Systems for Electric Vehicles, Polytechnic Institute of Brooklyn, New York, April 6–8, 1967.

27. Davy, H., *Ann. Phys.*, Vol. 8, p. 301, 1801.

28. Degobert, P., and O. Bloch, "Les Carbonates Alcalins Fondus comme Electrolyte de Pile à Combustible. Metaux Susceptibles de Servir d'Electrode," *Bull. Soc. Chim. France*, p. 1887, 1962.

29. Douglas, D. L., "Molten Alkali Carbonate Cells with Gas Diffusion Electrodes," in *Fuel Cells*, G. J. Young (ed.), Vol. 1, Reinhold, New York, 1960, p. 129.

30. Dugdale, I., "Fuel Cells," in *Direct Generation of Electricity*, K. H. Spring (ed.), Academic Press, New York, 1965.

31. Eisenberg, M., "Design and Scale-Up Considerations for Electrochemical Fuel Cells," in *Advances in Electrochemistry and Electrochemical Engineering*, W. Tobias (ed.), Vol. 2, Interscience, New York, 1962.

32. Eisenberg, M., *Ammonia Fuel Cells*, 18th Power Sources Conference Proceedings, PSC Publications Committee, Red Bank, N.J., 1964, p. 20.

33. Elmore, G. V., and H. A. Tanner, "Intermediate Temperature Fuel Cells," *J. Electrochem. Soc.*, Vol. 108, No. 7, p. 669, 1961.

34. Engle, M. L., "A Hydrocarbon Air Fuel Cell System for Military Application," *Chem. Eng. Progress, Symp. Series*, Vol. 63, No. 75, p. 41, 1967.

35. Fox, H. W., and R. Roberts, "Fuel Cells," *IRE Trans. Military Electronics*, Vol. MIL-6, p. 162, 1962.

36. Frysinnger, G. R., *Fuel Cell Battery Power Sources for Electric Cars*, Symposium on Power Systems for Electric Vehicles, Polytechnic Institute of Brooklyn, New York, April 6–8, 1967.

37. Gillibrand, M. I., and C. R. Lomax, "The Hydrazine Fuel Cell," in *Batteries*, D. H. Collins (ed.), Pergamon Press, New York, 1963, p. 221.

38. Giner, J., *Electrocatalyst Research*, 21st Power Sources Conference Proceedings, PSC Publications Committee, Red Bank, N.J., 1967, p. 10.

39. Glass, W. B., and G. H. Boyle, "Performance of Hydrogen-Bromine Fuel Cells," in *Fuel Cell Systems*, R. F. Gould (ed.), *Adv. in Chem. Series* 47, ACS, Washington, D.C., 1965, p. 203.

40. Goldstein, M., *Hydrox Fuel Cells and Their Application in Aircraft and Space Vehicles*, Aircraft Electrical Society Meeting, Feb. 1959.

41. Grove, W. R., "On a Gaseous Voltaic Battery," *Phil. Mag.*, Vol. 21, No. 11, p. 417, 1842.

42. Grubb, W. T., and L. W. Niedrach, "Batteries with Solid Ion Exchange Membrane Electrolytes, II. Low Temperature Hydrogen-Oxygen Fuel Cells," *J. Electrochem. Soc.*, Vol. 108, p. 131, 1960.

43. Grubb, W. T., and L. W. Niedrach, "Fuel Cells," in *Direct Energy Conversion*, G. W. Sutton (ed.), McGraw-Hill, New York, 1966,

44. Guillaume, J. P., "Quelques Applications Militaires des Piles à Combustible," in *Piles à Combustible*, Editions Technip, Paris, 1965, p. 437.

45. Haldeman, R. G., *Electrode-Matrix Materials*, 21st Power Sources Conference Proceedings, PSC Publications Committee, Red Bank, N.J., 1967, p. 1.

46. Hart, A. B., and J. H. Powell, "The Influence of Transport Processes in the Behavior of Gas Diffusion Electrodes," in *Batteries*, D. H. Collins (ed.), Pergamon Press, New York, 1963, p. 265.

47. Heath, C. E., et al., *Soluble Carbonaceous Fuel-Air Fuel Cell*, Esso Research and Engineering Co., Linden, N.J., ARPA Order 247-42, Task OST 760200471, Report No. 1, January 1–June 30, 1962.

48. Heath, C. E., and W. J. Sweeney, "Kinetics and Catalysis in Fuel Cells," in *Fuel Cells*, W. Mitchel (ed.), Academic Press, New York, 1963, p. 65.

49. Heath, C. E., and C. H. Worsham, "The Electrochemical Oxidation of Hydrocarbons in a Fuel Cell," in *Fuel Cells*, G. J. Young (ed.), Vol. 2, Reinhold, New York, 1963, p. 182.

50. Heath, C. E., *Methanol Fuel Cells*, 18th Power Sources Conference Proceedings, PSC Publications Committee, Red Bank, N.J., 1964, p. 33.

51. Heffner, W. H., A. C. Veverka, and G. T. Skoperdas, "Hydrogen-Generating Plant Based on Methanol Decomposition," in *Fuel Cell Systems*, R. F. Gould (ed.), Adv. in Chem. Series 47, 1965.

52. Holdsworth, N. D., and G. F. Eggelton, "Problems Associated with the Performance and Servicing of Secondary Batteries in Modern High Speed Aircraft," in *Batteries*, D. H. Collins (ed.), Macmillan, New York, 1963.

53. Howard, H. C., "Direct Generation of Electricity from Coal and Gas (Fuel Cells)," in *Chemistry of Coal Utilization*, H. H. Lowry (ed.), Wiley, New York, 1945.

54. Ives, D. J., and G. J. Janz, *Reference Electrodes*, Academic Press, New York, 1961.

55. Janz, G. J., and F. Saegusa, "Anodic Polarization Curves in Molten Carbonate Electrolysis," *J. Electrochem. Soc.*, Vol. 108, No. 7, p. 663, 1961.

56. Justi, E., et al., *High Drain Hydrogen Diffusion Electrodes Operating at Ambient Temperature and Low Pressure*, Academy of Sciences and Literature, Mainz, Komm-Verlag, Franz Steiner, Wiesbaden, Germany, 1959, p. 208.

57. Justi, E., and A. Winsel, *Kalte Verbrennung Fuel Cells*, Franz Steiner, Wiesbaden, Germany, 1962.

58. Justi, E. W., "Fuel Cell Research in Europe," *Proc. IEEE*, Vol. 51, No. 5, p. 781, 1963.

59. King, J., Jr., F. A. Ludwig, and J. J. Roulette, "General Evaluation of Chemicals for Regenerative Fuel Cells," in *Energy Conversion for Space Power*, N. W. Snyder (ed.), Academic Press, New York, 1961.

60. Kordesch, K. V., "The Hydrogen-Oxygen (Air) Fuel Cell with Carbon Electrodes," in *Fuel Cells*, G. J. Young (ed.), Vol. 1, Reinhold, New York, 1960, p. 11.

61. Kordesch, K. V., "Low Temperature Fuel Cells," *Proc. IEEE*, Vol. 51, No. 5, p. 806, 1963.

62. Kordesch, K. V., "Low Temperature H-O Fuel Cells," in *Fuel Cells*, W. Mitchell, Jr. (ed.), Academic Press, New York, 1963, p. 329.

63. Laroche, J., "Piles Hydrogène-Oxygène à Moyenne Temperature," in *Les Piles à Combustible*, Editions Technip, Paris, 1965, p. 361.

64. Liebhafsky, H. A., "The Fuel Cell and the Carnot Cycle," *J. Electrochem. Soc.*, Vol. 106, p. 1068, 1959.

65. Liebhafsky, H. A., and L. W. Niedrach, "Fuel Cells," *J. Franklin Institute*, Vol. 269, No. 4, p. 257, 1960.

66. Liebhafsky, H. A., and W. T. Grubb, Jr., "The Fuel Cell in Space," *J. Amer. Rocket Soc.*, p. 1183, Sept. 1961.

67. Liebhafsky, H. A., et al., "Current Density and Electrode Structure in Fuel Cells," in *Fuel Cell Systems*, R. F. Gould (ed.), Adv. in Chem. Series 47, ACS, Washington, D.C., 1965, p. 116.

68. Liebhafsky, H. A., "Fuel Cells and Fuel Batteries, An Engineering Value," *IEEE Spectrum*, Vol. 3, No. 12, p. 48, 1966.

69. Lindgren, N., "Electric Cars—Hope Springs Eternal," *IEEE Spectrum*, Vol. 4, No. 4, p. 49, 1967.

70. Lurie, R. M., C. Berger, and R. J. Shuman, "Ion Exchange Membranes in H-O Fuel Cells," in *Fuel Cells*, G. J. Young (ed.), Vol. 2, Reinhold, New York, 1960, p. 142.

71. Niedrach, L. W., and W. T. Grubb, "Ion Exchange Membrane Fuel Cells," in *Fuel Cells*, W. Mitchell, Jr. (ed.), Academic Press, New York, 1963, p. 253.

72. Peattie, C. G., et al., *Operating Characteristics of a High-Temperature Fuel Cell*, Spring Meeting of Electrochemical Society, Los Angeles, May 6–10, 1962.

73. Peattie, C. G., et al., *Performance Data for Molten-Electrolyte Fuel Cells Operating on*

Several Fuels, Pacific Energy Conversion Conference Proceedings, San Francisco, Aug. 12–16, 1962.

74. Peattie, C. G., et al., *Factors Involved in the Use of a High-Temperature Fuel Cell as a Space Power Source*, Paper No. 2565-52, ARS Space Power Systems Conference, Santa Monica, Cal., Sept. 25–28, 1962.

75. Peattie, C. G., "A Summary of Practical Fuel Cell Technology to 1963," *Proc. IEEE*, Vol. 51, No. 5, p. 795, 1963.

76. Phillips, G. A., "Status of Development and Future Prospects for the Ion-Exchange Membrane Fuel Cell," *Electrical Engineering*, Vol. 81, No. 3, p. 194, 1962.

77. Rideal, E. K., and U. R. Evans, "The Problem of Fuel Cells," *Trans. Faraday Soc.*, Vol. 17, p. 466, 1921.

78. Rossini, F. D., *Chemical Thermodynamics*, Wiley, New York, 1950.

79. Rossini, F. D., et al., *Selected Values of Chemical Thermodynamic Properties*, Circular of the National Bureau of Standards 500, Washington, D.C., 1952.

80. Ruetschi, P., J. G. Duddy, and D. T. Ferrell, "Alkaline Low Temperature Fuel Cells with Metal Electrodes," in *Batteries*, D. H. Collins (ed.), Pergamon Press, New York, 1963, p. 235.

81. Sandler, Y. L., "Chemical and Electrical Characteristics of the Alkali Carbonate High Temperature Fuel Cells," *J. Electrochem. Soc.*, Vol. 108, p. 1115, 1962.

82. Schanz, J. L., and E. K. Bullock, *Gemini Fuel Cell Power Source: First Spacecraft Application*, Paper No. 2561-62, ARS Space Power Systems Conference, Santa Monica, Cal., Sept. 25–28, 1968.

83. Schultz, E. B., et al., "Natural Gas Fuel Cells for Power Generation," *Am. Gas J.*, Vol. 188, p. 25, May 1961.

84. Shaw, R. H., and R. A. Thompson, *Hydrogen-Oxygen Fuel Cell Systems for Space Vehicles*, Paper No. 2560-62, ARS Space Power Systems Conference, Santa Monica, Cal., Sept. 25–28, 1962.

85. Tarmy, B. L., *Methanol Fuel Cells*, 16th Power Sources Conference Proceedings, PSC Publications Committee, Red Bank, N.J., 1962, p. 23.

86. Tomter, S. S., and A. P. Antony, "The Hydrazine Fuel Cell System," in *Fuel Cells*, Chem. Eng. Tech. Manual, Am. Inst. Chem. Engrs., New York, 1963, p. 22.

87. Truitt, J. K., "Engineering Aspects of Molten Carbonate Electric Generators Using Hydrocarbon Fuels," in *Fuel Cells*, Chem. Eng. Tech. Manual, Am. Inst. Chem. Engrs., New York, 1963, p. 1.

88. Vetter, J. J., *Elektrochemische Kinetik*, Springer-Verlag, Berlin, 1961.

89. Vielstich, W., *Methanol-Air Fuel Cells*, 4th International Symposium on Batteries, Brighton, England, Pergamon Press, New York, 1964.

90. Von Fredersdorff, C. G., *An Outline of the Economics of a Domestic Fuel Cell System*, Symposium on Recent Advances in Fuel Cells, Chicago, Sept. 3–8, 1961, Div. of Petroleum Chemistry, American Chemical Society, Washington, D.C., Vol. 6, No. 4-B, 1961, pp. B-35-B-48.

91. Warszawski, B., "Fonctionnement de la Pile Rédox en Milieu Tamponné," in *Les Piles à Combustible*, Editions Technip, Paris, 1965, p. 147.

92. Weissbart, J., and R. Ruka, "Solid Oxide Electrolyte Fuel Cells," in *Fuel Cells*, G. J. Young (ed.), Vol. 2, Reinhold, New York, 1960, p. 37.

93. Weissbart, J., and R. Ruka, "A Solid Electrolyte Fuel Cell," *J. Electrochem. Soc.*, Vol. 109, p. 723, 1962.

94. Wendell, M., "Oxidation States of the Elements and Their Potentials in Aqueous Solutions," Prentice-Hall, Englewood Cliffs, N.J., 1952.

95. Williams, K. R., J. W. Pearson, and W. J. Gressler, *Low Temperature Fuel Batteries*, 4th International Symposium on Batteries, Bournemouth, England, Oct. 1964.

96. Williams, K. R. (ed.), *An Introduction to Fuel Cells*, Elsevier, Amsterdam, 1966.

97. Winsel, A. W., "Statistisches Modell Einer Gas-Diffusionselektrode," *Adv. Energy Conversion*, Vol. 3, No. 4, p. 677, 1963.

98. Wynveen, R. A., "Preliminary Appraisal of the Ammonia Fuel Cell System," in *Fuel Cells*, G. J. Young (ed.), Vol. 2, Reinhold, New York, 1963, p. 153.

99. Yeager, J. F., "Fuel Cells as Energy Converters," in *Direct Conversion of Heat to Electricity*, J. Kaye and J. A. Welsh (eds.), Wiley, New York, 1960.

100. Young, G. J., and R. B. Rozelle, "Fuel Cells," *J. Chem. Educ.*, Vol. 36, p. 68, 1959.

101. Young, G. J., and R. B. Rozelle, "Catalysis Electrode Reactions," in *Fuel Cells*, G. J. Young (ed.), Vol. 1, Reinhold, New York, 1960, p. 23.

102. Young, G. J., and R. B. Rozelle, "Low Temperature Electrochemical Oxidation of Hydrocarbons," in *Fuel Cells*, G. J. Young (ed.), Vol. 2, Reinhold, New York, 1960, p. 216.

8 PHOTOVOLTAIC POWER GENERATION

8.1 Introduction

A photovoltaic power generator is a device which converts electromagnetic energy directly into electricity. The source of energy can be either radioactive or solar. In the first case, the wavelength of the radiation is of the order of 10^{-12} meter, whereas the wavelength of the solar radiation is of the order of 10^{-6} meter.

Photovoltaic cells are not subject to the Carnot limitation in the same manner as fuel cells. The conversion of energy is based on the photovoltaic effect, which was discovered in 1839 by Becquerel [2]. He noted that a difference of potential was developed when light was shone on an electrode immersed in an electrolytic solution. Four decades later, work on solid materials led to the discovery of the same effect in selenium and copper oxide by Adams [1], Schottky [76], and other early workers [36]. This led to the fabrication of photometers even before the turn of the century.

As for the other energy converters, work stopped for more than six decades, but resumed with the activities of the Bell Telephone Laboratories and the Radio Corporation of America in 1954 [2]. Simultaneously, research work in the field of semiconductors led to the fabrication of silicon and cadmium sulfide cells having efficiencies of about 5% (this figure was above the maximum 1% reached by earlier solar cells). This efficiency has greatly improved since then, reaching 16% in 1968. Solar cells based on the photovoltaic effect are now considered not only for space research, where their use in the space power systems has been necessary for the success of those systems, but also for medical and terrestrial uses [66, 68]. The power level ranges from around 0.1 mW for medical applications to the megawatt region for space vehicles.

The advantages of photovoltaic generators compared to other converters are numerous. They are simple to fabricate and to operate, they have a high power output per weight ratio, and a practically unlimited life. Furthermore, their present efficiency in converting solar energy is far higher than any other solar energy converter.

Nevertheless, progress in this field has been slowed by several problems leading to the high cost of a solar energy conversion system.

8.2 Radiation Energy

The energy $\mathscr{E}$ of one photon of light is proportional to the frequency v of the radiation,

$$\mathscr{E} = hv, \tag{8.1}$$

where vh is Planck's constant, $h = 6.624 \times 10^{-27}$ erg·sec. The energy quantum $\mathscr{E}$ represents the smallest amount of energy that can be obtained from the light. The total energy emitted can only be an integral multiple of the quantum expressed in Eq. (8.1).

If c is the speed of light in vacuum ($c = 2.998 \times 10^{10}$ cm/sec), then the frequency v can be written $v = c/\lambda$, where λ is the wavelength. Replacing v by its value in Eq. (8.1) and expressing $\mathscr{E}$ in electron-volts, $\mathscr{E} = 0.01238/\lambda$, where λ is in meters.

Solar energy is received in the form of a wide radiation spectrum including infrared, visible, and ultraviolet radiation. Outside the earth's atmosphere, the

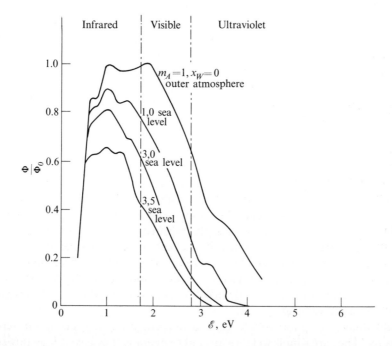

Figure 8.1. The solar spectrum intensity per unit wave number as a function of the electro-magnetic energy. Parameters are the ratio m_A of the illuminated to the effective area presented to the sun and the amount x_W of precipitable water in the atmosphere.

spectral distribution of the solar radiation can be accurately represented by the absorption of a blackbody with a power density equal to 0.135 W/cm² at 5900°K. When the radiation passes through the earth's atmosphere, a loss of radiated energy occurs due to the scattering and the absorption of the photons by the different materials existing in the atmosphere, such as air, clouds, water vapor, or dust. This leads to a change in the solar spectrum, which also depends on the effective area presented to the sun. Letting m_A be the ratio of the illuminated area to the effective area presented to the sun, $m_A = 1/\cos(\theta + \phi)$, where $(\theta + \phi)$ is the angle existing between the radiating beam and a vertical line at the point of the earth's surface receiving the radiation. If θ determines the latitude, ϕ will depend on the time of day and the season of the year.

Figure 8.1 illustrates the relative intensity Φ_s/Φ_0 of the photons as a function of the quantum energy; Φ_0 represents the maximum intensity attained. In Fig. 8.1, x_w represents the amount of precipitable water in the atmosphere expressed in meters. Note that the effect of the different scatterings and absorptions reduces the amount of ultraviolet and visible light of the spectrum.

The intensity Φ_s of the light is usually expressed in watts per square centimeter. It is equal to the product of the number n_{ph} of photons crossing a unit area perpendicular to the light beam per unit time, times the average energy $\bar{\mathscr{E}}$ of each photon; thus $\Phi = n_{ph}\bar{\mathscr{E}}$ or $\Phi_s = n_{ph}hc/\bar{\lambda}$, where $\bar{\lambda}$ is the average wavelength of the photon.

The average intensity received from the sun varies from a maximum of 0.135 W/cm² at the outer atmosphere to a maximum of 0.100 W/cm² at sea level when the sun is at the zenith.*

8.3 The Photovoltaic Effect

The pn junction. The photovoltaic effect can be described easily for a pn junction in a semiconductor material. In a piece of a pure intrinsic semiconductor such as silicon there are no free electrons at absolute zero, since each valence electron of the material atom is tied in a chemical bond. If this piece of silicon is doped with phosphorus or arsenic, there will be an excess of electrons leading to a n-type semiconductor. Phosphorus and arsenic have one more valence electron than silicon. Therefore, when a silicon atom is replaced in the lattice by a phosphorus atom, this excess electron will be free to move in the lattice. Similarly, if another piece of silicon is doped with boron, there will be a deficiency of electrons leading to a p-type semiconductor. Boron has one less valence electron than silicon, and this deficiency can be expressed in terms of an excess of holes or positive charges free to move in the lattice. The material used can also be any other semiconductor if its characteristics are favorable to the photovoltaic conversion.

If these two pieces of silicon are connected, a pn junction is obtained. The free electrons of the n-side will tend to flow to the p-side, and the holes of the p-side

---

* Sometimes solar radiation is measured in Langleys per minute. By definition 1 Langley = 1 calorie/cm².

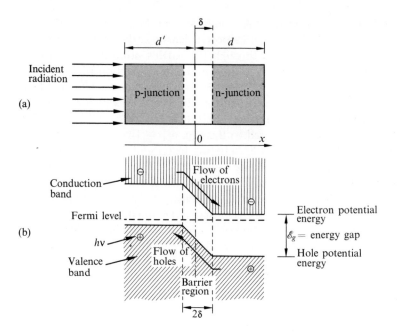

Figure 8.2. The operation of a photovoltaic junction: (a) schematic pn junction, (b) its corresponding energy level diagram.

will tend to flow to the n-region to compensate for their respective deficiencies. This diffusion will create an electric field $\mathbf{E}$ from the n-region to the p-region. This field will reach a maximum value at equilibrium proportional to an electric potential V_e: $V_e = V_p + V_n$, where V_p and V_n are the diffusion potentials for the holes and the electrons, respectively.

Figure 8.2 represents a pn junction and the distribution of the potential energy through the junction. The magnitude of the electric potential will increase upon increasing the number of the p-type and the n-type impurities in the p- and n-regions, respectively. This increase will reach equilibrium when balancing phenomena such as tunneling* through the potential barrier become significant.

When a photon of energy $h\nu$ enters the p-region, it is absorbed by an electron in the valence band. If $h\nu$ is larger than the energy gap $\mathscr{E}_g$ separating the valence band from the conduction band, the electron will be able to migrate to the n-region. Similarly, if $h\nu$ is larger than $\mathscr{E}_g$ in the n-region, the photon will be absorbed by a hole which will migrate to the p-region. This charge separation will create an electric field opposed to the electric field created by the diffusion. If the number of

---

* The probability of penetration of an electron into a region of negative kinetic energy is very small, but different from zero. The occurrence of this penetration in such a region (classically forbidden) is called the *tunnel effect* (or tunneling).

absorbed photons is large enough, these two fields will cancel each other, leading to an open circuit voltage between the two regions. If these electrons and holes are made to flow through an external load, electric energy will be obtained from the absorbed photons.

Note that only photons having energies larger than the energy gap of the material are useful. Consequently, as far as the open circuit voltage is concerned, the smaller the energy gap, the more energy can be converted. This will be limited by the effect of $\mathscr{E}_g$ on the output current, as will be discussed below.

The lifetime of holes and electrons should be large enough to enable them to migrate from one region to the other, before they can annihilate each other through recombination.

The performance characteristic [9, 36, 87]. Assume that the electrostatic field is confined to the area of the junction, the thickness of which is assumed to be very small compared to the dimensions of the p-region and the n-region and to the diffusion length of the minority carriers. Furthermore, the carrier densities are assumed to be small enough to allow the use of Boltzmann statistics instead of Fermi statistics.

The continuity equation for holes as excess minority carriers can be written, for the steady state, as

$$g(x) + \frac{p_n}{\tau_p} - \frac{p}{\tau_p} + \mathscr{D}_p \frac{\partial^2 p}{\partial x^2} = 0. \qquad (8.2)$$

The first term of this equation represents the generation rate of holes by the light quanta. The second term depicts the equilibrium thermal generation rate. The third term covers the nonequilibrium recombination rate and, finally, the last term expresses the net diffusion rate of the holes. In this equation, p is the nonequilibrium concentration of holes, p_n is the equilibrium concentration of holes in the n-material, τ_p is the lifetime of a hole, and $\mathscr{D}_p$ is the diffusion constant for holes.

Letting g_0 be the hole-electron pair generation rate at the junction due to the action of the photons, the generation rate $g(x)$ in the n-material can be expressed as

$$g(x) = g_0 \exp\left[-\alpha_\lambda(x - d)\right]. \qquad (8.3)$$

The junction is assumed to be a narrow band of width $2d$ centered at $x = 0$, as shown in Fig. 8.2; α_λ is the absorption coefficient for photons that have a wavelength λ. For $x < 0$, $g(x)$ will express the generation rate of holes in the p-material. To solve Eq. (8.2), use the following boundary conditions: at $x = 0$,

$$p(0) = p_n \exp\left(eV/kT\right), \qquad (8.4)$$

and at $x = d$, $dp/dx = 0$.

Equation (8.4) is obtained from the fact that the concentrations of holes on opposite sides of the barrier are related by

$$p_n = p_p \exp\left(-eV_c/kT\right),$$

where p_p is the concentration of holes in the p-region. When a voltage V is created under the effect of the impinging radiation, the quasi-equilibrium hole density can be written as

$$p(0) = p_p \exp\left[-e(V_c - V)/kT\right]$$

or

$$p(0) = p_n \exp(eV/kT).$$

Introducing Eq. (8.3) into Eq. (8.2) and solving the case where the irradiation is on the p-material with the junction near the surface, and for a finite n-material, one obtains the value of p. An equation similar to Eq. (8.2) will lead to the value of n.

The hole and electron current densities can then be readily calculated, since

$$J_p(0) = -e\mathscr{D}_p(\partial p/\partial x)_{x=0}, \tag{8.5}$$

and

$$J_n(0) = -e\mathscr{D}_n(\partial n/\partial x)_{x=0}. \tag{8.6}$$

For the total current $I = I_p(0) + I_n(0)$, or

$$I = eg_0\ell - eg\ell'[\exp(eV/kT) - 1], \tag{8.7}$$

where ℓ and ℓ' are coefficients having the dimensions of diffusion lengths, and g is a coefficient related to the thermal generation rates of both holes and electrons. The same result is obtained if the light irradiates the n-material instead of the p-material.

Equation (8.7) represents the performance characteristics of a photovoltaic cell under the assumptions stated above. Most of the losses are neglected. The results can be much improved by considering the effect of the electron and hole mobilities in Eqs. (8.5) and (8.6). This has been taken into account by Wolf [97] and other workers [17, 18]. The surface recombination effect in the boundary conditions and the resistivity of the cell have also been neglected. The voltage drop due to the resistivity can be readily taken into consideration by subtracting it from the calculated voltage in Eq. (8.7). Then,

$$I = eg_0\ell - eg\ell'\left\{\exp\left[\frac{e(V - R_sI)}{kT}\right] - 1\right\} + \frac{V - R_sI}{R_p}, \tag{8.8}$$

where R_sI is the voltage drop due to the internal series resistance R_s, and R_p is the internal parallel resistance of the cell.

When the voltage is expressed as a function of the current, and $(V - R_sI)/R_p$ is neglected, Eq. (8.8) leads to

$$V = \frac{kT}{e}\ln\left(1 - \frac{I}{eg_0\ell'} + \frac{g_0\ell}{g\ell'}\right) + R_sI. \tag{8.9}$$

The open circuit voltage V_0 can be found by setting $I = 0$ in Eq. (8.9). Thus,

$$V_0 = \frac{kT}{e}\ln\left(1 + \frac{g_0\ell}{g\ell'}\right). \tag{8.10}$$

The short circuit current I_s is obtained from Eq. (8.8) by setting $V = 0$. For small values of $R_s I_s$, approximately,

$$I_s = eg\ell. \qquad (8.11)$$

Also, note that

$$I_0 = eg\ell' \qquad (8.12)$$

is the reverse saturation current, often called the *dark current*, of the diode.

Introducing Eqs. (8.11) and (8.12) into Eqs. (8.9) and (8.10), respectively, yields

$$V = \frac{kT}{e} \ln \left(1 - \frac{I}{I_0} + \frac{I_s}{I_0} \right) + R_s I, \qquad (8.13)$$

and

$$V = \frac{kT}{e} \ln \left(1 + \frac{I_s}{I_0} \right). \qquad (8.14)$$

Often, under high light intensities, the ratio I_s/I_0 is much larger than unity. Taking account of this and introducing Eq. (8.14) into Eq. (8.13) yields

$$V_0 \cong (kT/e) \ln (I_s/I_0);$$

also,

$$\frac{V}{V_0} = 1 + \mathscr{C}_1 \ln \left(1 + \frac{I_0}{I_s} - \frac{I}{I_s} \right) + \mathscr{C}_2 \frac{I}{I_s}, \qquad (8.15)$$

where

$$\mathscr{C}_1 = kT/eV_0, \qquad (8.16)$$

and

$$\mathscr{C}_2 = R_s I_s / V_0. \qquad (8.17)$$

Figure 8.3 illustrates the ideal performance characteristic of a photovoltaic cell as expressed by Eq. (8.15) for $\mathscr{C}_2 = 0$; $\mathscr{C}_1$ was taken as parameter. Note that the best characteristics are given for small values of $\mathscr{C}_1$, i.e., for large open circuit voltages and low temperatures.

The power delivered to an external load will be $P = VI$. Thus,

$$P = V_0 I \left[1 + \mathscr{C}_1 \ln \left(1 + \frac{I_0}{I_s} - \frac{I}{I_s} \right) + \mathscr{C}_2 \frac{I}{I_s} \right]. \qquad (8.18)$$

The current I_m for maximum power output is obtained by setting $dP/dI = 0$; thus, if $\mathscr{C}_2 = 0$ is assumed

$$\frac{I_m}{I_s} = \frac{V_m}{V_0} \frac{\exp (1/\mathscr{C}_1)}{(V_m/V_0 + \mathscr{C}_1)[\exp (1/\mathscr{C}_1) - 1]}, \qquad (8.19)$$

where V_m is the output voltage corresponding to the maximum power. It is equal to

$$V_m = \left\{ 1 + \mathscr{C}_1 \ln \left[1 - \frac{I_m}{I_s} \exp \left(-\frac{1}{\mathscr{C}_1} \right) \right] \right\} V_0. \qquad (8.20)$$

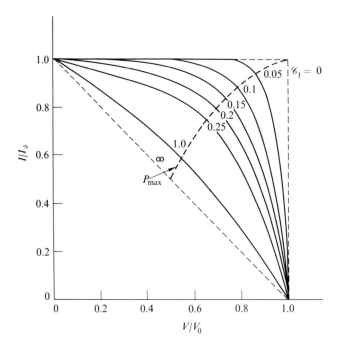

Figure 8.3. Theoretical performance characteristic of a photovoltaic cell as a function of $\mathscr{C}_1 = kT/eV$ at $\mathscr{C}_2 = 0$.

Therefore the maximum power output will be

$$P_m = \left\{ \frac{(V_m/V_0)^2 \exp(1/\mathscr{C}_1)}{(V_m/V_0 + \mathscr{C}_1)[\exp(1/\mathscr{C}_1) - 1]} \right\} I_s V_0. \tag{8.21}$$

The curve of maximum power is also shown in Fig. 8.3.

The equivalent circuit representing Eq. (8.7), taking into consideration Eqs. (8.11) and (8.12) and including the internal loss mechanisms due to the parallel and series resistances R_p and R_s, is shown in Fig. 8.4.

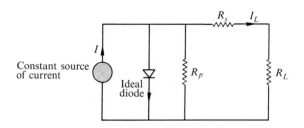

Figure 8.4. Equivalent circuit of a photovoltaic cell.

8.4 Different Types of Photovoltaic Cells

The different types of photovoltaic cells are usually classified according to the type of material used. They are all semiconductors and are chosen for their ability to convert electromagnetic energy into electrical energy. At the present state of the art, silicon photovoltaic cells are the most efficient, although, as predicted by Loferski [44], other materials may attain higher efficiencies. Such materials are indium phosphide (InP), gallium arsenide (GaAs), cadmium telluride (CdTe), aluminum antimonide (AlSb), and zinc telluride (ZnTe). The other materials for which maximum conversion efficiencies higher than 10% have been predicted are germanium (Ge), selenium (Se), gallium phosphide (GaP), and cadmium sulfide (CdS). Presently the most efficient solar cell is made up of silicon, attaining 16% efficiency, not far away from the 20% maximum efficiency predicted for this material. The maximum efficiency predicted for any material is 25% for AlSb at ambient temperatures.

Properties of a photovoltaic material. It is interesting to list, as was done for thermoelectric power generators, the qualities needed for a semiconductor to be used as a photovoltaic solar cell. To do this, it would be necessary to know the different events happening to the impinging radiation. Let J_i be the incident radiation current density, where $J_i = eN_g$ (A/cm^2), where N_g is the number of photons per square centimeter per second having energies larger than the energy gap $\mathscr{E}$ of the material. As J_i hits the front surface of the cell, a fraction $\mathscr{A}_r$ of the current is reflected and the remainder, $J_i(1 - \mathscr{A}_r)$, is transmitted. For a given material $\mathscr{A}_r$ should be as small as possible. If v_R is the refractive index of the material, then, for normally incident radiation,

$$\mathscr{A}_r = \left(\frac{v_R - 1}{v_R + 1}\right)^2.$$

Moss [55] related the index of refraction to the energy gap of the material through the following empirical expression:

$$\mathscr{E}_g v_R^4 = 172.$$

To have a small $\mathscr{A}_r$, v_R should be small and consequently $\mathscr{E}_g$ should be as large as possible. However, N_g depends on the energy gap $\mathscr{E}_g$, and since it is necessary to have N_g as large as possible, $\mathscr{E}_g$ should be small. An optimum should be reached. To avoid the dependence of $\mathscr{E}_g$ on $\mathscr{A}_r$, a surface coating is used by adopting one of several proposed methods [33]. A possibility is to reflect back the refracted radiation by a special construction for the surface of the cell.

The transmitted part of the current is exponentially attenuated by the p-region. This absorption is due to a decrease in the number of photons caused by the creation of electron-hole pairs. The constant of attenuation is α and it is approximately the same in the n-region, since α does not depend much on the dopant. The photons which cross the junction will continue to be attenuated at the same rate and eventually all will be absorbed if the n-region is thick enough. Minority carriers thus created by the absorbed photons should be able to cross the barrier of potential before being recombined. This requires that d'/ℓ be as small as possible, where ℓ is the diffusion length in the material and d' is the thickness of the p-region. Letting η_Q be the collection efficiency,

$$\eta_0 = \eta_Q[1 - \exp(-\alpha d')](1 - \mathscr{A}_r), \tag{8.22}$$

with η_0 being equal to $\eta_0 = I_s/I_i$. From Eq. (8.22), one sees that $d'\alpha$ should be as large as possible. However, d' is limited by the condition $(d'/\ell) < 1$ for the minority carriers to cross the potential barrier. Moss [57] has shown that when the electron and hole diffusion lengths are the same, the layer thickness d' should have an optimum value such that

$$\alpha d'[\exp(\alpha d') - 1] = \alpha\ell - 1, \tag{8.23}$$

where α should be as large as possible for the chosen semiconductor. Furthermore, ℓ is related to the minority lifetime τ by the relation

$$\tau = \ell^2/\mathscr{D}, \tag{8.24}$$

where $\mathscr{D}$ is the diffusion constant. Therefore, both the condition $(d'/\ell) < 1$, and the condition that τ should be as large as possible lead to desirable large values of ℓ. For large values of α, Eq. (8.23) can be written (assuming $\alpha\ell \gg 1$) as

$$\exp(\alpha d') = 1 + \ell/d'. \tag{8.25}$$

This equation leaves little choice for the value of d' for which an optimum is imposed by the values of ℓ and α. The optimum value of the p-layer thickness for different materials, ranges from one to several microns. Consequently, the technology of thin layers becomes extremely important for the fabrication of efficient photovoltaic cells.

In Eq. (8.24), $\mathscr{D}$ represents the diffusion constant, related to the mobility and the temperature T by the Einstein relation

$$\mathscr{D} = kT\ell/e. \tag{8.26}$$

Equation (8.24) indicates that $\mathscr{D}$ should be as small as possible; therefore both the temperature and the mobility should be small.

Thus a good photovoltaic material should have a large absorption coefficient α at a low temperature and the energy gap $\mathscr{E}_g$ should be small. However, too small values of the energy gap lead to a large intrinsic carrier concentration, as shown by the relation

$$n_i = \mathscr{A}T^{3/2} \exp(-\mathscr{E}_g/2kT), \tag{8.27}$$

where $\mathscr{A}$ is a constant depending on the effective mass of the carriers. It is approximately the same for all semiconductors, and equals $\mathscr{A} = 4.73 \times 10^{15} \ m^{-3}\cdot K^{-3/2}$. The reverse saturation current I_0 is related to the intrinsic carrier concentration

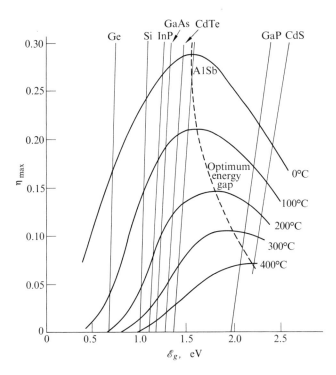

Figure 8.5. Variation of the maximum efficiency with the energy gap and the temperature. Doping level 10^{17} cm^{-3}. Solar spectrum above the atmosphere. The straight lines represent the temperature variation of the energy gaps for several materials. (Adapted from Wysocki and Rappaport [99].)

n_i by the relation

$$I_0 \cong \left(\frac{\ell_n}{\tau_n n_n} + \frac{\ell_p}{\tau_p p_p} \right) e n_i^2, \qquad (8.28)$$

where p_p is the equilibrium concentration of holes in the p-material.

An increase in n_i^2 will lead to an increase in I_0 which will lead to a decrease in the value of the current, as shown by Eq. (8.8), where $I_0 = e g \ell'$. This is undesirable, since it will lead to lower efficiencies. An optimum value for $\mathscr{E}_g$ should be attained (see below).

The variation of the maximum efficiency of a photovoltaic cell, as a function of the energy gap and the temperature, as obtained from results based on Wysocki and Rappaport's work [99], is shown in Fig. 8.5. Note that, as predicted, the efficiency decreases with increasing temperature whereas the optimum energy gap decreases with increasing temperature. This fact makes the use of elements of high energy gap attractive for high temperatures.

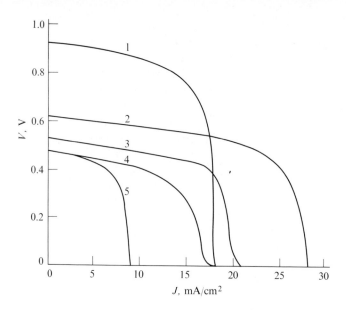

Figure 8.6. Performance characteristics of several photovoltaic cells. Curve 1: GaAs at 0°C under 100 mW/cm² illumination (after Gold [19]); curve 2: Si at 0°C under 100 mW/cm² illumination (after Gold [19]); curve 3: CdTe (crystal) under 97 mW/cm² illumination (after Cusano [10]); curve 4: CdTe (film) under 87 mW/cm² illumination (after Cusano [10]); curve 5: CdS under 79.2 mW/cm² illumination (after Schaeffer [75].

Silicon photovoltaic cells. By far, these are the most efficient solar cells at the current state of the art. Their efficiency now reaches 16%, and it can be raised to about 24% if work at low temperatures is made possible (see Fig. 8.5).

The energy gap of silicon is 1.12 eV at 0°C. The mobilities of the minority carriers are 1200 and 250 cm²/V·sec for electrons and holes, respectively, leading to a lifetime for holes equal to 10^{-5} sec. A typical performance characteristic of a silicon solar cell is shown in Fig. 8.6, curve 2, for an illumination of 100 mW/cm² at 0°C as reported by Gold [19]. In Fig. 8.7 the absorption coefficient of silicon is compared to those of other materials as a function of the incident energy $\mathscr{E}_i$. The fact that this curve varies relatively slowly with $\mathscr{E}_i$ makes the use of silicon for photovoltaic power generation attractive.

Silicon solar cells were the first to be used in artificial satellites [76, 89]. When these satellites moved through the Van Allen belts, the cells were subjected to bombardment by high-energy particles [32, 82]. This bombardment has the effect of degrading the lifetime of the minority carriers [13, 14] and thus damaging the solar cell. This effect can be minimized by using a very thin target material (of the order of 0.4 μ). It can also be reduced drastically if an n-on-p cell is used instead of a p-on-n cell. Because the current generated depends on the number of collected carriers created by the action of the photons deep within the cell, the carriers should have a long lifetime to be able to cross the large distances involved. The lifetime of the electrons in the p-material is more than three times greater than the lifetime of the holes in the n-material. This explains the demonstration of Mandelkorn [47] that n-on-p cells are more resistant to radiation bombardment. Furthermore, only the most energetic (>0.25 MeV) particles can create a defect in the p-material. The effect of particle bombardment can be reduced even further by using a cover glass of adequate thickness on the silicon cell.

The particle bombardment problem is insignificant in terrestrial applications. The high cost of the silicon solar cell per watt produced ($400/watt) is one of the most serious problems. This cost can be reduced greatly in three different ways:

1. By fabricating cells of higher efficiencies,

2. By reducing the amount of silicon used,

3. By perfecting the methods of silicon cell fabrication.

It is unlikely that greater improvements can be made in the direction of much higher efficiencies with silicon solar cells. On the other hand, it is now known that most of the material used in a cell does not contribute to the power conversion

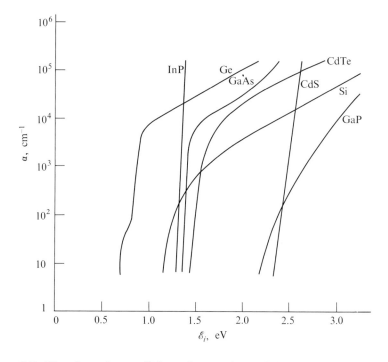

Figure 8.7. The absorption coefficient of some photovoltaic materials as a function of the incident energy. The slow variation of the absorption coefficient of silicon with the incident energy makes it one of the best materials for power generation.

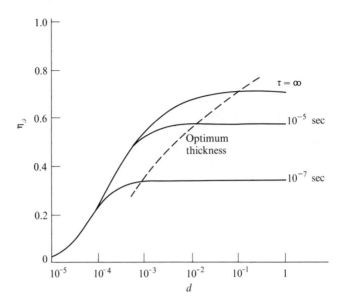

Figure 8.8. Relative number of carriers collected at a junction as a function of the cell thickness with the lifetime τ of the minority carriers as parameter. (From Elliott [15].)

process. As was shown theoretically by Elliott [15], there is minimum thickness for maximum efficiency, which depends on the minority carrier lifetime, τ. The number of carriers collected by the junction per number of photons in the sun's spectrum η_s is shown as a function of the cell thickness in Fig. 8.8, where τ is a parameter. This shows that the cell thickness can be reduced by at least two orders of magnitude without much loss in efficiency.

The use of much less material in silicon cells can also be achieved by using thin-film techniques. However, because of the now imperfect understanding of thin-film theory, the efficiencies of such cells are still exceptionally low.

Finally, silicon solar cells are usually prepared from wafers of single-crystal silicon. The whole process of manufacturing such cells is very expensive. Among the most promising new techniques is dendritic crystal growing, in which single silicon crystals are produced in a ribbon form.

Photovoltaic cells using other materials [27, 29, 35, 42]. Other materials have been considered for use in photovoltaic power generation. This is indicated in Fig. 8.5 where it is seen that silicon is, by no means, the most efficient photovoltaic converter in theory. The different energy gaps of the materials are represented, and it can be noted that the optimum energy gap increases from 1.55 eV at 0°C for AlSb to about 2.20 eV at 400°C for CdS.

Presently, the most efficient material after silicon is GaAs. Theoretically, GaAs should be more efficient than silicon, as shown in Fig. 8.5. The energy gap of GaAs is 1.34 eV and the mobilities of the minority carriers (3000 cm^2/V·sec for electrons and 600 cm^2/V·sec for holes) are much larger than those in silicon, leading to a much shorter lifetime (10^{-8} sec). Furthermore, as shown in Fig. 8.7, the absorption coefficient is such that most of the absorbed photons are absorbed in a very narrow region near the surface of incidence. This will greatly increase the internal resistance of the cell and consequently limit the overall efficiency. This effect can be reduced by applying a grid electrode on the incident surface. The effective incident surface will be somewhat reduced by the presence of the grid. The performance characteristic of a gallium arsenide photovoltaic cell at 0°C under 100 mW/cm^2 illumination is illustrated in Fig. 8.6. Finally, note that for space applications, GaAs seems more resistant to the particle bombardment effects. The technology of the GaAs solar cell is similar to that of silicon in that they are mostly obtained from a single-crystal material; however, special attention must be given to the optimization of the junction depth, due to the shorter lifetime of the minority carriers.

Cadmium telluride (CdTe) [34, 59] is another material which has received special attention as a photovoltaic semiconductor. Theoretically, it seems to be the most potentially efficient photovoltaic material, as shown in Fig. 8.5. Figure 8.7 shows the variation of the CdTe absorption coefficient. It appears from these two figures that CdTe is similar to GaAs in many respects, e.g., the energy gap at 0°C is equal to about 1.96 eV. As for the technology of this material, it has been directed toward both single crystals and thin films. Typical performance characteristics are shown in Fig. 8.6.

So far as the present state of the art is concerned, CdS is almost as efficient (about 6%) as CdTe. Theoretically, CdS seems more attractive for relatively high-temperature photovoltaic cells, because of its high energy gap (2.41 eV at 0°C), as shown in Fig. 8.5. As for CdTe, the technology efforts are directed toward both the single-crystal and the thin-film techniques. A performance characteristic of this type of cell, as reported by Schaeffer [75], is shown in Fig. 8.6; its absorption characteristic is shown in Fig. 8.7.

Other materials considered for photovoltaic use are named in Figs. 8.5 and 8.7. In order of increasing energy gaps, these are Ge, InP, AlSb, and GaP.

Multimaterial photovoltaic cells. It seems evident that there is no good reason for using the same material throughout a photovoltaic cell. The fact that materials of different energy gaps can be found or devised should be used to build a converter which matches more closely the energy spectrum of the sun. This is a concept which may lead to much higher efficiencies than those anticipated for single-material cells. Therefore, two or more layers using materials of different energy gaps can be utilized for the construction of a much more efficient solar energy converter. Theoretically, the number of layers can go to infinity, as was suggested by Jackson [25] in 1958 and by Wolf [95] in 1959. Jackson found that, for three layers, 60% of the energy of the sun could be transformed into electricity.

The problem of incomplete absorption of the solar energy can be solved by dividing the solar energy spectrum into different energy bands. The most suitable material could thus be used for each individual energy band. However, the generated current will be reduced for each layer with the decrease of the corresponding individual bandwidth. This suggests the existence of an optimum number of layers yielding the maximum electrical energy possible. Special consideration has to be given to the arrangement of the different layers. For instance, because each layer has a different energy gap and, therefore, a different output voltage, it seems that the only possible arrangement is a series connection of the different layers. Furthermore, the currents flowing should not destroy each other. For instance, for the three-layer cell it seems that the only possible workable arrangement is a "p-n-ohmic contact-metal-ohmic contact-p-n" series. Since the same current will be flowing in the different layers placed in series, it is most efficient when each layer absorbs the same number of photons from the sun, leading to the choice of the energy gap (and therefore the material) for each layer. For the merit of this method to be judged, more experimental work is needed in compound semiconductor technology.

Graded energy gap photovoltaic cells [92]. Instead of using a multilayer photovoltaic cell, it may be more interesting to use a single material which has a varying energy gap. If this variation of the energy gap is made to fit the solar spectrum exactly, it seems obvious that 100% solar energy conversion, or at least the electron-hole pair generation, will become possible.

Technically, it is possible to build semiconductors having graded energy gaps. However, this will create new problems, making the usefulness of such a method questionable; the gain in utilization factors is practically cancelled by the lower voltage corresponding to the smallest energy gap present. This loss is due to the fact that it is impossible to separate a small energy electron-hole pair by a junction of wider energy gap. This can happen only if the very improbable electron-electron interactions occur.

This method could be used just as an improvement of the collection efficiency. If one must improve this configuration by using ohmic contacts to separate the junction, one arrives at the multimaterial photovoltaic cell mentioned above. This separation becomes necessary, since a multiplicity of pn junctions in the same material is completely without effect for power conversion.

Multitransition photovoltaic cells. If energy levels in the forbidden gap are utilized so that transitions from photons with insufficient energy are made possible, these photons will lead to direct transitions from the valence band to the conduction band, thus yielding very high conversion efficiencies. For the photon to be able to interact with the electron while it is still trapped, the electron should stay in the transition gap long enough, thus a low trap level* is necessary. On the other hand, this trap level will lead to higher absorption coefficients for photons above the energy difference between the bottom of the conduction band and the trap level.

---

* A trap is an empty quantum state slightly below the conduction band. It is created by an impurity or a negative ion vacancy in the semiconductor crystal.

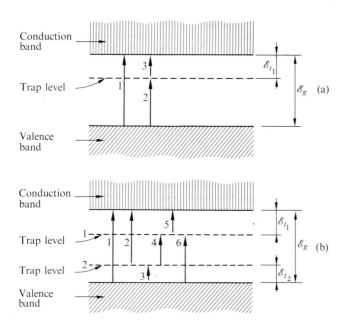

Figure 8.9. Trap levels for a better utilization of solar energy. (a) Single trap level, (b) two trap levels.

The introduction of a single trap level gives to the photon three possibilities for transition from the valence band to the conduction band; two trap levels will yield six possibilities, as shown in Fig. 8.9. It has been shown that the percentage of the solar energy utilized in pair generation is much higher with one or more trap levels than without any trap level.

The most suitable location for the trap for a material of a given energy gap can be obtained by assuming the following conditions: (1) that the density of trapping centers is such that a high absorption coefficient is obtained for transitions between the trap center and both bands, (2) that the number of filled traps is of the same order of magnitude as the number of empty traps under irradiation, (3) that the photon-electron interaction and atomic cross sections are equal, and (4) that most of the absorption takes place within one diffusion length from the pn junction. If it is further assumed that the number of photons available for transition from the valence band to one trap level and those available for transition from the trap level to the conduction band are equal, the generation rate of electron-hole pairs can be calculated. Using the same considerations as those taken for Fig. 8.5 in the case of a single pn junction, Fig. 8.10 is obtained. This figure shows the much higher maximum efficiencies for single and double trap levels at about 50°C as reported by Wolf [95].

The physical properties of trap levels should be investigated more thoroughly before the merit of this method can be fully appreciated. For instance, the re-combination properties of these levels should be better known. This and other

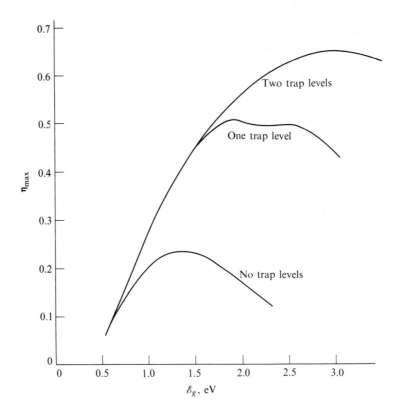

Figure 8.10. The maximum conversion efficiency as a function of the energy gap with and without trap levels. The existence of one or more trap levels greatly enhances the efficiency. (From Wolf [95].)

problems have been investigated by many workers since 1950 [20, 21, 80]. For more details, references should be consulted.

8.5 Collection of Solar Radiation

To produce larger amounts of energy from the sun, it seems necessary to use very large areas to be irradiated. For a solar constant of 0.1 W/cm^2 and an efficiency of 20%, it is necessary to cover 5000 m^2 to produce 1 MW of energy. Due to the high cost of photovoltaic semiconductors, this large area of photovoltaic cells will cost a prohibitive amount of money. To limit this, it is necessary to use reflectors and lenses to concentrate the solar energy. Special attention should be given to maintaining a low temperature, thus avoiding a degradation in efficiency. The internal resistance of the cell itself should also be greatly limited to avoid a loss in performance, due to the larger internal voltage drop.

On the other hand, arrangements should be taken to keep the solar panels oriented normally to the solar flux, in order to receive the largest amount of energy possible.

The systems of collecting solar radiation and of focusing can be classified into two important categories, refractive systems and reflective systems.

In the first category, the refractive behavior of some crystal subjected to Fresnel principles is utilized. This can lead to the use of small fixed lenses for low-energy generator use, as proposed by Chapin [6].

For high-energy generators, devices of the second category are more often considered. The reflecting areas are formed by mirrors made up of aluminum, aluminum compounds, or other reflective materials. These areas can be classified in many categories as far as the focal zone is concerned. Paraboloid mirrors having fixed forms were built for different terrestrial and space applications. For small mirrors, the surface is made up of one part; for large mirrors, many individual parts are used to form the overall paraboloid. The mirror can be made orientable to match the direction of the solar radiation. Mirrors of this type have been built [63] in Marseille, France, and Algiers, Algeria. The French mirror has a linear focus and a reflective area of 60 m^2.

The individual parts of the mirror can be made individually orientable, allowing the use of a fixed focus at which will be the photovoltaic generator.

Other types of mirrors being considered are of different forms: cylindrical, conical, or other possible configurations.

However, the main problem remains the heating of the photovoltaic cells due to the solar concentration. This heating should be limited greatly by adequate refrigeration methods and some optimal temperature has to be reached. This temperature is still extremely high (around 400°C). If this temperature is lowered to about 200°C, or eventually to about 100°C, a new era for photovoltaic power stations could be realized, especially if the different techniques described in Section 8.5 are improved. The use of thermoelectricity could also be considered to convert that part of solar radiation which cannot be converted by the photovoltaic effect.

Nevertheless, solar collector technology is still in its elementary empirical stage. Many requirements have to be met, and it is always dangerous to build a large mirror without a certain amount of experimental evidence being gathered. The different requirements, such as optical accuracy, low weight, high stiffness, and mechanical resistance to the environment, are often contradictory.

The efficiency of the collector as a function of temperature remains the most important characteristic. This efficiency decreases with increasing temperature. For a temperature of about 500°C, the best efficiency of a collector is about 92%, and the lowest mirror specific weight is about 1.5 kg/m^2 with a mirror diameter of about 10 m.

For a detailed study of the different solar energy collectors available and considered, McLelland's [49] paper should be consulted. The different technological aspects, including the different material and mirror considerations, are treated thoroughly.

8.6 Cell Fabrication

Since silicon is the most used material, its fabrication will be considered here. The fabrication of other cells generally follows the same pattern.

Silicon is obtained in polycrystalline chunk form from sand, which is mostly silicon dioxide. Although silicon is after oxygen the most abundant element on earth, the price of pure silicon is still very high (about $200/kg). This is mainly due to the high purity desirable in the material to be used in photovoltaic power generation.

This ultrapure silicon is placed with a certain amount of p-type silicon doped with boron or any other desired impurity in a quartz crucible. The ratio of the two materials depends on the desired final resistivity. The quartz crucible is then placed in a special furnace and crystals can be grown by some specialized method. Silicon is then brought slightly above its melting point and a single-crystal seed of the desired orientation is lowered into the melt. The melt is then continuously stirred slowly to promote homogeneity. A freezing surface exists at the surface of the seed, and a crystal with the preferred orientation of the seed grows as the seed is withdrawn with a speed slower than 10 cm/hr. The crystal obtained may be as long as 5 cm, with a weight of about 800 g.

The grown monocrystal is then annealed and checked for its conductivity. The initial seed is removed for reuse and cylindrical wafers of the crystal are obtained by the use of a diamond saw. After cleaning, these wafers are cemented to ceramic blocks and then cut into blanks slightly thicker than the finished cell. The blanks are cemented to steel plates and are highly polished on one side. They are placed in quartz boats and heated to approximately 900°C in a special furnace in the presence of a phosphorus compound (or any other desired n-type doping material). Phosphorus then diffuses into the silicon blank, forming an n-type material. The depth as well as the amount of phosphorus diffusion is regulated by controlling the temperature and the duration of the process. Thus a pn junction is obtained from the single crystal.

The diffusion layer is then eliminated by a sandblasting machine. After cleaning, the cells are placed in a vacuum evaporator where the grid and contact structures are applied. After being sintered in a reducing atmosphere at high temperature, the remaining diffusion layer is etched from the edges and the junction is cleaned. A silicon monoxide layer is then evaporated on the surface of the cells to decrease the losses by reflection. The cells are now ready to be used for photovoltaic power generation.

In the case of thin-film silicon cells, chemical methods are most often used for the deposition of the thin film. Pure hydrogen is passed through silicon tetrachloride $(SiCl_4)$, in a vessel. The hydrogen carries a certain amount of $SiCl_4$ vapor to a reaction furnace where a substrate is placed and as it reaches the substrate, the vapor is decomposed into a thin film of silicon, on the substrate, and chlorine, which is eliminated with the hydrogen from the furnace. To obtain an n-type or a p-type thin film, the $SiCl_4$ vapor is doped with phosphorus trichloride, boron tribromide, or any other desired vapor. The flow velocity of hydrogen, the vapor ratios, and the temperature of the furnace determine the quality of the thin film.

Thin-film [22] techniques are still in their infancy and the highest efficiencies of silicon thin-film cells rarely surpass 4%.

8.7 Power Output, Losses, and Efficiencies

Power output. The total power output of a photovoltaic cell has been given by Eq. (8.18). In the case where the parallel internal resistance is considered, the power generated becomes, from Eqs. (8.8), (8.11), (8.12), (8.16), and (8.17),

$$P = \frac{VI_s - VI_0\left[\exp\frac{1}{\mathscr{C}_1}\left(\frac{V}{V_0} - \mathscr{C}_2\frac{I}{I_s}\right) - 1\right] - \frac{V^2}{R_p}}{1 - \frac{\mathscr{C}_2}{R_p}\cdot\frac{V_0}{I_s}}.$$

For R_p going to infinity, this equation reduces to Eq. (8.18). If it is further assumed that $\mathscr{C}_2 = 0$,

$$\frac{P}{P_0} = \frac{V}{V_0}\left\{1 - \frac{\exp\left[(1/\mathscr{C}_1)(V/V_0)\right] - 1}{\exp(1/\mathscr{C}_1) - 1}\right\}, \tag{8.29}$$

where $P_0 = I_s V_0$. Equation (8.28) has been plotted in Fig. 8.11 with $\mathscr{C}_1$ as parameter. Once again the best characteristics are given for the smallest values of $\mathscr{C}_1$, i.e., for low temperatures. The effects of R_p and R_s will limit further the performances shown in Fig. 8.11. The maximum power output P_m is given by

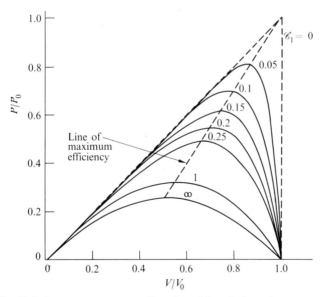

Figure 8.11. Relative power output as a function of the relative voltage output; $\mathscr{C}_1$ = parameter, $P_0 = I_s V_0$.

Eq. (8.20),

$$\frac{P_m}{P_0} = \frac{(V_m/V_0)^2 \exp{(1/\mathscr{C}_1)}}{(V/V_0 + \mathscr{C}_1)[\exp{(1/\mathscr{C}_1)} - 1]},$$

where V_m is the optimum voltage.

Input power. The input power is the electromagnetic power received from the sun, often measured as a function of the solar constant. The solar constant is, by definition, the total amount of energy received in 1 min from the sun on a 1-cm^2 surface perpendicular to the solar rays. This has been discussed in Section 8.2.

Losses. Several types of losses were treated under the heading of photovoltaic materials. These can be listed as follows:

1. Losses of collection of solar energy,
2. Reflection losses on the surface of the cells,
3. Losses due to incomplete absorption,
4. Losses in the photon energy necessary to create the electron-hole pairs,
5. Losses to the pn junction in the collection of the electron-hole pairs, lost in the diffusion process,
6. Losses due to the internal resistance of the cell.

Other losses may exist, but those listed above are of the greatest importance and much research work is being done to reduce them.

The losses of collection of solar energy are those due to the mirrors or lenses. These are generally small compared to the overall losses. More and more efficient mirrors and lenses are being built.

Reflection losses $\mathscr{A}_r$, treated in Section 8.4, have been greatly reduced, as was explained earlier. $\mathscr{A}_r$ is given by Eq. (8.15), and values as small as 6% have been reached, mostly by means of special coatings.

A part of the solar spectrum cannot be used. Only those photons having energies higher than the energy gap of the material could be absorbed. The problem of choosing the right value of $\mathscr{E}_g$ has been discussed in the preceding sections.

The photons having energies much higher than the energy gap of the material will create electron-hole pairs; however, the excess energy will be dissipated as heat and thus will be lost in the conversion process.

Photons crossing the semiconductor material will be attenuated with a constant of attenuation α as seen in Section 8.4. Due to the fact that α is finite, only a fraction of the photons will be absorbed.

The losses due to the internal parallel and series resistances are respectively V^2/R_p and $R_s I^2$. A part of these resistances is due to the electrical contacts. The techniques of applying ohmic contacts are being improved greatly; however, ohmic losses remain very important in thin-film cells.

Efficiencies. One of the most important ratios in the calculation of the overall efficiency of a photovoltaic cell is the voltage factor m_v.

The amount of energy utilized in the generation of electron-hole pairs is equal to the energy gap of the material. The largest obtainable voltage, i.e., the open circuit voltage is, however, always smaller than the gap. There are two important reasons for this. First, the maximum applicable forward voltage on a pn junction is determined by the difference in Fermi levels on both sides of the junction. These Fermi levels are functions of the impurity concentrations and temperatures in the n- and p-regions. They are normally located inside the forbidden gap in such a way that the barrier height becomes smaller than the barrier gap. Second, even this barrier height will not be reached by photons absorbed directly from the sun, as extremely high energy levels would be necessary. The voltage factor is defined by

$$m_v = V_0/V_g,$$

where V_g is the voltage gap corresponding to the energy gap $\mathscr{E}_g$ of an electron; $\mathscr{E}_g = eV_g$, and V_0 is given by Eq. (8.14). Thus,

$$m_v = \frac{kT}{\mathscr{E}_g} \ln\left(1 + \frac{I_s}{I_0}\right), \tag{8.30}$$

where I_0 is a function of the temperature and energy gap. Taking into account Eqs. (8.27) and (8.28), its value will be

$$I_0 = \mathscr{B}T^3 \exp\left(-\mathscr{E}_g/kT\right),$$

where $\mathscr{B}$ is a constant determined by the material properties and the junction configurations. It is equal to

$$\mathscr{B} = \left(\frac{\ell_n}{\tau_n n_n} + \frac{\ell_p}{\tau_p p_p}\right)e\mathscr{A}^2.$$

The short circuit current has been calculated by Wolf [95] and can be stated as the sum of the currents generated in the n- and the p-layers. It can be written

$$I_s = \int_0^\infty \left[I_n(\lambda) + I_p(\lambda)\right]\,d\lambda,$$

where λ is the wavelength of the radiation; $I_n(\lambda)$ and $I_p(\lambda)$ are complicated functions of the junction characteristics and the incident flux of photons.

The other important factor is the curve factor m_c, defined by the ratio

$$m_c = (V_m I_m)/(V_0 I_s).$$

Neglecting the internal resistances yields

$$m_c = \frac{V_m}{V_0}\left\{1 + \frac{I_0}{I_s}\left[1 - \exp\left(\frac{eV_m}{kT}\right)\right]\right\}. \tag{8.31}$$

The total output efficiency of a photovoltaic cell will be $\eta = P_{\text{out}}/P_{\text{in}}$, or taking into consideration Eq. (8.22),

$$\eta = m_v m_c \eta_Q[1 - \exp\left(-\alpha d'\right)](1 - \mathscr{A}_r).$$

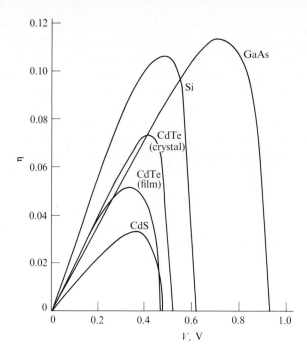

Figure 8.12. Measured efficiencies of several photovoltaic cells. (Adapted from Fig. 8.6.)

Since the factors η_Q, α, and $\mathscr{A}_r$ are functions of the wavelength,

$$\eta = m_v m_c \frac{e}{hc} \cdot \frac{\int_0^\infty [1 - \mathscr{A}_r(\lambda)]\{1 - \exp[-\alpha(\lambda)d']\}\eta_Q(\lambda)\mathscr{E}_i(\lambda)\lambda \, d\lambda}{\int \mathscr{E}_i(\lambda) \, d\lambda} \tag{8.32}$$

in the most general case.

Figure 8.12 shows the efficiencies of several photovoltaic cells as a function of the output voltage. These efficiencies correspond to the performance characteristics represented in Fig. 8.6. Note that the curve representing silicon is not the one giving the highest efficiency. As noted before, higher efficiencies are, however, obtainable with this material.

8.8 Photovoltaic Batteries and Systems [66, 77, 86]

Usually the power requirement is much larger than an individual cell can supply. Connections of individual cells in series and/or in parallel become necessary.

When the voltage requirements exceed the maximum output voltage of a cell, a connection of different individual cells in *series* becomes advisable. In this case, the voltage at any current level becomes equal to the sum of the individual voltages and the same current flows through all the cells. However, if the short circuit current is not the same for all the cells in series, the cells of lowest short circuit current will tend to limit the performance of the entire battery. This lowest current cell will be driven into the reverse voltage region by the other cells. Obviously, the short circuit current of the battery will be the point where the sum of the individual voltages equals zero.

When the current requirements exceed the maximum output current of an individual cell, it is a parallel connection of different cells which becomes necessary. The total output current is then equal to the sum of the individual currents and the same voltage output exists as for one cell. Here, the total performance will be limited by the cell having the lowest open circuit voltage.

Shadows should be absolutely avoided, as they are a source of important losses in the power produced by a photovoltaic battery. The effects of a shadow on a series battery and a parallel battery are different. For instance, if an individual cell falls completely under a shadow, its output may fall to zero and the output of the whole battery becomes approximately equal to zero if the shadowed cell is part of a series connection. If the shadowed cell is part of a parallel battery, the effect is less serious; it will simply cause a decrease in the total output current generated.

Other failures may become catastrophic for the operation of a solar battery. They can be divided into three distinct types:

1. A short circuit across a cell,

2. An open circuited cell (contact failure),

3. A cell shorting to the panel substrate.

The first failure causes, in the case of a series connection, a voltage drop equal to $V(n - 1)/n$, where n is the total number of cells. The present techniques of array fabrication are such that a failure of this type is very unlikely. The second failure is more serious since the entire battery becomes open. It is also the most likely to happen, due to high temperatures, meteorite impacts, or other reasons. To avoid this failure, it is necessary to build a matrixed array where each cell in a series string is connected to the corresponding cell in all other series strings. The third failure can be avoided by isolating the array substrate from the power supply system.

It is necessary to use matrixed arrays to avoid an open circuit in the system. Apart from the photovoltaic battery, the solar power generation system may also include concentrating mirrors and refrigerating arrangements to eliminate the heat produced by the losses. This heat may eventually be used to generate more electricity by another method.

8.9 Economic Aspects and Future Trends

Several factors have to be considered to make a method of generating electricity competitive. These factors are the power output per unit cost and per unit weight, the efficiency, the total power output, and the reliability. In the case of photo-voltaic converters, power output per unit area should also be taken into consideration.

The research conducted to improve these factors has led to more or less successful results. However, the biggest handicap to the use of photovoltaic batteries, outside space applications, is the prohibitive price of the cell materials. A great deal of work has to be done to improve the manufacturing methods and to reduce to the minimum, by the utilization of thin-film techniques, the amount of material used. The overall prices may eventually be decreased by using large area polycrystalline sheets, and studies have to be done both theoretically and experimentally, not only by engineers and physicists, but by metallurgists and chemists as well.

Photovoltaic batteries have already no competitor as converters of solar energy. Their efficiencies are by far the highest of any method of converting electromagnetic radiation, directly or indirectly, to electricity. It seems possible that in the near future, much higher efficiencies will be obtained. Solar energy is too important in some areas for men not to try to master it.

PROBLEMS

8.1 An extrinsic semiconducting slab is illuminated at room temperature by a radiation of wavelength λ_0. Assuming that pairs are optically generated in a uniform manner within the volume of the slab, write the general differential equation specifying the enhancement of the number of carriers due to optical generation.

8.2 A photosensitive semiconducting element having the dimensions shown in Fig. 8.13 is connected to a constant current source with ohmic contacts as indicated. (a) What is the resistance of the element when no light is incident? (b) What is the effect of an illumination at constant wavelength on the resistance? Find an expression in terms of the enhanced number of carriers.

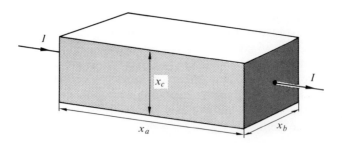

Figure 8.13. A photosensitive photoconducting element.

8.3 In Problem 8.1 assume that the incident light is time modulated at an angular frequency ω. Find the equivalent ac circuit of this photosensitive device.

8.4 Solve numerically Problems 8.2 and 8.3 for the following data:

$P_{\text{light}} = 1$ mW,	$\lambda_0 = 8000$ Å,
$P_{\text{light,max}} = 1$ mW,	$\omega = 1$ kHz,
$T = 30°C,$	$\tau_{\text{res}} = 150$ sec,
$\mathscr{E}_g = 0.7$ eV,	$t = 0.5$ mm,
$h = 8$ mm,	$d = 4$ mm,
$n_i = 2 \times 10^{13}$ particles/cm^3,	$n_p = 5 \times 10^{15}$ particles/cm^3,
$\mathscr{D}_n = 96$ cm^2/sec,	$\mathscr{D}_p = 45$ cm^2/sec,

where τ_{res} is the recombination time, n_p is the donor's density, and n_i is the intrinsic carrier concentration.

8.5 A pn junction silicon solar cell is illuminated as shown in Fig. 8.2(a). The highly conducting terminal of the p-surface is deposited in a manner which does not prevent the light from penetrating the semiconductor material. (a) If the light source is monochromatic, what frequency or wavelength should it have? What is the polarity of the solar cell voltage? (b) If the light source has a spectral density which is Gaussian, find the solution which would give the optimum center frequency for the distribution.

8.6 The solar cell of Problem 8.5(b) is loaded so that currents may flow. Find a relation of the carrier densities as a function of position and set up the one-dimensional problem which would permit the solution of the carrier flow.

8.7 Find the relation between the saturation current density versus the width of the energy gap for the following values:

$$T = 25°C,$$
$$\mathscr{E}_p = 250 \text{ cm}^2/\text{V·sec} = \text{mobility of holes},$$
$$\mathscr{E}_n = 1200 \text{ cm}^2/\text{V·sec} = \text{mobility of electrons},$$
$$\tau_p = 10^{-5} \text{ sec},$$
$$\tau_n = 10^{-7} \text{ sec},$$
$$n_n = 10^{17} \text{ particles/cm}^3,$$
$$p_p = 10^{19} \text{ particles/cm}^3.$$

8.8 Consider two pn junctions, one made up of silicon and the other of germanium. Assume that n and p are the same for the two materials and that for germanium.

$$\ell_p = 0.05 \text{ cm},$$
$$\tau_p = 80 \text{ sec},$$
$$\ell_n = 0.22 \text{ cm},$$
$$\tau_n = 500 \text{ sec}.$$

(a) Find the temperature at which the two pn junctions would have the same reverse saturation current. (b) Comment on your result. (Use Fig. 8.5.)

8.9 Large improvements of the silicon cell efficiency can be obtained either by reducing the cell thickness (Fig. 8.8) or by the application of a contact grid structure. Such grids consist of a certain number of very thin metal strips in contact with the p-layer. The effect of these grids is to reduce the series resistance of the p-layer. Calculate the optimum spacing and the optimum width of the grid lines as a function of solar cell characteristics near the maximum power point [95].

8.10 Study the effect of the series and the parallel resistances of a solar cell on the total power output. (a) Are there any optimum values for R_s and R_p? (b) What would be the ideal cell?

8.11 Design a photovoltaic pn junction cell producing 10 W of electric energy and receiving the solar spectrum at sea level under the best conditions (use Fig. 8.1). The material used is silicon with the characteristics given in Problem 8.7. Find (a) the dimensions of the cell, and (b) its efficiency.

8.12 Using the results of Problem 8.11, find the total illuminated area necessary for the production of 100 kWh/month. What would be the price of such installation, knowing that the density of silicon is 2.33 g/cm^3 and that its price is \$200/kg?

8.13 From Problem 8.12, one sees that the use of concentrators is an economic necessity. What should be the ratio of solar energy concentration to reduce the price of solar cells by a factor of 90%?

REFERENCES AND BIBLIOGRAPHY

1. Adams, W. G., and R. E. Day, "The Action of Light on Selenium," *Proc. Roy. Soc. (London)*, Vol. A25, p. 113, 1877.

2. Becquerel, E., "On Electric Effects under the Influence of Solar Radiation," *Acad. Sci. Compt. Rend. (Paris)*, Vol. 9, p. 561, 1839 (French).

3. Bube, R. H., "Photoconductivity of the Sulfide, Selenide, and Telluride of Zinc or Cadmium," *Proc. IRE*, Vol. 43, No. 12, p. 1837, 1955.

4. Chapin, D. M., C. S. Fuller, and G. L. Pearson, "A New Silicon p-n Junction Photocell for Converting Solar Radiation into Electrical Power," *J. Appl. Phys.*, Vol. 23, No. 5, 1954.

5. Chapin, D. M., "How to Make Solar Cells," *Electronics*, Vol. 33, No. 10, p. 89, March 1960.

6. Chapin, D. M., "The Direct Conversion of Solar Energy to Electrical Energy," in *Introduction to the Utilization of Solar Energy*, A. M. Zarem and D. D. Erway (eds.), McGraw-Hill, New York, 1963.

7. Cherry, W. R., "Solar Cells and the Applications Engineer," *Astronautics and Aerospace Engineering*, Vol. 1, No. 3, p. 54, 1964.

8. Chynoweth, A. G., and K. G. McKay, "Internal Field Emission in Silicon p-n Junctions," *Phys. Rev.*, Vol. 106, No. 5, p. 418, 1957.

9. Cummerow, R. L., "Photovoltaic Effect in p-n Junctions," *Phys. Rev.*, Vol. 95, No. 1, p. 16, 1954.

10. Cusano, D. A., *CdTe Solar Cells*, Photovoltaic Specialists Conference Transcript, Sec. A-3, DDC No. AD412819, July 1963.

11. Dale, B., and W. P. Smith, "Spectral Response of Solar Cells," *J. Appl. Phys.*, Vol. 32, No. 7, p. 1377, 1961.

12. Daniels, F., *Direct Use of the Sun's Energy*, Yale University Press, Princeton, N.J., 1964.

13. Denney, J. M., et al., *Estimate of Space Radiation Effects on Satellite Solar Cell Power Supplies*, Space Technology Laboratories, Los Angeles, Cal., Report No. EM1021 MR-13, Oct. 20, 1961.

14. Denney, J. M., R. G. Downing, and A. Grenall, "High Energy Proton Radiation Damage," in *Progress in Astronautics and Rocketry*, Vol. 3, Academic Press, New York, 1961, p. 363.

15. Elliott, J. F., "Home Generation of Power by Photovoltaic Conversion of Solar Energy," *Electrical Engineering*, Vol. 79, p. 735, 1960.

16. Elliott, J. F., V. F. Meikliham, and C. L. Kolbe, "Large Areas Solar Cells," in *Energy Conversion for Space Power*, N. W. Snyder (ed.), Academic Press, New York, 1961, p. 263.

17. Forlani, F., and N. Minnaja, "Photovoltaic Effect in Photoconductor-Dielectric Metal Sandwiches," *Physica Status Solidi*, Vol. 8, p. 177, 1965.

18. Gold, L., "A Theory of Photovoltaic Behavior in Semiconductors," *Electronics*, Vol. 19, p. 133, Aug. 1965.

19. Gold, R. D., *Current Status of GaAs Solar Cells*, Photovoltaic Specialists Conference Transcript, Sec. A-6, DDC No. AD412819, July 1963.

20. Haynes, J. R., and J. A. Hornbeck, "Trapping of Minority Carriers I, p-Type," *Phys. Rev.*, Vol. 97, No. 2, p. 311, 1955.

21. Haynes, J. R., and J. A. Hornbeck, "Trapping of Minority Carriers II, n-Type," *Phys. Rev.*, Vol. 100, No. 2, p. 606, 1955.

22. Heaps, J. D., *Thin Film Silicon Solar Cells*, Photovoltaic Specialists Conference Transcript, Sec. A-4, DDC No. AD412819, July 1963.

23. Hess, W. H., "Energetic Particles in the Inner Van Allen Belt," *Space Science Revs.*, Vol. 1, p. 273, 1962.

24. Iles, P. A., and B. Leibenhaut, "Diffusant Impurities-Concentration Profiles in Thin Film Layers on Silicon," *Solid State Electronics*, Vol. 5, p. 331, 1962.

25. Jackson, L. D., *Areas for Improvement of the Semiconductor Solar Energy Converter*, Transcript of the Conference on the Use of Solar Energy, University of Arizona Press, Tucson, 1958, p. 122.

26. Jain, G. C., and F. M. Stuber, "A Distributed Parameters Model for Solar Cells," *Adv. Energy Conversion*, Vol. 7, p. 167, 1967.

27. Jenny, D. A., J. J. Loferski, and P. Rappaport, "Photovoltaic Effect in GaAs p-n Junctions and Solar Energy Conversion," *Phys. Rev.*, Vol. 101, p. 1208, 1956.

28. Jordan, A. G., and A. G. Milnes, "Photoeffect on Diffused p-n Junctions with Internal Field Gradients," *IRE Trans. on Electron Devices*, Vol. ED-7, No. 10, p. 242, 1960.

29. Kallinkin, T., et al., "Preparation of Monocrystalline Layers of Cadmium Selenide," *Fiz. Tverdogo. Tela*, Vol. 3, p. 2640, 1961 (Russian).

30. Kaye, S., *Drift Field Solar Cells*, Photovoltaic Specialists Conference Transcript, Sec. B-14, DDC No. AD412819, July 1963.

31. Keating, P. N., "Photovoltaic Effect in Photoconductors," *J. Appl. Phys.*, Vol. 36, No. 2, p. 564, 1965.

32. Keller, J. W., et al., "A Study of the Effect of Geomagnetically Trapped Radiation of Unprotected Solar Cells," *Proc. IRE*, Vol. 50, No. 11, p. 2320, 1962.

33. Koltun, M. M., and T. M. Golovner, "Antireflection Coatings for Silicon Photocells," *Optics and Spectroscopy*, Vol. 11, No. 5, p. 347, 1966.

34. Laff, R. A., "Photoeffects in Lead Telluride p-n Junctions," *J. Appl. Phys.*, Vol. 36, No. 10, p. 3324, 1965.

35. Lamorte, M. I., "Internal Power Dissipation in Gallium Arsenide Solar Cells," *Adv. Energy Conversion*, Vol. 3, p. 557, 1963.

36. Lange, B., *Photoelements*, Reinhold, New York, 1938.

37. Lahovec, K., "The Photo-Voltaic Effect," *Phys. Rev.*, Vol. 74, No. 4, p. 463, 1948.

38. Loferski, J. J., "Theoretical Considerations Governing the Choice of the Optimum Semiconductor for Photovoltaic Solar Energy Conversion," *J. Appl. Phys.*, Vol. 27, No. 7, p. 777, 1956.

39. Loferski, J. J., and P. Rappaport, "The Effect of Radiation of Silicon Solar Energy Conversion," *RCA Rev.*, Vol. 115, p. 536, 1958.

40. Loferski, J. J., and P. Rappaport, "Electron Bombardment Induced Recombination Centers in Germanium," *J. Appl. Phys.*, Vol. 30, p. 1181, 1959.

41. Loferski, J. J., P. Rappaport, and J. J. Wysocki, *Recent Solar Converter Research*, 13th Power Sources Conference Proceedings, PSC Publications Committee, Red Bank, N.J., 1959.

42. Loferski, J. J., "Possibilities Afforded by Materials Other than Silicon in the Fabrication of Photovoltaic Cells," *Acta Electronica*, Vol. 5, p. 350, 1961.

43. Loferski, J. J., and J. J. Wysocki, "Spectral Response of Photovoltaic Cells," *RCA Rev.*, Vol. 22, p. 38, 1961.

44. Loferski, J. J., "Recent Research on Photovoltaic Solar Energy Converters," *Proc. IEEE*, Vol. 51, No. 5, p. 667, 1963.

45. Luft, W., "Effects of Electron Irradiation on n and p Silicon Solar Cells," *Adv. Energy Conversion*, Vol. 5, No. 1, p. 21, 1965.

46. Mandelkorn, J., et al., *A New Radiation-Resistant High-Efficiency Solar Cell*, USARDL Technical Report 2162, Oct. 1960.

47. Mandelkorn, J., et al., "Silicon n/p Solar Cells," *J. Electrochem. Soc.*, Vol. 109, p. 313, 1962.

48. Mayburgh, S., "Limitations on Lifetime in GaAs," *Solid-State Electronics*, Vol. 2, p. 195, 1961.

49. McClelland, D. H., "Solar Concentrator-Absorber," in *Energy Conversion Systems Reference Handbook*, Vol. 2, Sect. A, AD256701, Sept. 1960.

50. Menetrey, W. R., "Space Applications of Solar Energy," in *Introduction to the Utilization of Solar Energy*, A. M. Zarem and D. D. Erway (eds.), McGraw-Hill, New York, 1963.

51. Middleton, A. E., D. A. Gorski, and F. A. Shirland, "Vaporated CdS Film Photovoltaic Cells for Solar Energy Conversion," in *Progress in Astronautics and Rocketry*, Vol. 3, Academic Press, New York, 1961, p. 275.

52. Minden, H. T., "Intermetallic Semiconductors," *Semiconductor Products*, Vol. 2, No. 2, p. 30, 1959.

53. Moon, P., "Proposed Standard Solar-Radiation Curves for Engineering Use," *J. Franklin Institute*, Vol. 280, No. 11, p. 583, 1940.

54. Moss, H. I., "Large Area Thin-Film Photovoltaic Cells," *RCA Rev.*, Vol. 22, p. 29, 1961.

55. Moss, T. S., *Photoconductivity of the Elements*, Academic Press, New York, 1952, p. 244.

56. Moss, T. S., *Optical Properties of Semiconductors*, Academic Press, New York, 1959.

57. Moss, T. S., "The Potentialities of Silicon and Gallium Arsenide Solar Batteries," *Solid-State Electronics*, Vol. 2, p. 222, 1961.

58. Nathan, M. J., et al., "Stimulated Emission of Radiation from GaAs p-n Junctions," *Appl. Phys. Letters*, Vol. 1, p. 62, 1962.

59. Naumov, C. P., and O. V. Nikolaeva, "CdTe Photocells with Improved Efficiencies," *Fiz. Tverdogo. Tela*, Vol. 3, 1961, p. 3748, translation in *Soviet Physics—Solid State*, Vol. 3, 1962, p. 2718.

60. Novik, F. T., "A High Voltage Photo-EMF in 'Monocrystalline' Films of CdTe," *Fiz. Tverdogo. Tela*, Vol. 4, p. 3334, 1962.

61. Okamoto, H., "Crystallization of SiO_2 Prepared by Pyrolytic Decomposition of Silicon," *Japan. J. Appl. Phys.*, Vol. 7, No. 1, p. 82, 1968.

62. Perlman, S. S., "Heterojunctions Photovoltaic Cells," *Adv. Energy Conversion*, Vol. 4, No. 3, p. 187, 1964.

63. Perrot, M., and M. Touchais, "Rapport sur la Situation Actuelle de la Conversion Energétique Directe de l'Energie Solaire et Aperçus Futurs," *Adv. Energy Conversion*, Vol. 5, No. 4, p. 241, 1965.

64. Pfann, W. G., and W. Van Roosbroeck, "Radioactive and Photo-Electric p-n Junction Power Sources," *J. Appl. Phys.*, Vol. 25, No. 11, p. 1422, 1954.

65. Prince, M. B., "Silicon Solar Energy Converters," *J. Appl. Phys.*, Vol. 26, p. 534, 1955.

66. Prince, M. B., "Applications of Silicon Solar Cells for Space and Terrestrial Use," *Acta Electronica*, Vol. 5, p. 330, 1961.

67. Rappaport, P., J. J. Loferski, and E. G. Lindner, "The Electron-Voltaic Effect in Germanium and Silicon p-n Junctions," *RCA Rev.*, Vol. 17, p. 100, 1956.

68. Rappaport, P., "The Photovoltaic Effect and Its Utilization," *RCA Rev.*, Vol. 20, No. 3, p. 373, 1959.

69. Rappaport, P., and J. J. Wysocki, "The Photovoltaic Effect in GaAs, CdS, and Other Compound Semiconductors," *Acta Electronica*, Vol. 5, p. 364, 1961.

70. Reynolds, D. C., and S. J. Czyak, "Mechanisms for Photovoltaic and Photoconductivity Effects in Activated CdS Crystals," *Phys. Rev.*, Vol. 96, No. 12, p. 1705, 1954.

71. Reynolds, D. C., et al., "Photo-Voltaic Effect in CdS," *Phys. Rev.*, Vol. 96, p. 533, 1954.

72. Ross, B., and J. R. Madigan, "Thermal Generation of Recombination Centers in Silicon," *Phys. Rev.*, Vol. 108, No. 12, p. 1428, 1957.

73. Sah, C. J., R. N. Noyce, and W. Shockley, "Carrier Generation and Recombination in p-n Junctions and p-n Junction Characteristics," *Proc. IRE*, Vol. 45, No. 9, p. 1228, 1957.

74. Sawyer, D. E., and R. H. Rediker, "Narrow Base Germanium Photodiodes," *Proc. IRE*, Vol. 46, No. 6, p. 1122, 1958.

75. Schaeffer, J. C., *Thin Film CdS Front Wall Solar Cells*, Photovoltaic Specialists Conference Transcript, Sec. A-1 and 2, DDC No. AD412 819, July 1963.

76. Schottky, W., "Origin of Photoelectrons in Copper Oxide Photoelectric Cells," *Physik Zs.*, Vol. 31, p. 913, 1930.

77. Scott, W. C., "Space Electrical Power," *Astronautics and Aerospace Engineering*, Vol. 1, No. 5, p. 48, 1963.

78. Shockley, W., *Electrons and Holes in Semiconductors*, Van Nostrand, Princeton, N.J., 1950.

79. Shockley, W., and W. T. Read, "Statistics of Recombinations of Holes and Electrons," *Phys. Rev.*, Vol. 87, No. 9, p. 835, 1952.

80. Shockley, W., "Electrons, Holes, and Traps," *Proc. IRE*, Vol. 46, No. 6, p. 973, 1958.

81. Smits, F. M., K. D. Smits, and W. L. Brown, "Solar Cells for Communications Satellites in the Van Allen Belt," *J. Brit. Inst. of Radio Engineers*, Vol. 22, p. 161, 1961.

82. Smits, F. M., "The Degradation of Solar Cells under Van Allen Radiation," *IEEE Trans. on Nuclear Science*, Vol. NS-10, No. 1, p. 93, 1963.

83. Subashiev, J. K., and M. S. Sominskii, "Photovoltaic Effect and Solar Energy Conversion," in *Poluprovodniki v Nauke in Tekhnike, Moscow*, Vol. 2, p. 214, 1958 (Russian).

84. Tada, H. Y., "Theoretical Analysis of Transient Solar Cell Response and Minority Carrier Lifetime," *J. Appl. Phys.*, Vol. 37, No. 12, p. 4995, 1966.

85. Tallent, R. J., and H. Oman, "Solar Cell Performance with Concentrated Sunlight," *AIEE Transactions*, Vol. 81, Part II, Application and Industry, p. 30, March 1962.

86. Tarneja, K. S., *Dendritic Solar Cells and Array Investigation*, Contract AF33(615)-1049, DDC No. 451543, 1961.

87. Tauc, J., "Generation of an e.m.f. in Semiconductors with Non-equilibrium Current Carrier Concentrations," *J. Appl. Phys.*, Vol. 29, p. 308, 1957.

88. Van Allen, J. A., and L. A. Frank, "Radiation around the Earth to a Radial Distance of 107,400 km," *Nature*, Vol. 183, p. 430, 1959.

89. Vavilov, V. P., et al., "Silicon Solar Batteries as Electrical Power Supplies on Earth Satellites," *Uspekhi Fiz. Nauk*, Vol. 63, p. 123, 1957 (Russian).

90. Veloric, H. S., and M. B. Prince, "High Voltage Conductivity-Modulated Silicon Rectifier," *Bell System Tech. J.*, Vol. 96, No. 7, p. 975, 1957.

91. Vilms, J., and W. E. Spicer, "Quantum Efficiency and Radiative Lifetime in p-Type Gallium Arsenide," *J. Appl. Phys.*, Vol. 36, No. 9, p. 2815, 1965.

92. Webb, G. N., *Variable Energy Gap Devices*, Photovoltaic Specialists Conference Transcript, Sec. A-7, DDC, No. AD412 8199, July 1963.

93. Wernheim, G. K., "Electron Bombardment Damage in Silicon," *Phys. Rev.*, Vol. 110, p. 1272, 1958.

94. Wolf, M., *Design of Silicon Photovoltaic Cells for Special Applications*, AIEE-IRE Semiconductor Devices Conference, Boulder, Colo., July 15–17, 1957.

95. Wolf, M., "Limitations and Possibilities for Improvement of Photovoltaic Solar Energy Converters, Part I Considerations for Earth's Surface Operation," *Proc. IRE*, Vol. 48, p. 1246, 1960.

96. Wolf, M., and M. B. Prince, "Solid State Physics and its Applications in Electronics and Communications, Part B Semiconductors," *Proc. Congress in Brussels, Belgium*, June 2–7, 1958, Vol. 2, Academic Press, New York, 1960, p. 1180.

97. Wolf, M., "Drift Fields in Photovoltaic Solar Energy Converter Cells," *Proc. IEEE*, Vol. 51, No. 5, p. 674, 1963.

98. Wolf, M., and H. Rauschenbach, "Series Resistance Effects on Solar Cell Measurements," *Adv. Energy Conversion*, Vol. 3, No. 2, p. 445, 1963.

99. Wysocki, J. J., and P. Rappaport, "Effect of Temperature on Photovoltaic Solar Energy Conversion," *J. Appl. Phys.*, Vol. 31, p. 571, 1960.

100. Wysocki, J. J., "Photon Spectrum Outside the Earth's Atmosphere," *Solar Energy*, Vol. 6, p. 104, 1962.

9 ELECTROHYDRODYNAMIC POWER GENERATION

9.1 Introduction

Electrohydrodynamic (EHD) power generation, also called electrogasdynamic (EGD) generation, is based on the fact that when positive ions are transported against an electric field by a flow of a hot neutral gas or a belt, they will do electrical work if they are allowed to flow through a load. The difference between MHD and EGD generation is that, in this case, electromechanical coupling is provided by the effect of electric fields rather than magnetic fields. Since the hot "pushing" gas (or the belt) does work by making the positive ions move against the opposing electric field, the thermodynamic energy of the gas or the mechanical energy of the belt is converted directly into electricity.

Electrostatic phenomena were the first to be studied in the early beginnings of electrical science. During the seventeenth and eighteenth centuries, much work was done in the understanding of the forces existing between electric charges. By the end of the eighteenth century, Cavendish [46] was the first to study the capacitance of various capacitor geometries at the University of Cambridge in England. Influenced by his work, Coulomb [11] showed in 1785 that the force between two charges is proportional to the product of the charges and inversely proportional to the square of the distance between them. Later Faraday [20] introduced the idea of lines of force and fields to describe those coulombian forces. Fifty years later Maxwell [52] and Faraday [20] laid the basis for modern electromagnetic theory. Just before the turn of the century, Roentgen [2] discovered x-rays (1895), Becquerel [62] natural radioactivity (1896), and Thomson [83] the concept of the electron (1897). Since then a vast amount of experimental and theoretical work has followed in the field of atomic particle interactions.

The concept of an EGD generator was first announced by Chattock [8] in 1899, using a *corona wind* concept. Van de Graaff's [94] generator was developed in the 1930's. In this generator, the electric charges are mechanically transported by the use of a belt or a disc. Interest in EGD power generation entered a new era at the end of the 1950's following the development of a fossil fueled EGD generator by Gourdine Systems, Inc.,* under a contract with the U.S. Office of Coal Research. Extensive work is presently being carried out by other groups at Patterson Air Force Base [66–69], Curtiss Wright [26, 34] (N.J.), General Mills [71–81] (Minn.), and Marks Polarized Co. [48–51] (N.Y.).

Most of the EGD generators that have been built to date are high voltage sources. This is of great advantage in the use of EGD generators as power sources for klystrons, x-ray tubes, and other high voltage electron devices. If the idea of using EGD principles for power stations succeeds, it would eliminate boilers and steam turbines and thus end the dependence of power plants on water, which may be of great interest in arid zones. Furthermore, EGD generators have the advantage of being able to produce electricity of the voltage level of the transmission lines, thus eliminating the use of costly transformers.

9.2 Ion Mobility

The ionic mean free path λ_i is the average distance traveled by an ion before it suffers a collision with another particle. It can be expressed as the ratio of the average distance traveled in 1 second to the number of collisions per unit time, or $\lambda_i = 1/n_i Q$, where n_i is the density of the ions and Q is the collision cross section.

The motion of an ion under the effect of an electric field can be described by equating the force due to the electric field E to the Newtonian force of acceleration, $M_i \, dv_i/dt = q_i E$. When the ions are in a quiescent gas, they will interact with the neutral atoms of the gas through collisions and will lose part of their energy. Between collisions, the acceleration of the ion will be equal to

$$a_i = \frac{dv_i}{dt} = \frac{q_i E}{M_i}, \tag{9.1}$$

where q_i is the total charge of the ion. After each collision, the ion will lose part of its velocity, and, if a large number of collisions is considered, the average drift velocity can be calculated. It is proportional to the electric field strength and inversely proportional to the density of the target particles. The mobility ℓ_i is by definition the average drift velocity per unit electric field. Thus, $v_i = \ell_i E$, where the subscript i stands for *ion*. To calculate the value of this mobility, assume that the collisions are elastic, the masses of neutral particles and ions are equal, and the particle concentration is uniform.

Consider now that the neutral gas is moving against an electric field. The effect of this field will be to slow the ions and the amount of kinetic energy lost is transformed directly into electricity. Because of their larger mass only ions, not electrons, can yield useful work by collision with the neutral particles. In the following calculation, assume a reference system moving with the neutral gas.

---

*   "Physicist's Goal: to Make EGD a Common Household Word," *Product Engineering*, Aug. 1, 1966.

If, after each collision, the ion starts with zero velocity and a constant acceleration a_i, it will travel a distance d in a time t equal to

$$d = \tfrac{1}{2}a_i t^2,$$ (9.2)

and the average velocity of an ion between two collisions will be $\bar{v}_i = d/t = \tfrac{1}{2}a_i t$. Since the neutral particles are distributed at random in space, the average time $\bar{t}$ will be equal to the ratio of the free path x to the average random velocity $\bar{w}$,

$$\bar{t} = x/\bar{w}.$$ (9.3)

Introducing Eqs. (9.1) and (9.3) into Eq. (9.2) leads to $d = \tfrac{1}{2}q_i x^2/M_i \bar{w}^2$ and, on the average,

$$\bar{d} = \int_0^n d\,\frac{dn}{n},$$ (9.4)

where dn/n is the fraction of particles having free paths of lengths between x and $x + dx$. It is equal to

$$\frac{dn}{n} = \exp\left(\frac{-x}{\lambda}\right)\frac{dx}{\lambda}.$$

Replacing $\bar{d}$ and dn by their values in Eq. (9.4), and integrating from zero to infinity, yields

$$\bar{d} = q_i E \lambda^2/M_i \bar{w}^2.$$ (9.5)

The average drift velocity will be $\bar{v}_i = \bar{d}\bar{w}/\lambda$ or, introducing Eq. (9.5),

$$\bar{v} = q_i E \lambda/M_i \bar{w},$$

and the mobility becomes $\ell_i = q_i\lambda/M_i\bar{w}$. Compton and Langmuir [9] considered the case where the temperatures and the masses of the neutral particles and the ions are different. From their calculation, the mobility of the ions is

$$\ell_i = \frac{0.85 q_i \lambda_{\text{in}}}{M_i \bar{w}}\left(1 + \frac{T_n M_i}{T_i M_n}\right)^{1/2}.$$ (9.6)

This equation shows that for a given neutral gas at a given temperature, the mobility will decrease with an increase of the molecular weight of the ions. The result given by Eq. (9.6) has been checked satisfactorily by such early experimenters as Mitchell [54], Tyndall [93] and Thomson [84].

Up to now, a concentration constant throughout space has been considered, but this may not be true under the effect of an electric field. Writing the balance of forces in equilibrium on a unit volume of an ionic gas under the effect of an electric field in the z-direction, one has

$$-\frac{dp_i}{dz} + n_i q_i E = 0,$$ (9.7)

where p_i is the partial pressure of the ions at a point. The equation of state of an ionic perfect gas is $p_i = n_i k T_i$. Substituting in Eq. (9.7), for the case of a constant temperature, yields

$$-kT_i \frac{dn_i}{dz} + n_i q_i E = 0.$$

It is clear from this relation that the concentration gradient is equivalent to an electric field, the value of which is

$$E' = -\frac{kT_i}{n_i q_i} \frac{dn_i}{dz}. \qquad (9.8)$$

Thus, the total drift velocity of ions in a uniform field is

$$v_i = \ell_i (E + E'), \qquad (9.9)$$

or, replacing E' by its value in Eq. (9.8),

$$v_i = \ell_i \left(E - \frac{kT_i}{n_i q_i} \frac{dn_i}{dz} \right).$$

The velocity v_i is, therefore, the sum of the velocity due to the electric field and the velocity due to the concentration gradient dn_i/dz resulting from a diffusion of the ions. The value of the latter is

$$v_{iD} = -\frac{\ell_i k T_i}{n_i q_i} \frac{dn_i}{dz}.$$

The diffusion coefficient $\mathcal{D}_i$ is defined by the relation

$$v_{iD} = -\mathcal{D}_i \nabla n_i / n_i.$$

Thus $\mathcal{D}_i/\ell_i = kT_i/q_i$. This is called *Einstein's relation*. It is the basis of an experimental method devised by Townsend [86] by which the diffusion, mobility, and temperature of the ions can be obtained. When this is taken into account, the total drift velocity of Eq. (9.9) becomes

$$v_i = \ell_i E - \frac{\mathcal{D}_i}{n_i} \frac{dn_i}{dz}. \qquad (9.10)$$

9.3 EGD Equations

Mathematically, there is a close analogy between the dynamic behavior of a plasma in a magnetic field and that of a unipolar charged gas in an electric field. In MHD power generators, space charge effects were neglected since the plasma was considered to be electrically neutral. This assumption is far from true in EGD power generation. On the contrary, space charge phenomena here are of the greatest importance. The induced magnetic fields can, however, be neglected in EGD generation because of small electric currents involved. Since all the quantities considered will be related to ions, the subscript i will be omitted in the following relations.

In the derivation of the general EGD equations, assume a scalar pressure p_r and constant permittivity ε, permeability μ, and viscosity v. Gravitational and Coriolis forces and all magnetic fields will be neglected in the electric and hydrodynamic equations.

With these considerations taken into account, Maxwell's equations may be written as

$$\nabla \times \mathbf{E} = 0, \tag{9.11}$$

$$\nabla \cdot \mathbf{E} = \rho_e/\varepsilon, \tag{9.12}$$

$$\nabla \cdot \mathbf{J} = -\partial\rho_e/\partial t. \tag{9.13}$$

Also,

$$\mathbf{J} = \sigma(\mathbf{E} + v/\ell). \tag{9.14}$$

The conductivity σ is equal to

$$\sigma = \rho_e \ell, \tag{9.15}$$

and it is generally a space and time variable. The equation of motion in the most general case has been given by Eq. (4.16). If the magnetic field is neglected, this equation becomes

$$\rho \frac{D v}{Dt} = -\nabla p_r + \boldsymbol{\psi} + \rho_e \mathbf{E}, \tag{9.16}$$

where $\boldsymbol{\psi}$ is the pressure gradient due to viscosity and is given in Section 4.3. For a constant viscosity,

$$\boldsymbol{\psi} = v\,\nabla^2 v + \frac{v}{3}\,\nabla(\nabla \cdot v). \tag{9.17}$$

Replacing ρ_e by its value from Eq. (9.12) and introducing Eq. (9.17) in Eq. (9.16), one obtains for the equation of motion

$$\rho\left[\frac{\partial v}{\partial t} + (v \cdot \nabla)v\right] = -\nabla p_r + \varepsilon(\nabla \cdot \mathbf{E})\mathbf{E} + v\,\nabla^2 v + \frac{v}{3}\,\nabla(\nabla \cdot v). \tag{9.18}$$

Eliminating σ, ρ, and $\mathbf{J}$ from Eqs. (9.12), (9.13), (9.14), and (9.15), leads to

$$\partial\mathbf{E}/\partial t + v(\nabla \cdot \mathbf{E}) + \ell(\nabla \cdot \mathbf{E})\mathbf{E} = 0. \tag{9.19}$$

Actually, the left-hand side of Eq. (9.19) is equal to a constant. This constant can be assumed to be zero for a system without strong externally applied fields. In the presence of strong electric fields, the following more accurate relation should be used:

$$\nabla \cdot [\partial\mathbf{E}/\partial t + v(\nabla \cdot \mathbf{E}) + \ell(\nabla \cdot \mathbf{E})\mathbf{E}] = 0.$$

A scalar potential V is defined in terms of the electric field by Poisson's equation

$$\mathbf{E} = -\nabla V. \tag{9.20}$$

As in the case of MHD power generation, the equation of conservation of energy and the equation of continuity should be added to complete the system. If Coriolis forces are neglected, the equation of conservation of energy can be written as

$$\rho \frac{D}{Dt} (U + \tfrac{1}{2}v^2) = \mathbf{E} \cdot \mathbf{J} - \nabla \cdot (\rho v) + \nabla \cdot (\kappa \nabla T)$$

$$+ \tfrac{2}{3}v(\nabla \cdot v)^2 - v \nabla^2 v^2 + 2vv \nabla^2 v. \quad (9.21)$$

The equation of state of the ionic gas and the relations for the transport properties, as stated in Chapter 4, should be taken into consideration whenever necessary.

These are the most general electrogasdynamic equations. In the following sections, the viscosity terms are most often neglected. Other simplifications are used for the different geometries.

9.4 EGD Devices

When energy is supplied to a neutral gas to push ions against an electric field, electrical power can be generated and the device is an EGD generator.

If electrical energy is now supplied to a flow of ions in the presence of an insulating liquid or gas, the ion mobility will establish a strong coupling between electric and hydrodynamic systems, leading to the pumping of the fluid. In this case, the EGD device works as an ion drag pump. This system is often unstable and much work [71, 73] has been done to understand this instability and eventually to damp it to acceptable levels.

A multipoint drag pump [73] is shown in Fig. 9.1 in the schematic form reported by Stuetzer. A hydraulically permeable electrode creates unipolar ions

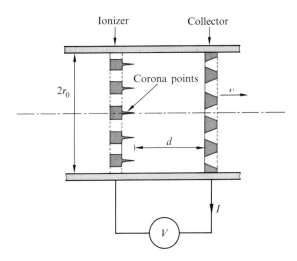

Figure 9.1. Ion drag pump. The ionizer creates unipolar ions which are accelerated by the applied voltage toward the collector. Both ionizer and collector are hydraulically permeable.

of mobility ℓ which are accelerated by an applied voltage V toward a hydraulically permeable collector situated at a distance d. If the fluid is initially at rest, a space-charge-limited ion current will be generated by this arrangement, the value of which has been found by Stuetzer to be

$$ I = \frac{9\pi}{8} \frac{\varepsilon_0 \ell r_0^2}{d^3} (V - V_0)^2, \qquad (9.22) $$

where r_0 is the radius of the pump and V_0 is the voltage at which space-charge-limited conduction starts. If the fluid to be pumped has a certain initial velocity, the value of the current will obviously be larger than that given by Eq. (9.22).

Frictional momentum transfer from the moving ions to the fluid was also calculated by Stuetzer. It was found [71] to create an overall pressure p_r equal to

$$ p_r = \frac{9}{8} \cdot \frac{\varepsilon}{d^2} (V - V_0)^2. \qquad (9.23) $$

Experimental pumps having from 1 to 100 corona points have been built at General Mills [73]. For instance, for an input voltage of about 20 kV and an input current of 6.5 μA, the pressure generated reached 1 psi in a 30-corona-point pump.

The same principle can be used for the cleaning of fluids. Such devices are commonly called EGD precipitators. Separation of suspended particles from a gas requires three important steps: (1) electrical charging of the particles in suspension, (2) collection of the charged particles in an electric field, and (3) their removal from the collectors to an external receptacle. The most widely used charging method utilizes the high voltage dc corona created in the EGD system. The problem created by this method is that often the electrodes generating the electric field are at the same time the collectors of the particulate matter. A typical EGD precipitator is composed of an ionizer and a collector. The ionizer has to supply unipolar ions which will create the necessary space charge. The fluid should be pumped through the ionizer with a velocity high enough to avoid the collection of the particles by the electrodes. A radial field is established by the collector, thus moving the particles to the outside walls. Other arrangements have been proposed by White and Penney [99].

EGD precipitation has important application capabilities in many fields, such as salt precipitation for desalination of sea-water and the cleaning of drinking water. In industrial areas, to avoid air pollution, separation of suspended particles from air becomes a must. Air pollution can be avoided at its source by controlling industrial emissions by adequate precipitation processes, and in many instances expensive material such as copper or gold can be recovered. The same precipitation principle can be applied to automobile exhausts. Air can also be cleaned in apartments and offices along the same lines; this is of greatest importance since, frequently, the pollution of the air can also come from natural sources (dust, humidity, pollen, odors, etc.). Air cleaning devices of this type are currently built for commercial use* and dust monitors are built by Gourdine Systems.† These latter devices serve to measure particulate matter in industrial processing exhausts.

Another important application of the EGD coupling of a corona discharge is the corona wind loudspeaker [65]. This device, invented in the early 1950's by David Tombs [85] in England, is based on the control of the wind produced by the discharge. The corona is created by a triode in which a ring is mounted coaxially about one electrode. The application of suitable external voltages on the electrodes controls the discharges. For instance, by applying an alternating voltage to the ring, a sound source results. A similar device has been proposed by Klein [40] in France and is called the Ionophone. The Ionophone is a point source of sound which functions on the same principle as the corona wind loudspeaker but has a power diode instead of a triode. The so-called electrostatic speakers can also be included in the same family of devices.

A host of other electric components are based on EGD principles. For instance, an EGD pump can be used as an element for mechanical and switching action. Several multipoint pumps, electrically and hydrodynamically in series, can be used to generate a pressure which will activate an outer mechanical switch. The choice of the right design and the appropriate fluids permits a wide variation of such parameters as the activating voltage and the response time of the switch.

The voltage-current characteristic of the pump being nonlinear, it can then be used as a voltage regulator for a given electronic current. When an EGD pump is combined with an EGD generator, a dc transformer can be obtained. The pump arrangement can also be used as an indicator similar to the dust monitor mentioned above.

Many other configurations of technical interest may be listed. Among these are a small motor, the "rotor", which is made of a circular disc supporting ionic emitters; elements energizing small hydraulic amplifiers; and elements for control in electronic circuits.

9.5 Different Types of EGD Power Generators

Today, the two important types in use for the generation of high potential power by electrostatic means are the *Van de Graaff generator* and the so-called *charged aerosol generator*. In the Van de Graaff device, ions are transported mechanically on an insulating belt from the corona discharges to a collector, against an externally applied electric field. In the aerosol EGD generators, they are transported by a gas flow.

Van de Graaff generator. As shown in Fig. 9.2, this device consists of a cylindrical high-voltage terminal and a rapidly moving belt which transports charges from lower spray points at the ground potential. Since the charging of the belt at its

---

*   Honeywell Merchandising Division. Minneapolis, Minn., Advertising leaflets, 1967.

†   Gourdine Systems, Inc., Livingston, N.J., Advertising leaflets, 1966.

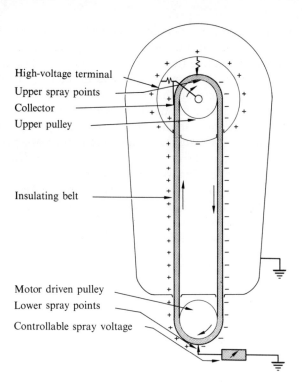

High-voltage terminal

Upper spray points

Collector

Upper pulley

Insulating belt

Motor driven pulley

Lower spray points

Controllable spray voltage

Figure 9.2. Diagram of Van de Graaff electrostatic belt generator. (From Trump [91].)

ground level and the removal of the charges at the high-voltage terminal are performed between field-free regions, the charging function remains independent of the terminal voltage. The maximum value of the latter depends only on its modulating capabilities. If I is the net current to the terminal and C_F is its capacitance to ground, the rate of rise of voltage V is $dV/dt = I/C_F$. This rise is often equal to 10^6 V/sec. If q is the charge on the terminal, the voltage of the generator is, at any instant, $V = q/C_F$. This is the equilibrium voltage at which the generator operates when the ionic current transferred by the belt is exactly equal to the current in the load. This can be controlled over a wide range by controlling the speed of the belt, the spray voltage, and the load current.

The ideal high-voltage terminal would be an isolated conducting sphere. However, for practical reasons, its shape is modified to provide a uniformly controlled electric field. Indeed, the shape of the terminal as well as that of the tank housing the generator greatly influence the maximum value attained by the terminal voltage.

To control the electric field around the terminal, intermediate metallic shields are often used, the potentials of which are provided by suitable connections to the generator column. It was found that the insulated voltage is optimum when the radii of these concentric cylinders follow the relation

$$r_1/r_2 = r_2/r_3 = \cdots = r_{n-1}/r_n,$$

where $r_1, r_2, r_3, \ldots, r_n$ are the radii of the electrode surfaces starting from the high-voltage terminal.

The charge conveyer is an endless flat belt made of good insulating material such as rubber fabric. The speed of its travel varies between 20 and 60 m/sec. A row of corona points extending across the width of the belt sprays ions radially on the belt at the ground level.

The electric power generated in the Van de Graaff device is converted from the mechanical work done by the belt to transport the ions from the ground to the terminal.

Aerosol EGD generators. The EGD generators using a gas flow as a charge conveyor are of several types and can be classified on a "geometric" basis. Three generator types are distinguished here, (1) the broad channel generators, (2) the dielectric, slender channel generators, and (3) the segmented electrode, slender channel generators. If the classification is now based on the way a liquid aerosol is formed and charged, the existence of three types may be noted: (1) In the *electrojet generator*, the flow is forced through a small hollow needle and subjected to an intense electric field as well as a pressure gradient. This type is illustrated in Fig. 9.3. (2) In the *condensation generator*, a thin tungsten wire is subjected to a strong electric field and exposed to a flow of expanding gas which contains a

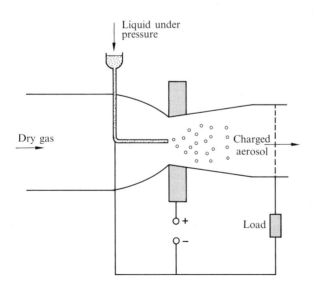

Figure 9.3. Electrojet generator. The flow is forced through a small hollow needle and subjected to an intense electric field and a large pressure gradient. (From Marks [50].)

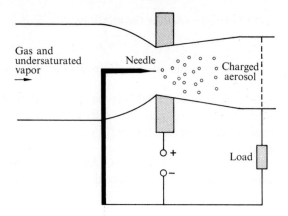

Figure 9.4. Condensation generator. The tungsten needle is exposed to a supercooled vapor flow in the presence of a strong electric field. (From Marks [50].)

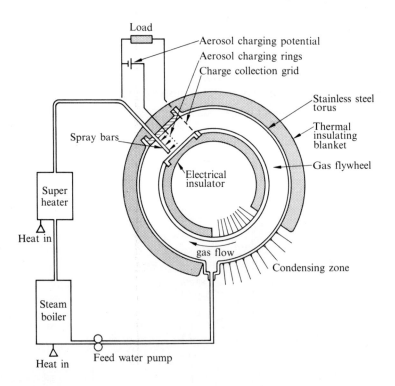

Figure 9.5. Gas flywheel EGD charged aerosol generator. The gas of low molecular weight flows at high pressure in the hollow torus and rotates in the same manner as a flywheel. (From Marks [50].)

supercooled vapor. The gas condenses on the individual ions, thus forming the charged aerosol as shown in Fig. 9.4. (3) The gas *flywheel generator*, illustrated in Fig. 9.5, is composed of a hollow torus in which a gas of low molecular weight and high electric breakdown strength flows at high pressure and rotates in the same manner as a flywheel: the gas is set into motion by the action of several internal jets.

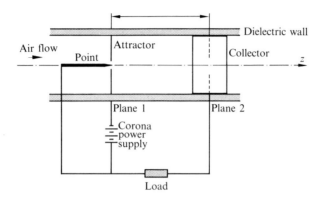

Figure 9.6. Schematic electrogasdynamic generator. The conversion region is situated between planes 1 and 2.

In calculations of the performance characteristics of the different types of generators, the geometric classification will be followed, since the mechanism by which the charged particles are introduced into the system is of no importance in understanding the basic principles of a theoretical analysis.

Broad channel EGD generators. Figure 9.6 shows a schematic view of an ion transport EGD generator. The conversion region consists of an insulating channel between the attractor and the collector, through which a neutral gas flows. The charges are injected by a conducting point at the same plane as the attractor, forming a corona discharge. Assume that all the charged particles are collected at the plane 2, and that the fluid is isotropic, nonviscous, non-heat-conducting, and in its steady state. In this case, the equation of continuity becomes

$$\rho v S = \dot{M} = \text{constant}, \tag{9.24}$$

where S is the cross section of the pipe and $\dot{M}$ is the mass flow per unit time. The equation of motion, given by Eq. (9.18), becomes

$$\rho v \frac{dv}{dz} + \frac{dp_r}{dz} = \rho_e E. \tag{9.25}$$

The equation of conservation of energy is obtained from Eq. (9.21):

$$\rho v \frac{d}{dz}(C_p T + \tfrac{1}{2}v^2) = EJ,$$

where C_p is the heat capacitance at constant pressure and T is the ionic temperature. The only Maxwell equations needed are Gauss' law and the generalized Ohm's law as represented, respectively, by Eqs. (9.12) and (9.14); they are, for this geometry,

$$dE/dz = \rho_e/\varepsilon_0, \tag{9.26}$$

and

$$J = \rho_e(v + \ell E). \tag{9.27}$$

To these equations, add the equation of state for a supposedly perfect ionic gas,

$$p_r = \rho \mathscr{R}T. \tag{9.28}$$

The electric field E is the sum of the space charge field and the field caused by the external load R_L. Assuming that the ionic mobility ℓ is a known constant, one has a set of six equations, Eqs. (9.24) through (9.28), with seven unknowns ρ_e, v, J, ρ, E, p_r, and T. One of these unknown variables must be specified to obtain a solution. For instance, assume that J is known. Then after some algebraic rearrangements,

$$\frac{1}{v}\frac{dv}{dz} = \frac{1}{(\mathscr{M}^2 - 1)}\frac{\rho_e E}{\gamma p_r}\left[1 - (\gamma - 1)\frac{\ell E}{v}\right], \tag{9.29}$$

and

$$\frac{1}{\mathscr{M}}\frac{d\mathscr{M}}{dz} = \frac{1}{(\mathscr{M}^2 - 1)}\frac{\rho_e E}{\gamma p}\left[\tfrac{1}{2}(\gamma + 1) - \tfrac{1}{2}(\gamma - 1)(1 + \gamma\mathscr{M}^2)\frac{\ell E}{v}\right], \tag{9.30}$$

where $\mathscr{M}$ is the Mach number, defined by $\mathscr{M} = (\rho v^2/\gamma p_r)^{1/2}$, and γ is the coefficient of adiabatic compression, defined by $\gamma = C_p/C_v$. For an easier determination of the EGD behavior, all the variables are nondimensionalized with respect to the initial conditions at plane 1, and the length ℓ of the conversion section, as follows:

$$\begin{aligned} z^* &= z/\ell, \\ \rho_e^* &= \rho_e/\rho_{e1}, \\ E^* &= E/E_1, \\ v^* &= v/v_1. \end{aligned} \tag{9.31}$$

Three nondimensional parameters appear; they are the interaction parameter,

$$N_1 = \rho_{e1}E_1\ell/\rho_1 v_1^2, \tag{9.32}$$

the ratio of drift velocity to flow velocity,

$$D_1 = \ell E_1/v_1, \tag{9.33}$$

and the ratio of charge density to the electric field,

$$K_1 = \rho_{e1}\ell/\varepsilon_0 E_1, \tag{9.34}$$

where the subscript 1 refers to the values at the inlet of the generator. Using Eqs. (9.31) through (9.34), one obtains, respectively, for Eqs. (9.29), (9.30), (9.26), and (9.27),

$$\frac{1}{v^*}\frac{dv^*}{dz^*} = \frac{\mathscr{M}^2}{\mathscr{M}^2 - 1} N_1 \frac{\rho_e^* E^*}{v^*}\left[1 - (\gamma - 1)D_1 \frac{E^*}{v^*}\right], \tag{9.35}$$

$$\frac{1}{\mathscr{M}}\frac{d\mathscr{M}}{dz^*} = \frac{\mathscr{M}^2}{\mathscr{M}^2 - 1} N_1 \frac{\rho_e^* E^*}{v^*}\left[\tfrac{1}{2}(\gamma + 1) - \tfrac{1}{2}(\gamma - 1)(1 + \gamma\mathscr{M}^2)D_1 \frac{E^*}{v^*}\right], \tag{9.36}$$

$$\frac{dE^*}{dz^*} = K_1\rho_e^*, \tag{9.37}$$

and

$$\rho_e^* = \frac{1 + D_1}{v^* + D_1 E^*}. \tag{9.38}$$

The four unknown v^*, $\mathscr{M}$, E^*, and ρ_e^* are determined as functions of z^*, N_1, D_1, K_1, and $\mathscr{M}_1$ by Eqs. (9.35) through (9.38). These equations have been studied by Khan [38], who computed the electric field distribution and found that E^* decreases with increasing z^*, $|K_1|$, and $|N_1|$ and with decreasing $|D_1|$.

From Eqs. (9.27), (9.32), (9.33), and (9.34), the current density J is found to be

$$J_1 = \left(\frac{\ell\rho_1 v^2}{\ell}\right)\left(\frac{1 + D_1}{D_1}\right)N_1, \tag{9.39}$$

or

$$J_1 = \left(\frac{\varepsilon_0 v_1^2}{\ell\ell}\right)D_1(1 + D_1)K_1. \tag{9.40}$$

Equation (9.39) gives the maximum current density for the limiting value of N_1. The current-voltage characteristic is obtained from Eqs. (9.35) through (9.38) and (9.40) by using a computer. The result as reported by Khan [38] is shown in Fig. 9.7, where $\tilde{J} = \ell\ell J_1/\varepsilon_0 v_1^2$ is a function of E^* with D_1 as a parameter. The two values of D_1 represent molecular and colloidal ion mobilities, respectively.

The study done by Smith [66] assumes a constant velocity operation instead of a constant current density. Smith found for the current I, as a function of the field induced by the external load, the expression

$$I = -\frac{E_L\ell}{R_L} - \frac{J_1^2\ell\ell^3}{12R_L\varepsilon_0^2(v_1 + \ell E_L)^3}. \tag{9.41}$$

Slender channel EGD generator. In the broad channel generator, multistaging is necessary, since the pressure drop per stage is limited by the breakdown field strength E_b of the dielectric between the electrodes. Indeed, by using Eqs. (9.25),

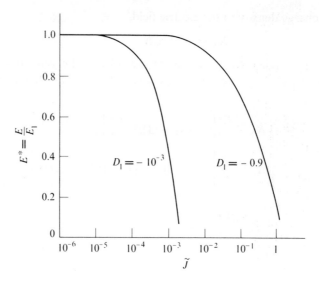

Figure 9.7. Theoretical current-voltage characteristics for two values of the ratio of drift velocity to flow velocity. (After Khan [38].)

(9.24), and (9.26), the value of the pressure drop is found to be

$$\Delta p_r = \tfrac{1}{2}\varepsilon_0(E_1^2 - E_2^2) - \dot{M}(v_1 - v_2)/S,$$

or

$$\Delta p_r \leqslant \tfrac{1}{2}\varepsilon_0 E_{1b}^2, \tag{9.42}$$

where $\dot{M}$ is the mass flow of the fluid.

It can be shown that for practical generators, thousands of stages may be required. If the number of stages can be limited, EGD power generation would be greatly simplified. This can be achieved if the pressure drop per stage is made independent of the breakdown strength between closely spaced electrodes. An obvious solution is the use of a slender geometry, in which the distance between the attractor and collector is large compared to the diameter of the channel.

There are two basic slender generators, the dielectric slender generator and the segmented electrode slender generator. In analyzing the dielectric model, the derivatives of the electric field in the axial direction are noted to be negligibly small compared to those in the radial direction. Thus Eq. (9.12) becomes, for the cylindrical geometry,

$$\frac{1}{r} \cdot \frac{\partial}{\partial r}(rE_r) = \frac{e}{\varepsilon_0} n(z), \tag{9.43}$$

where E_r is the radial component of the electric field and $n(z)$ is the average ion concentration at a given point z in the channel. After integration from 0 to r_0, where r_0 is the radius of the channel, the radial electric field at the wall $E^*(z)$ is found to be

$$E^*(z) = \frac{e}{\varepsilon_0} \cdot \frac{r_0}{2} n(z). \tag{9.44}$$

The decelerating force F'_z per unit volume exerted on the gas by the axial electric field E_z is $F'_z = eE_z n(z)$, and, replacing $n(z)$ by its value from Eq. (9.44),

$$F'_z = 2\varepsilon_0 EE^*/r_0.$$

If E_z and E^*_r are assumed to be constant, the total force F_z on the gas in the cylindrical channel of volume $\pi r_0^2 \ell$ is $F_z = \varepsilon_0 E_z E^*_r \cdot 2\pi r_0 \ell$, and the pressure drop for a single stage becomes $\Delta p_r = F_z/\pi r_0^2$ or $\Delta p_r = 2(\varepsilon_0 E_z E^*_r)\ell/r_0$. Since neither E_z nor E^*_r is limited by the breakdown field of the interelectrode space, the Maxwell shear stress $\varepsilon_0 E_z E^*_r$ can be made much larger than the Maxwell normal stress $\varepsilon_0 E_1^2$ of the broad channel case. Furthermore, the pressure drop is increased by the factor ℓ/r_0; E^*_r is limited only by the high electric field strength of the dielectric wall.

The problem with dielectric slender channels is that ions are deposited on the boundary layer of the channel walls and forced back to the attractor by the axial electric field E_z. This is a cause of loss which becomes completely unacceptable. This ion deposition can, however, be made to control the axial electric field, by the use of electrodes, leading to the concept of a segmented slender generator, as shown in Fig. 9.8. In this generator, the ions of the boundary layer are drained away by several spaced ring electrodes, and they are returned to the attractor through bias resistors chosen such that $R_1 < R_2 < \cdots < R_n$. This leads to a

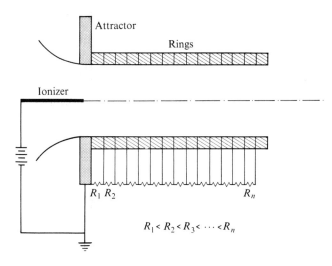

Figure 9.8. Segmented slender generator. The ring electrodes serve to drain the boundary layer ions and return them to the attractor through the resistors.

constant axial field E_z. When the radial velocity of the ions is neglected, Eq. (9.14) gives

$$J_r(r, z) = en\ell E_r(r, z),$$

and

$$J_z = en(v + \ell E_z), \tag{9.45}$$

where the subscripts r and z refer to the radial and the axial components, respectively. In the steady state, Eq. (9.13) yields

$$\frac{1}{r} \cdot \frac{\partial}{\partial r}(rJ_r) + \frac{\partial J_z}{\partial z} = 0. \tag{9.46}$$

Integrating Eq. (9.46) from 0 to r_0, replacing $E_r^*(z)$ by its value from Eq. (9.44), and using Eq. (9.45) leads to

$$\frac{n(z)}{n(0)} = \frac{1}{1 + z/\lambda}, \tag{9.47}$$

where

$$\lambda = \frac{\varepsilon_0(v + \ell E_z)^2}{\ell J_z(0)}. \tag{9.48}$$

The total force F_z on the ions is given by

$$F_z = \int_0^z \pi r_0^2 e E_z n(z)\, dz,$$

and, after integration,

$$F_z = \pi r_0^2 \varepsilon_0 E_z(E_z + v/\ell) \ln(1 + z/\lambda).$$

Therefore, the optimum force is given for an electric field equal to $E_z = -\frac{1}{2}v/\ell$. The resistance distribution R can be found by noting that

$$E_z = -\pi r_0^2 J_z \frac{dR}{dz}. \tag{9.49}$$

Taking account of Eqs. (9.45) and (9.47) leads to the following distribution for $R(z)$:

$$R(z) = -\frac{E_z}{I_z(0)} z\left(1 + \frac{z}{\lambda}\right). \tag{9.50}$$

9.6 EGD Materials

The most important elements in an EGD generator, as shown in Fig. 9.6, are the ionizer, the attractor, the collector, the walls, and the charge conveyer.

The ionizer material depends on whether the generator is of the electrojet type or the condensation type. In the electrojet type, a thin stream of water or other liquid is forced through a small needle into the moving working fluid and is subjected to the intense electric field of the attractor, thus ionizing the stream.

In the condensation type, the ionizer is usually made up of a tungsten wire, having a diameter of about 1 mm, exposed to the field of the attractor. The working fluid should then contain a supercooled vapor which will condense on the individual ions provided by the wire, thus producing a strongly charged corona.

The attractor, also called the *corona ring*, is of the same material as the corona point (tungsten). The walls should be made of a good insulating material. In the case of slender multielectrode channels, brass rings were used with relative success when they were coated with Glyptal* to keep the ions out of the boundary layer [34].

The function of the collector is to remove most of the ions from the working fluid. It should also provide an equipotential plane for the incoming ions. For this reason, grid-type electrodes made up of copper are often used, and some more sophisticated designs have been proposed by Smith [67, 68] and Stuetzer [74].

Except for the case of the Van de Graaff generator, where the charge conveyer is a solid belt, the working material in an EGD generator is liquid, gaseous, both liquid and gaseous, or a mixture of fluid with solid particulate matter such as dust. The belt of the Van de Graaff generator should be a good insulator which can resist mechanical stress. It is often made up of multi-ply rubber fabric.

Apart from vacuum, the insulating medium of greatest interest in electrostatics is provided by compressed gases. Figure 9.9 shows the dc breakdown voltage across several insulators as a function of the electrode separation. This figure shows that for an interelectrode distance $d < 3$ mm, vacuum is superior to all other insulators. This is not true when $d > 3$ mm, for which compressed gases become much more effective insulators.

It was shown by Paschen [56] that the breakdown voltage is proportional to the product of the interelectrode distance and the gas pressure for a given gas in a uniform electric field. Since Natterer's [55] work at the end of the last century, it was proved experimentally that gaseous compounds containing chlorine, fluorine, and other negative ions are better insulators than other gases such as air. These compounds are carbon tetrachloride (CCl_4), sulfur hexafluoride (SF_6), and especially Freon (dichlorodifluoromethane, CCl_2F_2), among many others.

In an EGD generator, it is essential that the ions have low mobility and high concentration. To achieve this, gases of low molecular weight are used as working fluids. It is found experimentally that both the output voltage and output current rise linearly with the gas density at constant temperature. In Marks' [49, 50] experiments, air having 70 % relative humidity of water vapor was used. In another experiment [50], the working fluid was air saturated with ethanol vapor. Because of its low molecular weight, helium has been proposed.

Stuetzer [71, 73, 74] considered working fluids such as kerosene and transformer oil. For kerosene, the ionic mobility was found to be 0.2 mm²/V-sec, leading to an output efficiency of about 10%.

---

* Glyptal is a tradename for some types of resins (alkyd) characterized by their flexibility, gloss, and good weathering properties.

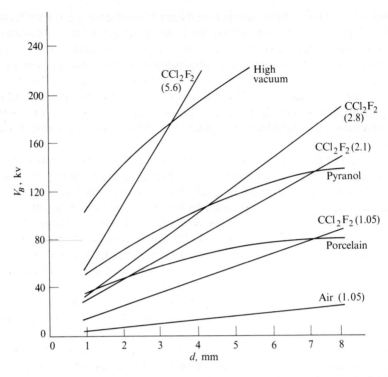

Figure 9.9. DC breakdown across several insulators under a uniform electric field. Figures in parentheses are pressures in kilograms per square centimeter. Freon under pressure is a better insulator than high vacuum for distances larger than 2 mm. (From Trump [91].)

Cox [12] produced charged colloidal particles in a high vacuum by the use of a supersonic metal vapor. The metal vapor condensed to form colloidal particles which have extremely small mobilities and very high kinetic energies.

Gourdine [101] proposed the addition of almost any kind of particulate matter, such as ashes directly from a mine, dust, or droplets, to the insulating gas to increase its ability to push the ions against the opposing electric field. He also proposed the use of the slender geometry [25], leading to relatively low-pressure operation of the working fluid.

To choose the best working fluid possible, much work should be done to understand the relationship between the size and the density of the particles and the mobility and the density of the ions. For any practical EGD generator to succeed, economic considerations should necessarily be taken into account.

9.7 Techniques for Ion Injection

There are three important methods which might be promising for ion injection in EGD generators: contact ionization, microwave ionization, and corona ionization.

Contact ionization is based on the principle that when a gas passes by a conductor of higher work function, ionization is produced by contact between the gas and the metal. This method is rarely considered in EGD, for two important reasons. First, to obtain enough ionization, a large surface to volume ratio for the duct is necessary, resulting in a decrease in efficiency. Second, most of the conductors useful for this application are expensive, at least at the present state of the art.

Microwave ionization utilizes a narrow microwave beam between two electrodes where a strong electric field exists. The electrons are thus trapped in a potential well, and the ions are extracted from the interelectrode region by the working fluid due to their lower mobility. This method is not easy to achieve because of the difficulties in separating the electrons from the ions before they recombine. If the microwave beam is strongly absorbed, the ions will remain near the surface on their way to the collector and there will be no volume charge transport. This will result in a large reduction in efficiency. If the beam is absorbed uniformly throughout the working fluid, then the ionization produced will be small compared to the microwave input energy. Microwave ionization will not be practical until these problems are understood and eventually solved.

By far, *corona ionization* [10, 87] is the most used method of ionization in EGD generators. Here, the neutral fluid is ionized by a discharge around a corona point, which traps the charges having polarity opposite that of the point. Those charges are attracted by a corona ring, thus allowing the removal of the ions by the flow of neutral gas. The corona point is thus the ionizer, and because its current is limited to about 10 μA, multipoint ionizers are often used.

In the electrojet method, the neutral fluid is supplied by a liquid under pressure, which is fed to the hollow needle which serves as the corona point. The electric field between the point and the corona ring induces a charge density on the surface of the liquid, forming highly charged droplets which are transported by the working fluid toward the attractor. Because of the nonuniformity of the droplet size, wetting of the generator walls becomes a serious problem.

To avoid the wetting, research work [49, 50] resulted in the condensation method, in which liquid is injected into the air stream (working fluid) in small amounts. The liquid vaporizes in the air, resulting in an air-vapor mixture. This mixture is expanded in a nozzle, resulting in a temperature drop of the supersonic air-vapor flow. The air approaches its saturation point when it reaches the corona region at the nozzle throat. The ions formed by the corona discharge act as condensation centers for the vapor-saturated air. The droplets caused by this method have the advantage of being too small to cause any wetting.

It is of interest to find a relationship between the mobility, the particle radius, the charge per particle, and the pressure of the gas. The slippage coefficient $\mathscr{C}_\alpha$ is defined as the ratio of the maximum ion slip velocity ℓE_b to the forward velocity v:

$$\mathscr{C}_\alpha = \ell E_b / v. \tag{9.51}$$

This quantity should be made negligibly small. From Paschen's [56] work, note that the maximum field strength E_b can be considered as a linear function of pressure, at least for $p_r < 20$ atm:

$$E_b = E_{b0}p_r/p_a, \tag{9.52}$$

where E_{b0} is the maximum field strength at the atmospheric pressure p_a. Introducing Eq. (9.52) into Eq. (9.51) yields

$$\ell = \mathscr{C}_a v p_a/E_{b0}p_r. \tag{9.53}$$

Now, if $\ell = \ell_0$ at atmospheric pressure, and with the assumption of a sonic flow ($v = v_s$) and a constant temperature, Eq. (9.53) becomes

$$\ell = \ell_0 p_a/p_r. \tag{9.54}$$

The equation of mobility for a droplet of radius r_d and viscosity v can be written* as

$$\ell = \left(\frac{Ne}{6\pi v r_d}\right)\left(1 + 0.87\frac{\lambda}{r_d}\right), \tag{9.55}$$

where N is the number of elementary charges per droplet and λ is the mean free path. As a function of the pressure, the mean free path is

$$\lambda = \lambda_0 p_r/p_a, \tag{9.56}$$

where λ_0 is the mean free path at atmospheric pressure. Combining Eqs. (9.54), (9.55), and (9.56) and eliminating λ and ℓ, one obtains, for the droplet radius,

$$r_d = \frac{pNe}{12\pi v \ell_0 p_a}\left[1 + \left(1 + \frac{20.88\pi v \ell_0 \lambda_0}{Ne}\right)^{1/2}\right].$$

Note from this relation that the radius of the droplets is linearly proportional to the pressure. If

$$\beta_c = \frac{Ne}{12\pi v \ell_0 \lambda_0}, \tag{9.57}$$

then

$$\frac{r_d}{\lambda_0} = \frac{p_r}{p_a}[1 + (1 + 1.74/\beta_c)^{1/2}]\beta_c. \tag{9.58}$$

The ratio $(r_d/\lambda_0)/(p_r/p_a)$ is plotted as a function of β_c in Fig. 9.10. For very small values of β_c, Eq. (9.58) becomes

$$\left(\frac{r_d}{\lambda_0}\right)_s = \frac{p_r}{p_a}(1.74\beta_c)^{1/2},$$

whereas, for large values of β_c,

$$\left(\frac{r_d}{\lambda_0}\right)_l = 2\frac{p_r}{p}\beta_c.$$

---

* See Problem 9.12.

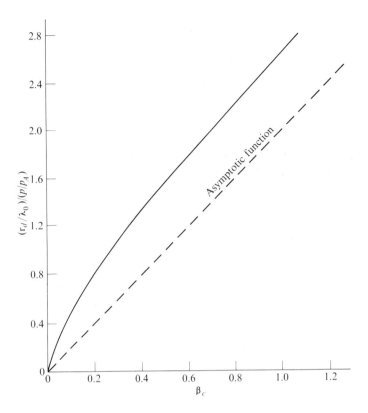

Figure 9.10. The ratio of the droplet radius to the pressure as a function of β_c.

9.8 Power Output, Losses, and Efficiency

Power output. The electrical power output of an EGD generator is given by

$$P = -\int_{\mathscr{V}} \mathbf{J} \cdot \mathbf{E} \, d\mathscr{V}, \qquad (9.59)$$

where $\mathscr{V}$ is the volume of the transport region. Following Smith's [66] work, assuming a constant velocity operation, Eq. (9.59) becomes

$$P = R_l I^2,$$

where R_l is the resistance of the load. Replacing $R_l I$ by its value from Eq. (9.41) yields

$$P = -IE_l \ell \left[1 + \frac{1}{12\varepsilon_0^2} \cdot \frac{J_1^2 \ell \ell^2}{E_L(v_1 + \ell E_l)^3} \right], \qquad (9.60)$$

putting

$$\mathscr{C}'_\alpha = \left(1 + \frac{\ell E_l}{v_1}\right) = 1 + \mathscr{C}_\alpha,$$

$$\mathscr{C}_v = \frac{J_1 \ell \ell}{2\varepsilon_0 \beta_c^2 v_1^2},$$

and

$$P_0 = 2\frac{\varepsilon_0 S v_1^2}{\ell^2},$$

where S is the cross section of the generator and where the subscript 1 refers to values at the inlet of the generator. For constant J, Eq. (9.60) becomes

$$\frac{P}{P_0} = \mathscr{C}_v \mathscr{C}'^2_\alpha [1 - (1 + \tfrac{1}{3}\mathscr{C}_v^2)\mathscr{C}'_\alpha]. \tag{9.61}$$

Since $\mathscr{C}_v$ is positive by definition, the net power output will be positive only if

$$0 < \mathscr{C}'_\alpha < (1 + \tfrac{1}{3}\mathscr{C}_v^2)^{-1}.$$

Furthermore, it can be noted that in order for the ions to be able to enter the transport region, $0 < \mathscr{C}_v < 1$. Equation (9.61) has been plotted in Fig. 9.11. Note

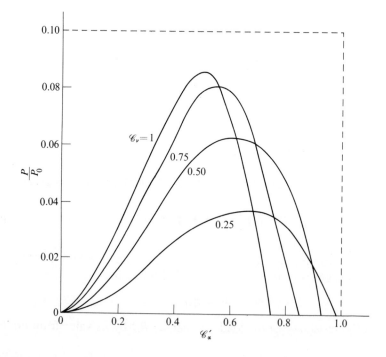

Figure 9.11. Dimensionless power output as function of parameters $\mathscr{C}'_\alpha$ and $\mathscr{C}_v$. (From Smith [66].)

that P/P_0 initially increases with increasing $\mathscr{C}'_\alpha$ and $\mathscr{C}_v$ until a maximum value is reached. This maximum of generated power is attained when

$$\mathscr{C}'_\alpha = \tfrac{2}{3}(1 + \tfrac{1}{3}\mathscr{C}^2_v)^{-1}.$$

For $\mathscr{C}_v = 1$, $\mathscr{C}'_\alpha$ is maximum when $\mathscr{C}'_\alpha = \tfrac{1}{2}$ and $P/P_0|_{max} = \tfrac{1}{12}$.

Losses. An important loss in an EGD generator is the frictional loss of the flowing working fluid. The frictional pressure drop is given by Darcy's equation [97],

$$\Delta p_f = \tfrac{1}{4}\mathscr{C}_f \ell \rho v^2/r_0,$$

where $\mathscr{C}_f$ is the friction coefficient of the gas with the wall, ℓ is the length of the duct, and r_0 is its radius. This pressure drop can be as high as 400 Newtons/m^2 for typical gas flows. The EGD body force pressure drop at a breakdown E^2_{1b} is given by Eq. (9.42). Therefore, the loss by friction is

$$\frac{\Delta p_f}{\Delta p_e} \geq \frac{\mathscr{C}_f \ell \rho v^2}{2\varepsilon_0 E^2_{1b} r_0}.$$

This value can be made as low as 10%. Friction losses are also very important at the electrodes, the structures of which should be studied if these losses are to be avoided. Airfoil arrays have been proposed [50] as the best structures possible.

Ion slip* due to Joulean heating is another source of loss. To avoid it, it was proposed [50] to use converters operating in the low subsonic flow regions. Ion loss to the walls is extremely important, especially in slender geometries. Under the action of the radial electric field, ions drift to the walls, are trapped by the boundary layer, and then drift back upstream.

Other losses, such as those due to the viscosity of the working fluid, exist, but they are almost negligible compared to the losses mentioned above.

Efficiency. The electrical efficiency η_e of EGD generators can be defined as the ratio of the electrical power output to the power required to move the gas against the opposing electric field. The electric power output is given by Eq. (9.59), whereas the power required to move the gas is

$$P_{in} = -\int_\mathcal{V} v_1 \frac{dp_r}{dz} d\mathcal{V}.$$

Calculating η_e in terms of the dimensionless units $\mathscr{C}'_\alpha$ and $\mathscr{C}_v$ leads to

$$\eta_e = \frac{1 - \mathscr{C}'_\alpha(1 + \mathscr{C}^2_v/3)}{(1/\mathscr{C}'_\alpha - 1)}. \qquad (9.62)$$

This efficiency can be shown to be practically equal to the equivalent turbine

---

* Under the action of the applied axial field, the ions tend to slip back upstream when Joulean heating is important. This effect is known as *ion slip*.

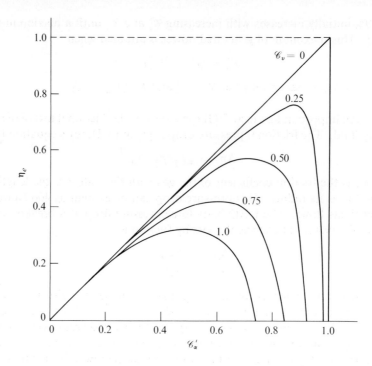

Figure 9.12. Efficiency as a function of parameters $\mathscr{C}'_\alpha$ and $\mathscr{C}_v$. (From Smith [66].)

efficiency of the generator. Equation (9.62) is plotted in Fig. 9.12. Under maximum power conditions, it is equal to 33%.

Another parameter is of great interest in EGD generators. This is the pressure ratio η_p, which is equal to the ratio of the pressure at the entrance of the transport region to the pressure at the exit of the transport region. It is found to be, at the maximum,

$$\eta_p = \frac{1}{1 - \frac{1}{2}(\varepsilon_0 v_1^2/\ell^2 p_1)}.$$

The ideal thermal efficiency for the case of an ideal Brayton cycle operating with η_p is

$$\eta_t = 1 - \eta_p^{(1-\gamma)/\gamma}, \tag{9.63}$$

where γ is the ratio of the heat capacitances.

Experimentally, the most interesting results are those which were reported by Gourdine and Malcolm [28]. These results are shown in Fig. 9.13. In these experiments the power output is found as a function of the load for dry air and for water-saturated air. The working fluid, which is at 215°C and a vapor pressure of 250 psi, expands from a boiler in a supersonic nozzle. It is seen that the mobility of the ions is reduced by about two orders of magnitude by the use of the saturated air.

Other experiments on EGD generators are less conclusive, although they prove convincingly the feasibility of this method of power generation. Such results were reported by Smith and Fried [68], who have shown that the ratio of the electrical power output to the electrical power input of a generator tended toward values larger than unity for high flow rates.

9.9 EGD Generators and Systems

It can be seen from Eq. (9.63) that the ideal thermal efficiency of an EGD generator would be high only if the pressure ratio is large. To raise the efficiency to reasonably high levels, many methods have been proposed.

The use of multistage generators is one of these methods. It was shown [66] that staging of several generators is experimentally feasible. However, to obtain thermal efficiency of 10%, about 100 stages would be required. It is doubtful that such a generator would be interesting or even feasible because of the many problems created by the ion injection requirements.

Another method is the use of high-pressure generators. In this case, an efficiency (ideal) of 10% is predicted [66] for a pressure of 100 atm.

If these two methods are used simultaneously, much higher efficiencies could be obtained. Efficiencies between 60 and 70% have been predicted by Marks [50] for generators using multiloop cycles of the flywheel type (see Fig. 9.5).

Two types of EGD systems can be distinguished: an open-circuit system and a closed-circuit system. The system being developed by the Foster Wheeler Corporation and Gourdine Systems, Inc. [17] is of the open type, as shown in

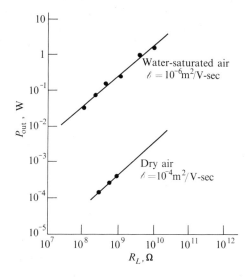

Figure 9.13. Experimental EGD generator power load characteristics: $v_1 = 300$ m/sec, $\ell = 0.03$ m, $r_0 = 0.001$ m, $I_{cor} = 50\ \mu A$. (From Gourdine and Malcolm [28].)

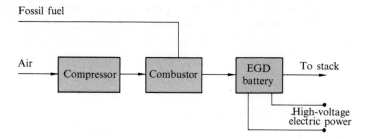

Figure 9.14. Open-circuit EGD system.

Fig. 9.14. In this system, the working fluid is generated by a mixture of air and pulverized coal which is burned at a rate of 68 kg/hr at 30 atm. The resulting hot gases of combustion are then made to expand in a series-parallel system of staged EGD tubes. At the entrance of the tubes, the gas is ionized by a corona discharge. The walls of the tube are made of beryllium oxide, which is a dielectric ceramic material. Yeaple [101] reported an overall efficiency of 5 % for such a system using a single EGD stage. Work is under way to minimize the amount of friction loss as well as to improve the design of the overall configuration of the tubes, the nozzle, and the electrodes.

Nuclear energy can be used as the heat source of an EGD closed-cycle system. Such a system is much more practical than an MHD nuclear system, since high temperatures are not required. Figure 9.15 shows this system, in which the nuclear reactor is cooled by the working gas of the EGD generator at high pressure.

In both these cycles, the problems of corrosion and electrical breakdown are particularly acute, especially in the ionizer-attractor region. Because of this

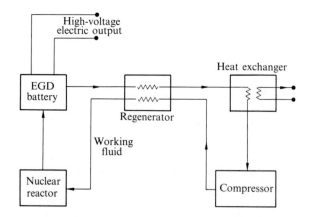

Figure 9.15. Closed-circuit EGD system.

serious problem, thought is being given to developing other means of ionization. Another problem in the fossil fueled open-cycle systems is the contamination of the walls by dust. This problem is under study [35, 50], although its effect is not yet clear.

9.10 Applications and Future Trends

When the different technological problems are dealt with satisfactorily, EGD conversion will show many advantages which are unmatched by the other methods of energy conversion. Because of the fact that power can be extracted over a large range of temperatures (from about 250 to 2500°K), EGD systems can operate as independent power plants. Since there will be no need for heat exchangers, the need for water is minimized, making such systems of great value for arid regions where finding water is a major problem in itself and where, because of this, huge deposits of fossil fuel still wait to be tapped.

Because EGD systems generate high-voltage power at the transmission line voltage level, further savings can be made in the cost of the installation by the elimination of unnecessary transformers. This, of course, will lead to a simpler system and will avoid further losses in the process of transformation from a low to a high potential.

EGD power plants can even generate high alternating voltage directly from a fossil fuel or nuclear source of energy. Indeed, the output voltage form depends on the variations in the ionization rates, and if appropriately modulated ionization sources are used, ac power would be generated.

The Foster Wheeler Corporation's open-cycle experiment will be followed by a large scale project with the General Public Utilities Corporation. In this project [101], a pilot plant will be built at the Whippany, N.J., station using oil as fossil fuel and generating 0.1 MW of power at 140 kV. Work is being done [6] by the same company on EGD channels using coal as fuel. Since most present-day power plants have capacities reaching the 500-MW level, the Whippany project with its 0.1 MW seems very modest. However, if it succeeds, there will be no obstacle to building larger plants. It is hoped by the people supporting these projects that the current cost of about $100/kW capacity of a big coal burning plant could be reduced by about a fifth if work on EGD plants succeeds.

EGD power generation has not yet entered the commercial field, although Gourdine Systems, Inc.,* is currently marketing a 100-kV EGD generator for experimental purposes.

As a source of high voltage, EGD generators could have some other very important applications. As pointed out by Bennett [5], if deuterons or protons are accelerated to about 1000 kV, neutrons could be produced by their collision with suitably chosen target materials. If such a beam of neutrons is injected into a sub-critical reactor, a chain reaction could be initiated, leading to a much higher

---

* Gourdine Systems, Inc., Livingston, N.J., Advertising leaflet, 1967.

efficiency for the nuclear reactor. Such a reactor would be much lighter than the existing ones, thus opening new horizons for nuclear energy, especially in the aircraft propulsion field.

Similar applications of high potentials could be used in fusion research and in radiography. High voltages can also have a host of less important applications, such as portable sources for local irradiation of plastic coatings for polymerization without heat, for electron beam welding, and for power sources for klystrons, x-ray tubes, and other similar devices.

In conclusion, in spite of all the great difficulties and challenges involved, the potential use of EGD power generation for practical large scale power stations seems most promising. The most pressing problems which need to be solved are those related to the corrosion of electrodes, high-voltage insulation, and the economical production of practical EGD batteries. However, these problems cannot be discouraging compared to the prospects of this method of power generation.

PROBLEMS

9.1 Study the effect of a magnetic field on ionic mobility.

9.2 Equation (9.7) is a unidimensional form of the equation of motion. The latter can be written in vectorial form as

$$\nabla p_i + \rho_e \mathbf{E} = 0.$$

(a) Compare this relation to Eq. (9.16). What terms were neglected to arrive to this simplified form? Justify these simplifications. (b) Solve the above equation for a cylindrical geometry with the assumption that there is an angular symmetry.

9.3 What is the temperature of an argon ion having a mobility of 1.2×10^{-4} m²/V·sec and a diffusion velocity of 100 m/sec. Further, assume that the ionic density is given by the relation $n_i = 0.7 \times 10^{14}/z$ particles/m³ and that $\mathscr{D}_i = 10^4 z$ m²/sec, where z is a coordinate.

9.4 From Poisson's law prove Eq. (9.20). Assume that the mobility is constant.

9.5 From the equation of motion derive Eq. (9.21).

9.6 (a) What is the current produced in a Van de Graaff generator having the following characteristics: width of belt $= 0.5$ m, speed of belt $= 60$ m/sec, maximum charge density of belt surface $= 6.0 \times 10^{-5}$ Cb/m², and field intensity $= 30$ MV/m?

9.7 (a) Calculate the output voltage, output power, and the capacitance to the ground of the generator of Problem 9.6, knowing that the distance between the lower and the upper spray points is 2 m. (b) Assuming that the only power lost is that necessary to drive the belt, estimate the efficiency of such a generator.

9.8 Prove Eq. (9.41).

9.9 A slender channel EGD generator has the following dimensions: $\ell = 4$ cm and $r = 0.4$ cm. An electric field $E = 10^7$ V/m is applied at a 45° angle from the axial direction (in average). (a) Find the pressure drop in this slender channel. (b) The gas in the channel is a sonic flow of argon having the following characteristics: $\ell = 1.2 \times 10^{-4}$ m²/V·sec, $v = 480$ m/sec, and $n = 2.2 \times 10^{16}$ ions/m³. What is the total average current density?

9.10 In Problem 9.9, an average value of $E_r(r, z)$ has been taken. (a) Find the variation of the angle existing between E_r and E_z as a function of r and z. (b) What is the value of the current density at the load of the slender channel generator described above? Take into consideration the variation of the radial component of the electric field.

9.11 Find the performance characteristic $V(J)$ of a slender channel generator.

9.12 Equation (9.55) has been obtained from the equation of motion of a droplet by taking into consideration the effect of viscosity. (a) Write the relevant equation of motion. (b) Prove Eq. (9.55).

9.13 The viscous drag D_v of a gas on a particle is given by

$$D_v = \frac{6\pi \upsilon r \nu}{1 + 0.87\lambda/r},$$

where $\lambda = 0.1 \mu$ for air at normal temperature and pressure. Find the radius of a droplet having a viscosity $\upsilon = 1.81 \times 10^{-6}$ poise, a velocity $\nu = 15$ cm/sec, and a mobility

$$\ell = 6 \times 10^{-4} \text{ m/sec.}$$

9.14 Show in an elementary calculation that the maximum power output per unit volume that can be produced by an EGD channel is

$$P_{max}/\mathscr{V} = \tfrac{1}{2}q_1^2\upsilon^2/\ell.$$

9.15 Design an EGD generator using as working fluid water-saturated air and producing 5 W output power. The ion mobility is equal to 10^6 m^2/V·sec, the ion number density is 10^{17} particles/m^3, the corona current is 5×10^{-5} A, and the corona voltage is 6 kV. (a) Calculate the highest value possible of the output voltage, (b) the dimensions of the channel (if necessary use more than one stage), and (c) the efficiency of the generator.

REFERENCES AND BIBLIOGRAPHY

1. Armstrong, W. G., "On the Efficiency of Steam as a Means of Producing Electricity, and on a Curious Action of a Jet of Steam Upon a Ball," *Phil. Mag.*, Series 3, Vol. 22, Jan. 1843.

2. Barker, G. F., *Roentgen Rays*, American Book Co., New York, 1899.

3. Becker, E. W., K. Bier, and W. Henkes, "Strahlen aus Kondensierten Atomen und Molekeln im Hochvakuum," *Zs. für Physik*, Vol. 146, p. 33, 1956.

4. Becker, E. W., R. Klingelhofer, and P. Lohse, "Strahlen aus Zondensiertem Helium im Hochvakuum," *Zs. Naturforsch.*, Vol. 16a, p. 1259, 1961.

5. Bennett, W. E., "The Generation of Direct Current at High Potentials," *Res. Appl. Ind.*, Vol. 12, No. 12, p. 455, 1959.

6. Bishop, J. E., "A Tiny Firm Devises Possibly Cheaper Way to Generate Electricity," *The Wall Street Journal*, March 2, 1966.

7. Charlton, E. E., and F. S. Cooper, "Dielectric Strength of Insulating Fluids," *General Electric Rev.*, Vol. 40, No. 9, p. 438, 1937.

8. Chattock, A. P., "On the Velocity and Mass of the Ions in the Electric Wind in Air," *Phil. Mag.*, Vol. 48, p. 401, 1899.

9. Compton, K. T., and I. Langmuir, "Electrical Discharges in Gases, Part I. Survey of Fundamental Processes," *Rev. Modern Phys.*, Vol. 2, No. 2, p. 123, 1930.

10. Cooperman, P., *The Effects of Gas Temperature, Pressure, and Composition on the Electrical Characteristics of the Corona Discharge*, Research Corps. Technical Report, 1953.

11. Coulomb, C. A., "Seven Papers on the Discovery of the Inverse Square Law of Electrostatics," *Memoires de l'Académie Royale des Sciences*, Paris, 1785–1789.

12. Cox, A. L., *Condensation Colloid Propellants for Electrostatic Propulsion*, 3rd Symposium on Advanced Concepts, Cincinnati, Ohio, Oct. 1962.

13. Cox, A. L., "Colloidal Electrohydrodynamic Energy Converter," *J. AIAA*, Vol. 1, No. 11, p. 2491, 1963.

14. Cox, A. L., and S. Harrison, *The Controlled Growth Colloidal Ion Source*, AIAA Electric Propulsion Conference, Colorado Springs, Col., March 1963.

15. Cox, A. L., *High Voltage Colloidal Energy Converter*, Symposium on Electrostatic Energy Conversion, Interagency Advanced Power Group PIC-ELE 209/1, Philadelphia, April 1963.

16. Craggs, I. D., and I. N. Merk, *High Voltage Laboratory Technique*, Butterworths, London, 1954.

17. Daman, E. L., and M. C. Gourdine, *Electrogasdynamic Power Generation*, Paper SM-74/197, Foster Wheeler Corp., Livingston, N.J., May 1966.

18. Dowdell, R., et al., *The Case for Electrohydrodynamic Energy Conversion*, Marks Polarized Corporation Technical Paper, July 1963.

19. Drozin, V. G., and V. K. LaMer, "The Determination of the Particle Size Distribution of Aerosols by Precipitation of Charged Particles," *J. Colloid Science*, Vol. 14, p. 74, 1959.

20. Faraday, M., *Experimental Researches in Electricity*, Dover, New York, 1965.

21. Flodin, C. R., "Specification and Selection of Dust and Fume Collection Equipment," *Consulting Engineer*, Vol. 16, p. 118, May 1961.

22. Gourdine, M. C., *Electrogasdynamic Channel Flow*, Jet Propulsion Laboratory, California Technical, Technical Release 34-5, Feb. 1960.

23. Gourdine, M. C., *Power Generation by Means of the Electric Wind*, Jet Propulsion Laboratory, TR 32-6, April 1960.

24. Gourdine, M. C., *One-Dimensional Electrogasdynamics*, Plasmadyne Corp. Report PLR-76, Aug. 1960.

25. Gourdine, M. C., *Slender Channel EGD Power Generation*, Symposium on Electrostatic Energy Conversion, Interagency Advanced Power Group, PIC-ELE 209/1, Philadelphia, April 1963.

26. Gourdine, M. C., E. Barreto, and M. P. Khan, *On the Performance of Electrogasdynamic Generators*, 5th Symposium on the Engineering Aspects of MHD, Cambridge, Mass., April 1–2, 1964.

27. Gourdine, M. C., and D. H. Malcolm, *Prospect for Electrogasdynamic Power*, International Conference on Energetics, Rochester, N.Y., 1965.

28. Gourdine, M. C., and D. H. Malcolm, *Feasibility of an EGD High Voltage Power*

Source, 19th Power Sources Conference Proceedings, PSC Publications Committee, Red Bank, N.J., May 1965.

29. Green, H. L., and W. H. Lane, *Particulate Clouds, Dusts, Smokes and Mists*, Van Nostrand, Princeton, N.J., 1957.

30. Harvey, D. J., *An Aerodynamic Study of the Electric Wind*, California Technical, Guffenhein Aero Laboratory Report Misc. No. 2, 1957.

31. Hemeon, W. C., "Gas Cleaning Efficiency Requirements for Different Pollutents," *J. Air Pollution Control Assoc.*, Vol. 12, No. 3, p. 105, 1962.

32. Howell, A. H., "Breakdown Studies in Compressed Gases," *AIEE Trans.*, Vol. 58, p. 193, 1939.

33. Jorgenson, G. V., and E. Will, "Improved Ion Drag Pump," *Rev. Sci. Instruments*, Vol. 33, No. 1, p. 55, 1962.

34. Kahn, B., and M. C. Gourdine, *A Basic Study of Slender Channel Electrogasdynamics*, ARL 63-203 Aerospace Research Laboratories, USAF Office of Aerospace Research, Nov. 1963.

35. Kahn, B., and M. C. Gourdine, "Electrogasdynamic Power Generation," *J. AIAA*, Vol. 2, No. 8, p. 1423, 1964.

36. Kahn, B., *Continuation of the Basic Study of Slender Channel Electrodynamics*, ARL 65-4, Aerospace Research Laboratories, USAF Office of Aerospace Research, Jan. 1965.

37. Kantrowitz, A., and J. Grey, "A High Intensity Source for the Molecular Beam I, Theoretical," *Rev. Sci. Instruments*, Vol. 22, p. 328, 1959.

38. Khan, M. P., *High-Voltage Electrogasdynamic Generators*, ASME paper 64-WA/ENER-11, Sept. 1964.

39. Kistiakowsky, G. B., and W. R. Shichten, "A High Intensity Source for the Molecular Beam Part II, Experimental," *Rev. Sci. Instruments*, Vol. 2, p. 333, 1951.

40. Klein, S., "L'Ioniphone," *L'Onde Electrique (France)*, Vol. 32, p. 314, July 1952.

41. Knoennschild, E., "Problems of Energy Conversion by Ion Convection," *Raumfahrtforschung*, Jan.–March 1965.

42. Krohn, V. E., Jr., *Glycerol Droplets for Electrostatic Propulsion*, ARS Preprint 2398-62, March 1962.

43. Kunkel, W. B., and J. W. Hansen, "A Dust Electricity Analyzer," *Rev. Sci. Instruments*, Vol. 21, No. 4, p. 308, 1950.

44. Lawson, M., E. Von Chain, and H. F. Wattendorf, *Performance Potentialities of Direct Energy Processes Between Electrostatic and Field Dynamic Energy*, ARL-178, Aeronautical Research Laboratories, USAF Office of Aerospace Research, Dec. 1961.

45. Little, A., "Practical Aspects of Electrostatic Precipitative Operation Experiments on a Pilot Plant," *Trans. Inst. Chem. Eng.*, Vol. 34, p. 259, 1956.

46. Mackenzie, A. S., *The Laws of Gravitation*, American Book Co., New York, 1900.

47. Malyshev, I. F., F. G. Zheleznikov, and G. Y. Roshal, *Electrostaticheskiye Generatory*, Moscow, 1959.

48. Marks, A. M., *Heat-Electrical Power Conversion Through the Medium of a Charged Aerosol*, U.S. Patent 2638555, 1953.

49. Marks, A. M., *The Charged Aerosol Generator*, Symposium on Electrostatic Energy Conversion, Interagency Advanced Power Group PIC-ELE 209/1, Philadelphia, April 1963.

50. Marks, A. M., *State of the Art Paper on the Charged Aerosol EHD Generator*, prepared for Executive Office of the President, Office of Science and Technology, Marks Polarized Corp., New York, Aug. 1963.

51. Marks, A. M., E. Barreto, and C. K. Chu, "Charged Aerosol Energy Converter," *J. AIAA*, Vol. 2, No. 1, p. 45, 1964.

52. Maxwell, J. C., *A Treatise on Electricity and Magnetism*, Dover, New York, 1954.

53. Merriman, M. R., *Investigations on Particle-Type Electrostatic Generators*, Symposium on Electrostatic Energy Conversion, Interagency Advanced Power Group PIC-ELE 209/1, Philadelphia, April 1963.

54. Mitchel, J. H., and K. E. Ridler, "The Speed of Positive Ions in Nitrogen," *Proc. Roy. Soc. (London)*, Vol. 146A, p. 911, 1934.

55. Natterer, A., *Wied. Ann. der Phys.*, Vol. 38, p. 63, 1889.

56. Paschen, F., "Uber die Zum Funkenübergang in Luft, Wasserstoff, Und Kohlensaure bei Verschiedenen Drucken Erferderliche Potentialdiferenz," *Wied. Ann. der Phys.*, Vol. 37, p. 69, 1889.

57. Pauthenier, M., and M. Moreau-Hanot, "La Charge des Particules Sphériques Dans un Champ Ionisé," *J. Physique et Radium*, Vol. 3, No. 12, p. 590, 1932.

58. Pauthenier, M., *Lois de Charge des Particules Sphériques Conductrices dans un Champ Electrique Di-Ionisé*, International Colloquium on the Physics of Electrostatic Forces and Their Application, CNRS, Paris, 1961.

59. Penney, G. W., "A New Electrostatic Precipitator," *Electrical Engineering*, Vol. 56, No. 1, p. 159, 1937.

60. Penney, G. W., and R. E. Matick, *A Probe Method for Measuring Potentials in DC Corona*, AIEE Reprint, 1957.

61. Pirovar, L. I., and V. M. Tubayev, "An Investigation of the Electric Strength of Compressed Gases in Weakly Unbalanced Fields by Means of an Electrostatic Generator," *Tekhnicheskiy Otchet FTI AN USSR*, 1956.

62. Rayleigh, R. J., *The Becquerel Rays and the Properties of Radium*, Arnold, London, 1906.

63. Schumb, W. C., "Preparation and Properties of SF_6," *Phys. Rev.*, Vol. 69, Series 2, p. 692, 1946.

64. Servinov, A. H., "A High Frequency Ion Source," *Tekhnicheskiy Otchet FTI AN USSR*, 1956.

65. Shirley, G., "The Corona Wind Loudspeaker," *J. Audio Engineering Soc.*, Vol. 5, No. 1, p. 23, 1957.

66. Smith, J. M., *Theoretical Study of the Electrohydrodynamic Generator*, General Electric TIS Report R61SD192, 1961.

67. Smith, J. M., et al., *Study of Electrical Energy Conversion Systems*, General Electric Aeronautical Systems Division TR 61-379, 1961.

68. Smith, J. M., and W. Fried, *Electrohydrodynamic, Experimental Studies*, General Electric TIS Report R62SD27, March 1962.

69. Smith, J. M., *Performance Criterion for EHD Generators*, Symposium on Electrostatic Energy Conversion, Interagency Advanced Power Group PIC-ELE 209/1, Philadelphia, April 1963.

70. Stairmand, C. J., *The Design and Performance of Modern Gas Cleaning Equipment*, presented to the Institute of Fuel, London, Nov. 1955.

71. Stuetzer, O. M., "Ion Drag Pressure Generation," *J. Appl. Phys.*, Vol. 30, No. 7, p. 984, 1959.

72. Stuetzer, O. M., "Instability of Certain Electrohydrodynamic Systems," *Phys. Fluids*, No. 2, p. 642, 1959.

73. Stuetzer, O. M., "Ion Drag Pumps," *J. Appl. Phys.*, Vol. 31, No. 1, p. 136, 1960.

74. Stuetzer, O. M., "Ion Transport High Voltage Generators," *Rev. Sci. Instruments*, Vol. 32, No. 1, p. 16, 1961.

75. Stuetzer, O. M., "Electrohydrodynamic Components," *IRE Trans. on Component Parts*, Vol. CP-8, p. 57, 1961.

76. Stuetzer, O. M., "Apparent Viscosity of a Charged Fluid," *Phys. Fluids*, Vol. 4, No. 10, p. 1226, 1961.

77. Stuetzer, O. M., "Magnetohydrodynamics and Electrohydrodynamics," *Phys. Fluids*, Vol. 5, No. 5, p. 534, 1962.

78. Stuetzer, O. M., "Electrohydrodynamic Precipitator," *Rev. Sci. Instruments*, Vol. 33, No. 11, p. 1171, 1962.

79. Stuetzer, O. M., *Fundamentals of EHD Power Generation*, Symposium on Electrostatic Energy Conversion, Interagency Advanced Power Group PIC-ELE 209/1, Philadelphia, April 1963.

80. Stuetzer, O. M., "Pressure Analysis of Conduction in Liquids," *Phys. Fluids*, Vol. 6, No. 2, p. 190, 1963.

81. Stuetzer, O. M., "Gas Bubbles in a Charged Liquid," *J. Appl. Phys.*, Vol. 34, No. 4, p. 958, 1963.

82. Thomas, J. B., and E. Wong, "Experimental Study of DC Corona at High Temperatures and Pressures," *J. Appl. Phys.*, Vol. 29, No. 8, p. 1226, 1958.

83. Thomson, J. J., "The Electron in Chemistry," *J. Franklin Institute*, 1923.

84. Thomson, J. S., and J. P. Thomson, *Conduction of Electricity Through Gases*, Cambridge University Press, New York, 1928 and 1933 (2 volumes).

85. Tombs, D. M., "Corona Wind Loudspeaker," *Nature*, Vol. 176, p. 923, 1955.

86. Townsend, J. S., *Electricity in Gases*, Oxford University Press, New York, 1914.

87. Trichel, G. W., "The Mechanism of the Negative Point-to-Plane Corona Onset," *Phys. Rev.*, Vol. 54, p. 1078, 1938.

88. Trump, J. G., *Vacuum Electrostatic Engineering*, Doctorate Thesis, M.I.T., Cambridge, Mass., 1933.

89. Trump, J. G., F. J. Safford, and R. W. Cloud, "DC Breakdown of Air and of Freon in a Uniform Field at High Pressures," *AIEE Trans.*, Vol. 60, p. 132, March 1941.

90. Trump, J. G., and R. J. Van de Graaff, "The Insulation of High Voltage in Vacuum," *J. Appl. Phys.*, Vol. 18, No. 3, p. 327, 1947.

91. Trump, J. G., *Electrostatic Sources of Electric Power*, Conference on Energy Sources, New York, Jan. 29, 1947.

92. Trump, J. G., "Electrostatic Sources of Power," *AIEE Trans.*, Vol. 70, p. 1021, 1951.

93. Tyndall, A. M., and C. F. Powell, "The Mobility of Ions in Pure Gases," *Proc. Roy. Soc. (London)*, Vol. 129A, p. 162, 1930.

94. Van de Graaff, R. J., "A 1,500,000 Volt Electrostatic Generator," *Phys. Rev.*, Vol. 38, p. 1919, 1931.

95. Van de Graaff, R. J., K. T. Compton, and L. C. Van Atta, "Electrostatic Production of High Voltages for Nuclear Investigations," *Phys. Rev.*, Vol. 43, p. 149, 1933.

96. Vollrath, R. E., "A High Voltage Direct Current Generator," *Phys. Rev.*, Vol. 42, No. 10, p. 298, 1932.

97. Welsh, E. M., "Electrogasdynamic Energy Conversion," *IEEE Spectrum*, Vol. 4, No. 12, p. 57, 1967.

98. Whitby, K. T., "Generator for Producing High Concentration of Small Ions," *Rev. Sci. Instruments*, Vol. 32, No. 12, p. 1351, 1961.

99. White, H. J., and G. W. Penney, *Electrical Precipitation Fundamentals*, Penn State University Press, College Park, Pa., 1961.

100. White, H. J., *Industrial Electrostatic Precipitation*, Addison-Wesley, Reading, Mass., 1963.

101. Yeaple, F. D., "Electrogasdynamics, a Bold New Power Source," *Prod. Eng.*, Vol. 37, No. 5, p. 90, 1966.

102. Zapata, R. N., H. H. Parker, and J. H. Bodine, "Performance of a Supersonic Molecular Beam," in *Rarefied Gas Dynamics*, L. Talbot (ed.), Academic Press, New York, 1961, p. 61.

10 PIEZOELECTRIC POWER GENERATION

10.1 Introduction

Electricity can be generated directly from mechanical energy by means of generators based on the piezoelectric effect. The piezoelectric generator is a relative newcomer among direct energy converters, and, like all the other converter types, it can also work in the reverse mode by converting electricity directly into mechanical energy.

The piezoelectric effect was discovered in 1880 by the brothers Pierre and Jacques Curie [24], while they were studying the effect of an electric field on the symmetry of crystals such as quartz (SiO_2), Rochelle salt, and tourmaline. They found that in asymmetrical crystals possessing a polar axis, the effect of a compression parallel to the polar axis was to polarize the crystal; i.e., positive electric charges were generated on one side of the crystal and negative charges on the other side. The reversible nature of this effect was then predicted on a thermodynamic basis by Lippmann [66] and experimentally verified by the Curies. They found that if a stress produces a polarization in a certain direction, an electric field applied in this direction will give rise to a stress directly proportional to the field and changing sign with it. This makes the piezoelectric effect distinctly different from the electrostriction and magnetostriction effects in which the stress is proportional to the square of the electric field and consequently is independent of its sign.

Between the Curies' discovery and World War I, many scientists were interested in this phenomenon, and by their work they laid the theoretical basis of its understanding. The word "piezoelectricity" itself was coined by Hankel [42] from the Greek word "piezo," which means "pressure." Lord Kelvin [51] and Pockels [78] contributed greatly to piezoelectric molecular theory and Duhem [95] laid the basis of piezoelectric principles. It was, however, Voigt [95] who developed the work of all his predecessors into a coherent theory.

Interest in some piezoelectric applications was triggered between the two world wars by Langevin's [61] work on piezoelectric crystals, leading to many piezoelectric resonator and oscillator systems. Several experimental studies were also conducted on piezoelectric materials during that period by Born and Gibbs.

During and after World War II, many commercial and military applications based on the piezoelectric effect, such as sonar (*S*ound *N*avigation and *R*anging) for the detection of submarines, were developed. More recently, in the 1960's, interest has grown in the use of piezoelectricity to power artificial internal organs for man.

Some secondary effects were also noticed in piezoelectric materials. Generally, the dielectric constant of a piezoelectric crystal depends on the direction of the applied field. Thus, when a field is applied, a piezoelectric polarization will be located in the direction of that field, giving an apparently larger value to the dielectric constant.

Some materials are electrically polarized when heated; they are said to be *pyroelectric*. Pyroelectricity is closely related to piezoelectricity and occurs simultaneously with it in most of the relevant materials. It should then be taken into consideration with all the other secondary effects in any quantitative study of piezoelectricity.

10.2 Crystallography

An ideal crystal consists of identical unit cells forming a crystal lattice. The unit cells are parallelepiped in form, identically situated throughout the lattice, and their edges are parallel to the crystallographic axes. The overall crystal is anisotropic, i.e., its physical characteristics vary from one direction to the other. Real crystals are somewhat different from the ideal picture due to the phenomena of dislocations and other imperfections in the crystal structure.

The Miller system. Crystals are classified on a geometrical basis. The crystallographic system can be considered as a collection of points which are the intersection of three systems of parallel planes. These planes are at equal distances from each other, thus forming in the crystal a series of elementary and identical parallelepipedic simple lattices. The intersection points mentioned above lie only at the edges of these simple lattices.

There are evidently many ways of determining simple lattices, depending on the choice of the parallel planes. It is, however, more convenient to choose a large lattice, in which case the intersection points are situated in the center of the faces, the center of the parallelepiped, and on its edges. These large lattices are called *multiple lattices*.

Both simple and multiple lattices are determined by the axial ratio $a:b:c$ and the angles between the three axes of intersection represented by the vectors a, b, and c. If O is taken as the origin of the system as shown in Fig. 10.1, and axes Ox, Oy, and Oz are taken parallel to a, b, and c, respectively, it can be shown that the points of the lattice are on the planes that cut the axes at the coordinates a/h, b/k, c/l, respectively where h, k, and l are integers including zero. In the Miller system, the reference face has the $(1, 1, 1)$ index and the general formula

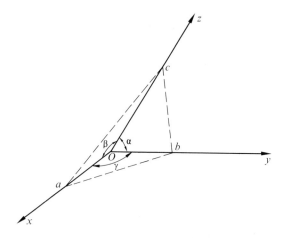

Figure 10.1. Crystallographic system of reference.

is (h, k, l) for any face corresponding to the axes Ox, Oy, and Oz, respectively. Then h, k, and l are called *Miller indices*.

Generally, crystals possess a certain number of inherent symmetrical properties. These properties are studied by considering the following possible elements of symmetry:

1. Center of symmetry,
2. Planes of symmetry,
3. Axis of symmetry.

A crystal as well as any one of its physical properties may be symmetrical with respect to one or a combination of several of these elements of symmetry. If symmetry with respect to a point is considered, it can be shown that there are 32 groups of crystals having different centers. These crystals are divided into 32 classes, each corresponding to a different group.

The 32 crystal classes have been conventionally divided into seven systems. Each crystal belongs to one of the seven crystal systems:

1. The *cubic system*, in which there are three orthogonal two- or fourfold axes a_1, a_2, and a_3 of equal length,
2. The *tetragonal system*, in which there are equal orthogonal axes a_1 and a_2 different from c,
3. The *rhombic system*, in which there are three unequal axes a, b, and c,
4. The *monoclinic system*, in which the axes are unequal in length, the b-axis

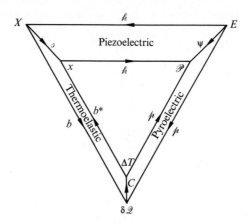

Figure 10.2. Relationships among elastic, dielectric, and thermal phenomena.

being perpendicular to the a- and c- axes which are not orthogonal to each other,

5. The *triclinic system*, in which the three axes a, b, and c are unequal and oblique,
6. The *hexagonal system*, where the c-axis is of sixfold symmetry,
7. The *trigonal system*, which has the appearance of a three-sided pyramid at each crystal edge.

Crystal properties.* The relationships among elastic, dielectric, and thermal phenomena are illustrated in Fig. 10.2. From linear theory, an electric field E will cause a piezoelectric stress X such that

$$X = hE, \tag{10.1}$$

where h is the piezoelectric stress coefficient. Similarly, a polarization $\mathcal{P}$ will be created by a strain x,

$$\mathcal{P} = hx. \tag{10.2}$$

In the same way,

$$\delta \mathcal{Q} = pE, \tag{10.3}$$

and

$$\mathcal{P} = p\,\Delta T \tag{10.4}$$

for the pyroelectric effect, where p is the pyroelectric constant, T is the temperature, and $\mathcal{Q}$ is the heat. Thus

$$\delta \mathcal{Q} = bX$$

and

$$x = b^*\,\Delta T,$$

---

* The units of the several quantities in this section, as in all the others are in the MKSC system. The unit of each quantity is reported in the glossary.

where b and b^* are the coefficients of expansion for the thermoelastic effect. By analogy,

$$x = \delta X, \qquad \mathscr{P} = \chi E, \qquad \Delta T = C\, \delta \mathcal{Q},$$

where χ is the dielectric susceptibility and C is the specific heat.

Each of the straight lines in Fig. 10.2 represents a primary effect, but there are other possible paths which represent the secondary effects. In thermodynamics, one can distinguish C_p at constant pressure and C_v at constant volume. Similarly, in the solid, heat capacitances at constant stress c_X and at constant strain c_x may be distinguished. Therefore, to measure c_X, X and E should be kept constant, and to measure c_x, x and $\mathscr{P}$ should be constant. The compliance coefficient δ depends on the electrical and thermal state of the crystal.

The secondary piezoelectric effect takes place when a piezoelectric contribution to the observed expansion occurs. For instance, a temperature gradient ΔT in a crystal creates a polarization $\mathscr{P}$ which, in turn, induces an electric field E (pyroelectric effect). This induced electric field results in a stress X leading to a strain x (reverse piezoelectric effect). Thus, the reverse piezoelectric effect is, in this example, a secondary effect resulting from the field induced through a primary effect (here pyroelectric).

Thermodynamic theory has been generalized to the solid state by Duhem and Voigt [95]. When the free energy is expressed in terms of strains, it is called the *first thermodynamic potential* ξ. When it is expressed in terms of stresses, it is called the *second thermodynamic potential* ζ.

Shearing strains and stresses are important phenomena in piezoelectricity. A simple shear involves the rotation of the body about an axis perpendicular to the plane of the shear. A pure shear involves no rotation of the body as an entity. It can thus be considered as the superposition of two simple shears. These two cases are represented in Figs. 10.3 and 10.4, respectively. Other phenomena such as torsion and compression are of lesser importance in the study of the piezoelectric effect.

In many applications, the phenomenon of piezoelectric crystal vibration is of great importance. There are three broad classes of modes of vibration in a parallelepiped; (1) compressional and shear modes, (2) flexural modes, and (3) torsional modes. To these one must add the different overtones possible. Also, there are modes coupled between two or more of the other modes. The coupling is often due to frictional forces, inertia, or elastic coupling. The theories of vibrations in solids are rather complicated and often lack generality. Precise descriptions of compressional, shear, flexural, and torsional vibrations in simple crystals can be found in the literature [10, 15, 74].

10.3 Piezoelectric Theory

What makes the 20 piezoelectric classes of crystals different from the other 12 classes is that the piezoelectric crystals possess a structural bias that determines

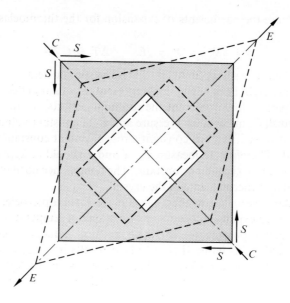

Figure 10.3. Pure shear S is equivalent to combined compression C and extension E.

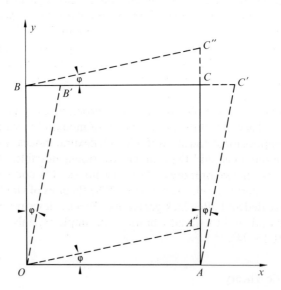

Figure 10.4. A simple shear; the angle ϕ measures the shearing strain. Plane xy is here the plane of shear.

whether a certain surface of the crystal will be negatively or positively charged upon compression, a characteristic that the other crystals do not possess. The 20 piezoelectric classes are, however, very different in their piezoelectricity. In only one class (asymmetric triclinic) will a random stress (compression, extension, shearing) produce a polarization. For all others, only certain types of stresses having some particular relations with the crystal axes of symmetry will polarize the crystal.

Since there are three electric polarization components and six stress components, there will be 18 piezoelectric constants which are independent relations between the electrical and the mechanical states of the crystal. For instance, if an electric field is applied to an asymmetric triclinic crystal, each one of the three components of the field will excite the six independent components (directions and angles) of the stress. Sometimes secondary effects will also be excited due to the fields induced by the stress.

In the piezoelectric theory developed by Voigt [95], it is assumed that the piezoelectric forces are body forces and that all conditions are homogeneous. The problem is complicated by the boundary conditions whenever they exist. The fundamental piezoelectric equations will be written in terms of all the 18 piezoelectric constants. Some of these constants may be zero or may be equal in certain types of crystals. The number of independent constants decreases with increasing symmetry.

In the following discussion, pyroelectric effects will be disregarded and only isothermal processes will be outlined. In the piezoelectric crystals, the generalized Hooke's law can be used, since the linearity of the characteristic variation is convincingly proved experimentally. The first or second thermodynamic potentials can be expanded in products of the components of strain or stress. For small stresses, Hooke's law can be applied and only the quadratic components need to be kept. Considering X and $\mathscr{P}$ as generalized forces and x and E as generalized displacements, one has

$$
\zeta = -\frac{1}{2}\sum_{h=1}^{6}\sum_{i=1}^{6}\sigma_{hi}^{*}x_{h}x_{i} + \frac{1}{2}\sum_{k=1}^{3}\sum_{m=1}^{3}\chi_{km}^{x}E_{k}E_{m} + \sum_{m=1}^{3}\sum_{h=1}^{3}e_{mh}E_{m}x_{h}
$$
$$
+ \frac{C(\Delta T)^{2}}{T} + \Delta T\sum_{h=1}^{6}q_{h}x_{h} + \Delta T\sum_{m=1}^{3}p_{m}E_{m}, \tag{10.5}
$$

and

$$
\zeta = \frac{1}{2}\sum_{h=1}^{6}\sum_{i=1}^{6}\sigma_{hi}X_{h}X_{i} + \frac{1}{2}\sum_{k=1}^{3}\sum_{m=1}^{3}\chi_{km}^{X}E_{k}E_{m} + \sum_{m=1}^{3}\sum_{k=1}^{3}\delta_{mh}E_{m}X_{h}
$$
$$
+ \frac{1}{2}\frac{C(\Delta T)^{2}}{T} - \Delta T\sum_{h=1}^{6}b_{h}X_{h} + \Delta T\sum_{m=1}^{3}p_{m}E_{m}, \tag{10.6}
$$

where the coefficients σ^{*} and σ are the elastic stiffness and the elastic compliance, respectively, χ^{x} and χ^{X} are the dielectric susceptibilities at constant strain x and at constant stress X, respectively, e and δ are the piezoelectric stress and strain coefficients, q is the heat coefficient, b^* is the coefficient of expansion, μ is the pyroelectric constant, and C is the heat capacitance. The six terms in each equation represent the energy expressed in terms of the elastic, dielectric, piezoelectric, thermal, thermoelastic, and pyroelectric properties of the material by considering the six stress components and the three electric polarization components. When the constant temperature is taken into consideration, Eqs. (10.5) and (10.6) become

$$\xi = -\frac{1}{2}\sum_{h=1}^{6}\sum_{i=1}^{6}\delta_{hi}^{*}x_h x_i + \sum_{m=1}^{3}\sum_{h=1}^{3}e_{mh}E_m x_h + \frac{1}{2}\sum_{k=1}^{3}\sum_{m=1}^{3}\chi_{km}^{x}E_k E_m, \qquad (10.7)$$

and

$$\zeta = \frac{1}{2}\sum_{h=1}^{6}\sum_{i=1}^{6}\delta_{hi}X_h X_i + \sum_{m=1}^{3}\sum_{h=1}^{3}\delta_{mh}E_m X_h + \frac{1}{2}\sum_{k=1}^{3}\sum_{m=1}^{3}\chi_{km}^{X}E_k E_m. \qquad (10.8)$$

When a piezoelectric crystal is placed in an electric field and is at the same time subjected to a mechanical stress, there will be a polarization and a strain present in the crystal. For small variations, the change in internal energy U will be given by an exact differential such as

$$dU = \mathscr{P}\,dE - x\,dX. \qquad (10.9)$$

For a reversible process,

$$\left(\frac{\partial \mathscr{P}}{\partial x}\right)_E = -\left(\frac{\partial X}{\partial E}\right)_x = h,$$

$$\left(\frac{\partial \mathscr{P}}{\partial X}\right)_E = \left(\frac{\partial x}{\partial E}\right)_X = -\delta.$$

Taking the derivatives of Eqs. (10.7) and (10.8) with respect to the electric field leads to

$$\partial\xi/\partial E_m = \sum_{k=1}^{3}\chi_{km}^{x}E_k + \sum_{h=1}^{6}e_{mh}x_h = \mathscr{P}_m, \qquad (10.10)$$

and

$$\partial\zeta/\partial E_m = \sum_{k=1}^{3}\chi_{km}^{X}E_k + \sum_{h=1}^{6}\delta_{mh}X_h = \mathscr{P}_m. \qquad (10.11)$$

For the converse effect,

$$\partial\xi/\partial x_h = \sum_{i=1}^{6}\delta_{hi}^{*E}x_i + \sum_{m=1}^{3}e_{mh}E_m = X_h,$$

and

$$\partial\zeta/\partial X_h = \sum_{i=1}^{6}\delta_{hi}^{E}X_i + \sum_{m=1}^{3}\delta_{mh}E_m = x_h.$$

This treatment is valid only for piezoelectric crystals which are not ferro-electric. For those which are ferroelectric, the use of the polarization $\mathscr{P}$ or the displacement D is more valid, as will be seen later in the polarization theory developed by Mason [69].

As an example, take the case where there is no variation in temperature and where the electric field has only one component associated with a single strain and a single stress. Then Eqs. (10.5) and (10.6) become

$$\xi = -\tfrac{1}{2}\jmath^{*E}x^2 + \tfrac{1}{2}\chi^x E^2 + eEx,$$
$$\zeta = \tfrac{1}{2}\jmath^E X^2 + \tfrac{1}{2}\chi^x E^2 + \delta EX,$$

and, after taking the derivative with respect to the electric field,

$$\partial\xi/\partial E = \chi^x E + ex = \mathscr{P}, \tag{10.12}$$
$$\partial\zeta/\partial E = \chi^x E + \delta X = \mathscr{P}. \tag{10.13}$$

In Eq. (10.12), as in Eq. (10.10), note that the polarization is the sum of dielectric and piezoelectric contributions. The elastic constants $\jmath^*$ and $\jmath$ are related by the relations

$$\sum_{h=1}^{6} \jmath^*_{ih}\jmath_{ih} = 1,$$

and

$$\sum_{h=1}^{6} \jmath^*_{ih}\jmath_{kh} = 0 \qquad (k \neq i).$$

The δ's and e's are related to the elastic constants, according to

$$e_{mh} = \sum_{i=1}^{6} \delta_{mi}\jmath^{*E}_{ih},$$

$$e_{kh} = \sum_{m=1}^{3} b^*_{mk}\psi^x_{mk},$$

and

$$\delta_{mh} = \sum_{i=1}^{6} e_{mi}\jmath^E_{ih},$$

$$\delta_{mh} = \sum_{m=1}^{3} b_{mh}\chi^X_{mh},$$

which leads to the following expression for the polarization:

$$\mathscr{P}_m = \sum_{i=1}^{6}\sum_{h=1}^{6} \delta_{mi}\jmath^{*E}_{ih}X_h.$$

It can be proved [15, Sec. 2.7] that this polarization has the same direction whether the strain and polarization are due to a mechanical force or to a field.

Four types of piezoelectric effects can be distinguished as shown by the general matrix of Fig. 10.5, reported by Cady [15]: the longitudinal compressional effect L, the transverse compressional effect T, the longitudinal shear effect L, and the transverse shear effect T.

	Strain	1	2	3	4	5	6
Polarization		x_x	y_y	z_z	y_z	z_x	x_y
1	$\mathscr{P}_x$	L	T	T	L_s	T_s	T_s
2	$\mathscr{P}_y$	T	L	T	T_s	L_s	T_s
3	$\mathscr{P}_z$	T	T	L	T_s	T_s	L_s

Figure 10.5. General piezoelectric matrix. (From Cady [15], Vol. 1, p. 193 in Dover Edition, 1964.) This matrix gives the relations between the $\mathscr{P}$'s and x's, y's, and z's, respectively. For instance, if the transverse shear effect of the upper right-hand corner of the figure is considered, the polarization is then $\mathscr{P}_x = h_{16}x_y$, and L_s at the lower right-hand corner represents $\mathscr{P}_z = h_{36}x_y$, etc.

The quadratic* electrostrictive effect is often present with piezoelectric phenomena, but it is always negligibly small and it will not be considered.

10.4 Polarization Theory

Voigt's [95] theory, in spite of its relative mathematical simplicity and, therefore its usefulness, unfortunately fails to give an accurate description of some piezoelectric crystals such as Rochelle salt.† In the Voigt theory, it is postulated that the piezoelectric strain is linearly proportional to the electric field. Mason [69], inspired by his experimental results, proposed that the stress be proportional to the polarization, thus creating a new theory.

Under the assumption that all the processes are isothermal and reversible, the exact derivatives of the thermodynamic potentials ξ and ζ can be written in terms of the electromechanical quantities:

$$d\xi_{\mathscr{P},X} = \mathscr{P}\, dE - X\, dx, \tag{10.14}$$

$$d\xi_{E,X} = E\, d\mathscr{P} + X\, dx, \tag{10.15}$$

$$d\zeta_{E,x} = E\, d\mathscr{P} - x\, dX, \tag{10.16}$$

$$d\zeta_{\mathscr{P},x} = \mathscr{P}\, dE + x\, dX, \tag{10.17}$$

---

*   Proportional to the square of the electric field.

†   Rochelle salt is a crystalline salt $KNaC_4H_4O_6 \cdot 4H_2O$ called also potassium sodium tartrate and seignette salt.

where the subscripts stand for the corresponding constant quantities, $\xi_{\mathscr{P},x}$ is similar to the internal energy, $\xi_{E,x}$ to the Helmholtz free energy, $\zeta_{E,x}$ to the Gibbs free energy, and $\zeta_{\mathscr{P},x}$ to the enthalpy of a reversible thermodynamic system. The quantities $\mathscr{P}$, E, x, and X correspond to the temperature, entropy, pressure, and volume, respectively. From the exact derivatives of Eqs. (10.14) through (10.17),

$$\hbar = \left(\frac{\partial \mathscr{P}}{\partial x}\right)_E = -\left(\frac{\partial X}{\partial E}\right)_x, \tag{10.18}$$

$$b^* = -\left(\frac{\partial E}{\partial x}\right)_{\mathscr{P}} = -\left(\frac{\partial X}{\partial \mathscr{P}}\right)_x, \tag{10.19}$$

$$b = \left(\frac{\partial E}{\partial X}\right)_{\mathscr{P}} = \left(\frac{\partial x}{\partial \mathscr{P}}\right)_X, \tag{10.20}$$

$$\delta = \left(\frac{\partial \mathscr{P}}{\partial X}\right)_E = \left(\frac{\partial x}{\partial E}\right)_X. \tag{10.21}$$

Equations (10.18) and (10.21) were used in Voigt's theory, and are the basis of the polarization theory in which the potentials are expressed in terms of polarization instead of field. Similar to Eqs. (10.7) and (10.8),

$$\xi = -\frac{1}{2}\sum_{h=1}^{6}\sum_{i=1}^{6}\sigma_{hi}^{*\mathscr{P}}x_h x_i - \frac{1}{2}\sum_{k=1}^{3}\sum_{m=1}^{3}\chi_{km}^{x}\mathscr{P}_k\mathscr{P}_m + \sum_{m=1}^{3}\sum_{h=1}^{6}b_{mh}^{*}\mathscr{P}_m x_h,$$

$$\zeta = \frac{1}{2}\sum_{h=1}^{6}\sum_{i=1}^{6}\sigma_{hi}^{\mathscr{P}}X_h X_i - \frac{1}{2}\sum_{k=1}^{3}\sum_{m=1}^{3}\chi_{km}^{X}\mathscr{P}_k\mathscr{P}_m + \sum_{m=1}^{3}\sum_{h=1}^{6}b_{mh}\mathscr{P}_m X_h.$$

From these can be derived the fundamental piezoelectric equations in the polarization theory,

$$\partial\xi/\partial\mathscr{P}_m = -\sum_{k=1}^{3}\chi_{km}^{x}\mathscr{P}_k + \sum_{h=1}^{6}b_{mh}^{*}x_h = -E_m, \tag{10.22}$$

$$\partial\zeta/\partial\mathscr{P}_m = -\sum_{k=1}^{3}\chi_{km}^{X}\mathscr{P}_k + \sum_{h=1}^{6}b_{mh}X_h = -E_m, \tag{10.23}$$

and

$$\partial\xi/\partial x_h = -\sum_{i=1}^{6}\sigma_{hi}^{*\mathscr{P}}x_i + \sum_{m=1}^{3}b_{mh}^{*}\mathscr{P}_m = -X_h, \tag{10.24}$$

$$\partial\zeta/\partial X_h = \sum_{i=1}^{6}\sigma_{hi}^{\mathscr{P}}X_i + \sum_{m=1}^{3}b_{mh}\mathscr{P}_m = x_h. \tag{10.25}$$

In matrix form, these equations become, respectively,

$$E = \chi^x\mathscr{P} - b^*x,$$
$$E = \chi^X\mathscr{P} - bX,$$
$$X = \sigma^{*\mathscr{P}}x - b_t^*\mathscr{P},$$
$$x = \sigma^{\mathscr{P}}X + b_t\mathscr{P},$$

in which b_t^* and b_t stand for transposed matrices. In Eqs. (10.22) and (10.23), E_m is the electric field required to produce a given polarization when a given strain and respectively a given stress are present. In Eq. (10.24), the externally applied stress X_h is equal to the sum of both the effects of polarization and of strain. In Eq. (10.25), the applied strain x_h is the sum of the effects of stress and polarization.

Other theories have been developed by making different assumptions [12, 51]. Nevertheless, Voigt's theory is the most used, although many modern authors prefer the polarization theory of Mason [69] or its equivalent, the D-theory. More recently, some workers have tried to seek a better understanding of the piezoelectric phenomenon in the light of atomic theory. It is obvious that the best theory is that which is most supported by the experimental evidence. Because of the presence of secondary effects, it is often difficult to determine experimentally the different electromechanical coefficients.

10.5 Piezoelectric Resonators and Oscillators

A *piezoresonator* is a piezoelectric elastic crystal which is excited at resonance under the effect of an applied electric field oscillating at the proper frequency. The resonator is said to be *composite* when both piezoelectric and nonpiezoelectric materials are present. By a proper choice of the electrode locations, the field can be applied in such direction as to excite the desired mode of vibration through the action of the converse piezoelectric effect. When the crystal vibrates under the effect of this field, the periodic deformation will create a periodic charge polarization through the direct piezoelectric effect similar to the counter emf created in a synchronous motor by the driving circuit.

The phase angle between the applied electric field and the resonator-reaction field is determined by the frequency of the applied field, the frequency of the reaction, and the losses which are the only "load" on the resonator.

The crystal can work as a "motor" by converting electrical energy into mechanical energy, as in the case of undersea ultrasonic signaling, or it can work as a generator, as will be seen below. The crystal is called a *transducer* in these two cases.

The deformation of the crystal is maximum at the frequency of resonance. This maximum depends on the losses as well as on the inertia and elastic compliance of the crystal. Electrically, the piezoelectric resonator is often described by an LRC network having the same resonant frequency and losses as the crystal. An LRC network equivalent to a piezoresonator is shown in Fig. 10.6. The resonator vibrates near the harmonic h. The parameters R_h, L_h, C_h, and C_1 are called the *electric constants* of the resonator, and it can be shown [99] that they are the same whether or not there is an air gap accompanying the crystal (Fig. 10.6). Only their values change.

Piezoresonators have extremely sharp resonances and, depending on their geometries, they can vibrate at extremely high frequencies with excellent elastic, mechanical, and thermal qualities.

When an amplifying circuit oscillates only when a piezoresonator is inserted in it, the system is called a *piezooscillator*, the frequency of which is determined by the vibrational modes of the crystal. Sometimes the crystal is used in connection with an oscillator to stabilize its frequency within a narrow range. In this case it is called a *crystal stabilizer*.

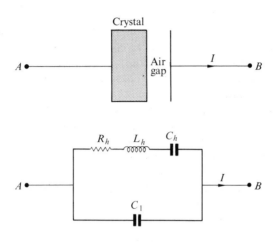

Figure 10.6. Equivalent circuit of a piezoresonator.

One of the most important applications of piezoelectric materials is their use as crystal filters. It was Cady [17] who first proposed the use of piezoelectricity in coupling between two different circuits, with the piezoelectric material thus working as a filter. The crystal will transmit energy only near its resonant frequency. The advantage of such a crystal filter is that its characteristics are almost unaffected by changes in temperature, and its bandwidth is extremely narrow. It can be shown that the bandwidth of the filter is determined by the ratio C_1/C_h; the higher this ratio, the narrower the band. Thickness vibrations are used for very high frequencies (>0.5 MHz), compressional vibrations for frequencies in the 50 to 500-kHz range, and flexural vibrations are used for lower frequencies. Crystals can work as bandpass filters or, if connected properly, as high-pass or low-pass elements [68]. They are used in many commercial applications in radio and in acoustics.

Many meters utilize the piezoelectric effect of quartz, Rochelle salt, or tourmaline to measure pressure in many applications: measurement of the pressure of explosions, of internal combustion engines, of firearms, of cutting tools, and of blood along with other biological applications. Others measure vibrations and stresses in machinery.

Nonresonant piezoelectric transducers are used in many electromechanical and electroacoustic devices. These transducers can operate as motors or as generators. In the latter case, the load impedance is usually very large and the output voltage of the generator has the advantage of being independent of temperature. Such devices are used in record players, tape recorders, microphones, earphones, oscilloscopes, and meters. They are also used in loudspeakers and amplifiers.

Ultrasonics [81, 82] is another important field of application for piezoelectricity. Although the study of ultrasonics began in the last century, it was only in the 1920's that Langevin [61] conceived the idea of using the piezoelectric property of quartz to emit undamped acoustic waves when the crystal is connected to a high-frequency generator. From this basic idea, Langevin was able to produce an ultrasonic beam of high intensity to detect and locate undersea objects by means of echoes. This method is very valuable in surveying the bottoms of the seas. It can also be used by a vessel in motion to avoid reefs and wrecks. Due to extensive work during World War II, piezoelectricity became very helpful in the detection of enemy submarines (by sonar). With a suitable choice of frequency and geometry of the crystal, a cone of sound with any desired amount of divergence can be produced. The distance through which the signals can be sent is limited by the damping of sound waves in water. It can be easily shown [61] that this distance is a function of the frequency. For instance, with a frequency of 20 kHz, the wave will travel 140 miles before its intensity is reduced to about 1 % of its value at the source. For 1 MHz this distance is reduced to about 100 m. With the combination of high frequency and high intensity, ultrasonic waves can cause cavitation* in a liquid. It is utilized in the production of finely dispersed emulsions, in the degasing of liquids, in the dissipation of fogs, in the destruction of bacteria, and in many other biological and metallurgical applications.

Transducers are used in the storage or memory cells in radar and computers. They are employed as ultrasonic interferometers and light relays. Two relatively new applications are in television reception and horology [35].

The modulation of a light beam in a television set makes use of an ultrasonic light modulator conceived first in the mid-1930's by Jeffree [49] (Scophony supersonic television systems).

One of the most important qualities of piezoelectric resonators is their high Q_m. One of the highest figures (55×10^6) was recorded by White [101] with an AT quartz plate† at 4.2°K. This led to the use of quartz in high-precision clocks. For instance, the primary timekeepers in the U.S. and Britain are based on a resonating monitor driven by a quartz oscillator.

Note that the frequency range of piezoelectric transducers is extremely large. It varies from about 1 kHz to about 10,000 MHz at a 4.2°K temperature as reported by Jacobsen [45].

---

* Cavitation is the liberation of bubbles of absorbed gases from a liquid.

† An AT quartz plate is a plate cut from the natural crystal perpendicularly to the x-axis and at 35°15′ from the z-axis (see Fig. 10.11).

10.6 Piezoelectric Generators

Piezoelectric generators would differ by the type of the piezoelectric material used, as will be seen below. They will also be made of different individual cells which can be put in parallel, in series, or both, to yield the desired output voltage and current. In this section, the performance characteristic of an individual piezoelectric generating cell will be studied by the use of the piezoelectric theory and by using the equivalent electrical circuit.

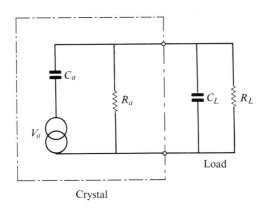

Figure 10.7. Equivalent circuit of a piezoelectric generator. Usually ωC_L is much larger than R_L.

The piezoelectric generator can be visualized as a parallel plate capacitor to which a charge is applied by the effect of an external stress. The equivalent circuit is illustrated in Fig. 10.7 which shows the generator represented by an internal capacitance C_a, an internal resistance R_a, and a voltage source V_a. The load is assumed to consist of a capacitance C_L in parallel with a resistance R_L.

The input mechanical energy in the form of compression is given by

$$\mathscr{E}_{in} = \tfrac{1}{2}F\,\Delta\ell, \tag{10.26}$$

where $\Delta\ell$ is the average distance of motion of the crystal. When Young's modulus Y (in Newtons/m^2) is introduced, Eq. (10.26) becomes

$$\mathscr{E}_{in} = \tfrac{1}{2}F^2\ell/YS, \tag{10.27}$$

where ℓ is the thickness of the crystal (assumed to be a parallelepiped) and S is the cross section of the crystal perpendicular to the direction of the applied force. The energy output is equal to the energy stored in the capacitance of the piezoelectric crystal:

$$\mathscr{E}_{out} = \tfrac{1}{2}q^2/C_a, \tag{10.28}$$

where q is the electric charge. As a function of the permittivity $\varepsilon_0\varepsilon_r$ of the material, the capacitance C_a becomes

$$C_a = \varepsilon_0 \varepsilon_r S/\ell. \tag{10.29}$$

If the effect of resistances R_L and R_a is neglected by assuming that they are very large, the electrical energy would be equal to the input mechanical energy. Thus, equating Eq. (10.29) to Eq. (10.28) and taking into consideration the fact that

$$q = C_a V_a, \tag{10.30}$$

one obtains for the open-circuit voltage V_a;

$$V_a = \frac{F\ell}{S} (Y\varepsilon_0 \varepsilon_r)^{-1/2}.$$

When δ_{33}, the piezoelectric electromechanical constant for the electric effect in the longitudinal direction only ($\delta_{33} = F/q$), and g_{33}, the piezoelectric generator coefficient ($g_{33} = \delta_{33}/\varepsilon_0 \varepsilon_r$) are introduced, then $V_a = g_{33} F\ell/S$, and the output voltage V_L on the capacitive load will be

$$V_L = \frac{C_a V_a}{C_L + C_a}$$

or

$$V_L = \frac{\delta_{33} F}{C_a + C_L}. \tag{10.31}$$

This is a linear function of the mechanical load. When resistances R_a and R_L are taken into consideration, Eq. (10.31) becomes

$$V_L = \frac{\delta_{33} F}{C_a + C_L + (1/j\omega)(1/R_a + 1/R_L)}. \tag{10.32}$$

The magnitude of this is

$$V_L = \frac{\delta_{33} F/C_a}{[(1 + C_L/C_a)^2 + (R_a + R_L)^2/\omega^2 C_a^2 R_a^2 R_L^2]^{1/2}}.$$

The experimental $V_L(F)$ characteristic [63] is shown in Fig. 10.8 for a lead-zirconate-titanate-5 (PZT-5) piezoelectric material having a capacitance equal to 190 pF. This characteristic is linear only for large values of F, most probably because of the variation of δ_{33} with F.

Consider now the case of Rochelle salt, where only the x-direction is considered, where ξ and ζ are in terms of strains and stresses confined to shears in the yz-plane, and Y_z is the externally applied stress whereas y_z is the strain. When the polarization theory is applied, Eqs. (10.3) and (10.4) become

$$\xi = -\tfrac{1}{2}\delta_{44}^* y_z^2 - \tfrac{1}{2}\chi^x(\mathscr{P}^{yEO})^2 - \tfrac{1}{4}\mathscr{B}(\mathscr{P}^{yEO})^4 + b\mathscr{P}^{yEO} y_z, \tag{10.33}$$

$$\zeta = \tfrac{1}{2}\delta_{44} Y_z^2 - \tfrac{1}{2}\chi^X(\mathscr{P}^{YEO})^2 - \tfrac{1}{4}\mathscr{B}(\mathscr{P}^{YEO})^4 + b\mathscr{P}^{YEO} Y_z, \tag{10.34}$$

where $\mathscr{P}^{EO} = \mathscr{P}^O + \mathscr{P}^E$, $\mathscr{P}^O$ being the spontaneous polarization, and $\mathscr{P}^E$ is the part of the field due to E. The superscripts y and Y denote that the polarization

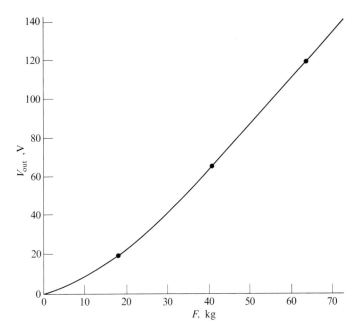

Figure 10.8. Experimental output voltage as a function of the input force for PZT (4100-5) and $C_F = 190$ pF. (Adapted from Lehmann and Stirnkob [64].)

is due to strain or stress, respectively; $\mathscr{B}$ is an experimentally determined factor. It can be shown [15, Chap. 23] that the application of a stress Y_z is similar to the application of a biasing field $b_{14}Y_z$. The electric field is obtained by taking the derivative of Eq. (10.34) with respect to the polarization $\mathscr{P}$,

$$-\partial\zeta/\partial\mathscr{P} = E = -b_{14}Y_z + \chi^X\mathscr{P}^{YEO} + \mathscr{B}(\mathscr{P}^{YEO})^3,$$

and, since there is no applied electric field, this equation becomes

$$Y_z = \frac{\chi^X}{b_{14}}\mathscr{P}^{YO} + \frac{\mathscr{B}}{b_{14}}(\mathscr{P}^{YO})^3.$$

This equation is plotted in Fig. 10.9 for Rochelle salt at two different temperatures (15 and 31.5°C), b_{14} being equal to 0.192 m²/C. It can be shown that this figure is similar to the experimental result shown in Fig. 10.8. The polarization is found to increase with increasing stress and with decreasing temperature. Similar curves can also be drawn for negative values of the polarization. Finally, note that the value of b_{14}, as well as those of χ^X and $\mathscr{B}$, is experimental.

10.7 Piezoelectric Materials

The piezoelectric quality of a material can be measured in terms of the coefficient δ. Indeed, it can be seen that the components of the piezoelectric polarization in a

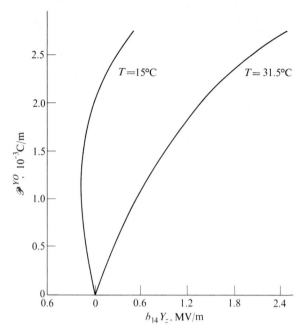

Figure 10.9. Theoretical polarization curves for Rochelle salt, in the case of a nonclamped crystal (free crystal) at two different temperatures. (Adapted from Cady [15, Vol. 2, Fig. 138].)

crystal under uniform hydrostatic pressure Π_p are

$$-\mathscr{P}_1 = (\delta_{11} + \delta_{12} + \delta_{13})\Pi_p,$$
$$-\mathscr{P}_2 = (\delta_{21} + \delta_{22} + \delta_{23})\Pi_p,$$
$$-\mathscr{P}_3 = (\delta_{31} + \delta_{32} + \delta_{33})\Pi_p.$$

The usefulness of a material for power generation is often determined by the electromechanical coupling coefficient k, defined by

$$k^2 = \frac{\xi_{in}}{\xi_{out}} = \frac{\text{energy input}}{\text{energy output}}.$$

Most of the piezoelectric materials belong to one of the 20 piezoelectric classes of single crystals mentioned above. Others, manufactured more recently, are ceramics. Today, more than 1000 materials are recognized to be piezoelectric. Only about 100 of them have been investigated experimentally. The 20 piezoelectric classes are all devoid of a center of symmetry and are either *hemihedral* or *tetratohedral*.* Those which might be useful for power generation should be

---

* From Greek hemi = half; tetrato = fourth; hedra = surface. A hemihedral crystal has half the faces required by complete symmetry. A tetratohedral crystal has one-quarter of the faces required by complete symmetry.

useful in the L and T modes (see Section 10.3). These classes are in the triclinic system (class 1), the monoclinic system (3, 4), the rhombic system (7), the tetragonal system (9, 10, 14), the trigonal system (16, 18, 19), the hexagonal system (21, 22, 23, 24, 26), and the cubic system (28, 31).*

Table 10.1

Properties of some piezoelectric materials

Material*	k_{33}	$\varepsilon_{r_{33}}$	δ_{33}, 10^{-12} C/N	g_{33}, 10^{-3} Vm/N	Y_{33}, 10^{10} N/m²
Quartz [88]	0.10	4	2	58	7
Rochelle salt [88]	0.90	350	150	90	2
ADP [88]	0.38	15	24	180	2
Lead-titanate-zirconate [63]	0.68	800	220	26	8.0
PZT-4 [63]	0.70	1,300	285	24.9	6.6
PZT-5 [63]	0.71	1,500	374	24.8	5.3
Ceramic B [63]	0.48	1,200	149	14.0	11.1
BaTiO$_3$ [23]	0.50	1,700	190	12.0	11.2
PbNb$_2$O$_6$ [88]	0.42	280	75	40.0	3.3

* Often the figures are calculated from values given in the cited literature and adapted to the MKSC system.

In spite of the importance of single crystals, ceramics are becoming more and more used in piezoelectric conversion. This is because the longitudinal and transverse piezoelectric effects are the most commonly employed. The latter involve the constants

$$\delta_1 = (\delta_{11} + \delta_{12} + \delta_{13}),$$
$$\delta_2 = (\delta_{21} + \delta_{22} + \delta_{23}),$$
$$\delta_3 = (\delta_{31} + \delta_{32} + \delta_{33}),$$

which happen to be very large in ceramics. Some properties of piezoelectric materials are shown in Table 10.1.

Very few materials in the triclinic system, which is the most complicated with its 18 independent piezoelectric constants, have been studied.

Several materials of class 3 of the monoclinic system have been investigated. Tartaric acid [HOOC(CHOH)$_2$COOH] has strong piezoelectric properties with $\delta_3 = 4.78 \times 10^{-12}$ C/N [26] but, because of its easy cleavage, it has found few

---

* There are no L and T modes in the three remaining classes.

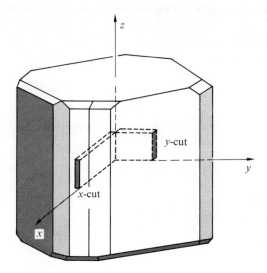

Figure 10.10. Artificial Rochelle salt crystal [18]. The two most used cuts are shown in the figure.

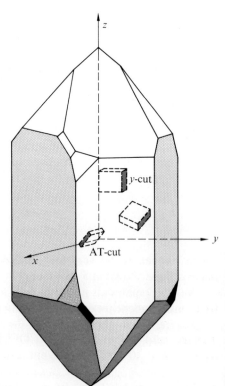

Figure 10.11. Natural quartz crystal [18]. The three most common cuts are shown by small slabs in the figure.

applications. Cane sugar is not very useful because of its rather weak piezoelectric properties ($\delta_3 = -1.13 \times 10^{-12}$ C/N) [44]. Lithium sulfate monohydrate ($Li_2SO_4 \cdot H_2O$) has received considerable attention because of its low dielectric constant, its high electromechanical coupling factor, and its tightly bound crystals. It was studied by Mason [69] and Bechmann [4], who found that $\delta_2 = 14.4 \times 10^{-12}$ C/N. Rochelle salt [28, 58, 86] is of class 3 only in the temperature range from -18 to $+24°C$, the limits being determined by the Curie points, which are the temperatures at which the ferromagnetism of the crystal disappears (see Chapter 11). Outside these limits, the crystal is rhombic of class 6 and is of little use in power conversion.

Rochelle salt ($KNaC_4H_4O_6 \cdot 4H_2O$), a crystal of which is represented in Fig. 10.10, is also pyroelectric and has a large k, as can be seen from Table 10.1. In spite of this, it is only rarely used in power generation applications. This is because of its limited useful temperature range, its low permissible stress level, and its solubility in water. The heavy water variety is characterized by temperature invariance in the useful temperature range. Other monoclinic piezoelectric crystals are beet and milk sugar, experimented with in some resonators; brushite (calcium diorthophosphate, $CaHPO_4 \cdot 2H_2O$); and potassium and sodium tartrates.

The only tetragonal crystal of piezoelectric importance is barium antimonyl tartrate $[Ba(SbO)_2(C_4H_4O_6)_2 \cdot H_2O]$ of class 10. It has a field to strain ratio equal to $\delta_{33} = 3.67 \times 10^{-12}$ C/N [92].

The most important crystal of the trigonal system is quartz [85], shown in Fig. 10.11. Because of its low value of k, quartz is not especially useful as a generator material, but, because of its excellent immunity to aging and changes in temperature, it is used for resonant mechanoelectrical frequency standards. Other crystals of class 18 are benzil* and rubidium tartrate. Tourmaline of class 19 has historic importance but it is not very practical because of its low δ_3 ($\delta_3 = 2.43 \times 10^{-12}$ C/N).

Other crystals of some piezoelectric importance are the ammonium dihydrogen phosphate (ADP), shown in Fig. 10.12, lithium trisodium molybdate, and ethylene diamine tartrate (EDT).

The aim of work started in Japan in 1952 [87] and continued in this country† was done to obtain piezoelectric materials having better temperature and time (aging) stability and better values for the electromechanical constants, the permittivity ε_r, and the mechanical resonance Q_m. This work was directly related to the study of ferroelectric crystals such as certain tartrates (especially Rochelle salt) and the alkali metal dihydrogen phosphates. Two important groups of these ferroelectrics have been discovered more recently; the perovskites and materials similar to the guanadine aluminum sulfate hexahydrate (GASH).

---

*   Yellow crystal $C_6H_5COCOC_6H_5$ made by oxidizing benzoin, not to be confused with benzyl, which is the radical $C_6H_5CH_2$.

†   See Refs. 22 through 25, 45, and 94 of Chapter 11.

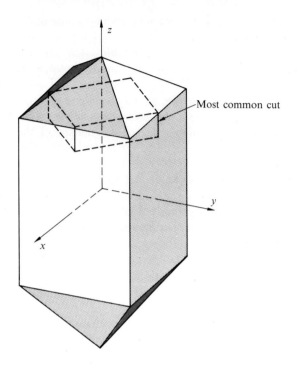

Figure 10.12. Artificial crystal of ammonium dihydrogen phosphate (ADP) [18].

The materials of the perovskite family are titanates of barium [84], calcium, and lead, as well as lead zirconate and lead titanate zirconate (PZT) [56, 57]. Useful elements with the most favorable properties for different applications have been patented in the U.S. by Clevite as the PZT series and in the United Kingdom by Brush Clevite as the LZ and LZ_a series.

$BaTiO_3$ (ceramic B) has been grown in the form of single crystals since 1947 [21]. Its single domain is piezoelectric and its high spontaneous polarization ($\mathscr{P}_0 = 0.0028$ C/m) makes it useful in memory applications. Barium titanate is trigonal rhombohedral for temperatures below $-80°$C, orthorhombic for temperatures from -80 to $+5°$C, tetragonal for temperatures from 5 to 120°C (120°C is the Curie point), and cubic for higher temperatures. At each transformation, a spontaneous strain takes place which can be used for piezoelectric power conversion. This crystal has been studied most thoroughly in the temperature range from 5 to 120°C.

Polycrystalline aggregates of the perovskite type have good piezoelectric properties when prepared by ceramic methods. They have the same symmetry as the hexagonal system (class 26) and can have a spontaneous polarization of the order of 3 mC/m. Ceramics have the advantages of a more stable polarization than single crystals and can be molded in any desirable form.

The dielectric, elastic, and piezoelectric properties of barium titanate ceramics change greatly with additions to improve some desired characteristics. The same statement can be made about lead zirconate titanate ceramics, in which only trace quantities of additives are needed. These additives are strontium salts and rare earth oxides which do not enter into the solid solution. As an example, the effect of the addition of $PbZrO_3$ on the coupling coefficient and permittivity of lead zirconate titanate is shown in Fig. 10.13. Both k_{33} and ε_r have a maximum value corresponding to a certain molecular percentage of the additive.

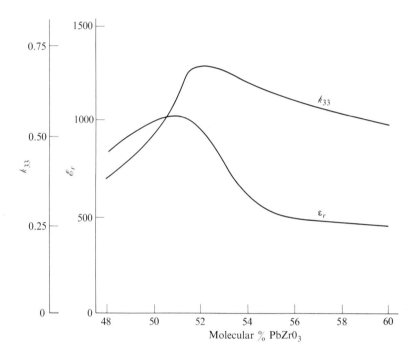

Figure 10.13. Effect of composition of lead zirconate titanate ceramic on the coupling coefficient and permittivity. Note that there is an optimum composition for both functions. (From Crawford [23].)

The advantages of the lead zirconate titanate [$Pb(Zr-Ti)O_3$] ceramic [87] are due to its very high Curie points (300 and 400°C) and larger piezoelectric effects than for barium titanate. The high Curie points are surpassed only by lead metaniobate (570°C), which is of a great interest in high-temperature piezoelectric applications.

There are three important types of PZT materials: PZT-4 is characterized by a large mechanical Q_m and is often used in transducer applications working in the "motor" mode,* PZT-5 is better suited for generator applications because of its low Q_m, whereas PZT-6 has been developed to yield optimum stability at the resonant frequency. Similar to the PZT series, and corresponding to them, are the British LZ series, LZ-4a, LZ-5a, and LZ-6, respectively.

It was found experimentally [64] that the material stability† increases under stress for PZT-4, and decreases for PZT-5, making it unsuitable for stresses above 3000 psi. The exposure to any new stress always begins a new aging cycle, but at relatively low stresses, no perceptible deterioration in the PZT-4 activity was measured under a long-time stress. However, at high compressive stresses, severe degradations have been observed by Krueger and Berlincourt [55]. At 200°C and 1500 psi, they observed a 70% decrease in δ for PZT-4, and 50% for PZT-5, but at 70°C and 4000 psi the decrease was no larger than 3%.

The piezoelectric and electric properties of a material vary with temperature. Figures 10.14 and 10.15 show that g and k decrease and δ increases for an increasing

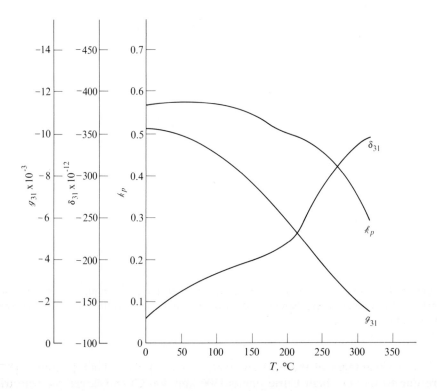

Figure 10.14. Piezoelectric characteristics of LZ-4a ceramic as a function of temperature. δ_{31} is given in C/N and g_{31} is given in Vm/N. (From Crawford [23].)

---

* Converting electrical energy into mechanical energy.

† Stability with respect to the output voltage (stable output).

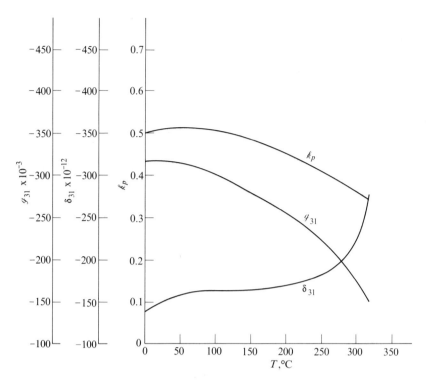

Figure 10.15. Piezoelectric characteristics of LZ-5a ceramic as a function of temperature. δ_{31} is given in C/N and g_{31} is given in Vm/N. (From Crawford [23].)

temperature for LZ-4a and LZ-5a, respectively. Therefore, an optimum working temperature must be chosen for each particular material.

10.8 Power Output, Losses, and Efficiency

Under the effect of compression, a piezoelectric crystal stores energy in a mechanical and an electrical form. The first is due to the elastic property of the crystal and the second to its piezoelectric behavior. The behavior of the piezoelectric generator itself will depend on whether it is short circuited or open circuited. When the crystal is short circuited, it is known to lend itself better to compression and other mechanical stresses, as in the case of the piezoelectric slabs cemented back to back and called *bimorphs*. These are used in many electroacoustic transducer applications. The relation between Young's modulus for the short-circuit case and that for the open-circuit case is given by

$$Y_{sc}/Y_{0c} = 1 - k^2, \qquad (10.35)$$

where k is the electromechanical coupling coefficient. It can also be proved that

$$\varepsilon_{rc}/\varepsilon_{ru} = 1 - k^2, \qquad (10.36)$$

where ε_{r_c} is the dielectric constant of a material clamped to avoid deformation and ε_{r_u} is that of the mechanically unstressed material.

The power output is directly related to variations in internal energy of the crystal. Using the polarization theory of Section 10.4 and introducing the electric displacement D leads to

$$D = \mathscr{P} + \varepsilon_0 E.$$

In terms of the components at constant electric displacement D and at constant stress X, the total differential of the strain x and that of the electric field E are

$$dx = \left(\frac{\partial x}{\partial X}\right)_D dX + \left(\frac{\partial x}{\partial D}\right)_x dD, \tag{10.37}$$

$$dE = \left(\frac{\partial E}{\partial x}\right)_D dx + \left(\frac{\partial E}{\partial D}\right)_x dD. \tag{10.38}$$

The change in internal energy under adiabatic conditions is given by an equation similar to Eq. (10.15):

$$d\xi_{E,X} = E \, dD + X \, dx. \tag{10.39}$$

Since this has to be an exact differential for a conservative system, there exist relations similar to Eqs. (10.18) to (10.21):

$$-d \equiv (\partial E/\partial x)_D,$$

$$g_t \equiv (\partial x/\partial D)_x,$$

$$1/\varepsilon^* \equiv (E/D)_x,$$

$$\jmath^D \equiv (x/X)_D.$$

Thus, substituting in Eqs. (10.37) and (10.38), one obtains, after integration,

$$x = \jmath^D X + g_t D \tag{10.40}$$

and

$$E = -dx + D/\varepsilon^*, \tag{10.41}$$

where ε^* is the permittivity at constant strain, g_t is the transverse generator constant, and $\jmath^D$ is the ratio of strain to stress at constant electric displacement.

Under open-circuit conditions, $D = 0$ and,

$$X_{0c} = x/\jmath^D.$$

Under short-circuit conditions, $E = 0$, and Eqs. (10.38) and (10.41) lead to

$$X_{sc} = (1 - \varepsilon^* d \, g_t)x/\jmath^D. \tag{10.42}$$

The ratio of the short-circuit stress to the open-circuit stress is, therefore,

$$X_{sc}/X_{0c} = (1 - \varepsilon^* d \, g_t).$$

A comparison of this equation to Eq. (10.35) leads to

$$k^2 = \varepsilon^* d \, g_t,$$

which is the square of the electromechanical coupling coefficient. If a mechanical stress is applied without an electric field, the change in energy will be obtained from Eq. (10.39):

$$d\xi_X = X \, dx.$$

After replacing X by its value in Eq. (10.42) and x by its value in Eq. (10.41) (for $E = 0$), one obtains, after integration,

$$\xi_X = \frac{D}{2\varepsilon^x}\left(\frac{1 - \mathit{k}^2}{\mathit{k}^2}\right),$$

where ξ_X represents the original input mechanical energy for a nonelectrostrictive material. The stored electrical energy is equal to $\mathscr{E}_d = \mathit{k}^2 \xi_X$, or

$$\mathscr{E}_d = (1 - \mathit{k}^2)\tfrac{1}{2}D^2/\varepsilon^x.$$

Therefore, only a fraction k^2 of the input mechanical energy is converted into electricity, and k^2 represents the square of the efficiency.

The relations between the different piezoelectric constants can be found readily. In the most general case, they are written in matrix notation:

$$b^* = \varepsilon^x d/\chi^x, \qquad b = \varepsilon^x g/\chi^x,$$
$$\delta = \mathit{s}^E e = \varepsilon^X g, \qquad e = \mathit{s}^{*E}\delta = \varepsilon^x \hbar.$$

There are two types of losses which are of interest in piezoelectric power generation, electrical losses and mechanical losses. The electrical losses are due to Joulean heating and to hysteresis. These losses increase rapidly with increasing temperature. They are produced in the form of heat, and arrangements should be taken to remove the heat thus produced. Mechanical losses are often a function of the state of polarization and stress.

The addition of impurities tends to limit the effect of aging and to give thermal stability to the material. However, losses under low stress in the modified material are considerably larger than those in the unmodified material.

10.9 Bioenergetics

The clinical use of implantable pacemakers for human organs, the utilization of stimulating electronic devices in physiological research, and the need of energy for artificial organs (especially the heart) are forcing physicians and engineers to find generators of energy which can be safe, reliable, long lived, and able to produce from a few microwatts to a few watts of energy. In general, the study of energy transformation in living organisms is referred to as *bioenergetics*. One of the bioenergetics areas of study dealing with direct energy conversion has been seen in Chapter 7; i.e., the biochemical fuel cell, where microbiological processes are utilized for the conversion of chemical energy into electrical energy. Another area is concerned with implantable power sources and is of biomedical interest. The implantable sources may either derive their energy from the body itself or be autonomous units.

The physician is often interested in collecting *in vivo* data from an unrestrained subject. For this, implanted telemeters with a power of the order of a fraction of a watt are needed.

Another important application is the artificial heart. A healthy human adult heart develops an average of 8 W of mechanical power and it is estimated that the internal power source used should furnish at least 35 W for normal functioning of an artificial heart. The power source should work unattended for the entire lifetime of the individual.

A third application is electrical pacing (stimulation) of the natural organs, such as the gastrointestinal tract, the bladder, and the heart itself, which do not function properly for one reason or another. The power requirements of those devices vary from a few microwatts to several hundred microwatts. It is hoped that by increasing the efficiency of the stimulating electrodes, the required stimulating energy might be reduced by a factor of 20. Present day pacemakers work on conventional electrochemical batteries and are not completely satisfactory because of their short lifetimes (no more than 2 years).

In spite of the intense research in this field, it is only recently that some success has been obtained toward the development of safe and long-lasting internal power sources. The different methods proposed are: (1) direct use of muscle power, (2) implantable radioisotope energy sources, (3) power generating implantable electrodes, and (4) piezoelectric energy conversion.

The direct use of muscle power to drive a blood pump and thus "boost" the heart has been investigated by Kusserow [59] *in vivo* on a dog. However, the practicability of this method in humans is questionable. Cellular damage was observed after several hours of stimulation, and the periodic stimulation of a muscle diverted from some other function might be uncomfortable for the subject.

The use of an implanted radioisotope source driving a miniature Rankine cycle reciprocating engine or a thermoelectric generator is indeed extremely attractive. Lindgren [65] has reported work in this field at Thermo-Electron. Several problems need to be solved, especially the danger of irradiation damage, but it seems that this is a promising source of power for the artificial heart as well as for pacemakers and implanted telemeters.

Power generating implantable electrodes use similar approaches to biochemical or redox fuel cells. In the first case, the power source may use some metabolites as fuel and the oxygen dissolved in the living tissues as oxidizer. Many experiments were carried out on dogs and rats using regular and hybrid fuel cells [80, 83]. The future developments of this attractive method are directed toward the evaluation of the long-term effects on the power output. Much work should be done before human use can be tried, although experiments on rabbits, which are much more sensitive than man to foreign bodies, are encouraging.

Piezoelectricity attracted the attention of many workers developing the artificial heart. Loehr [67] and his co-workers developed a bimorph piezoelectric pump with electricity as the source of energy. They expect to increase its pumping capability to 5 l/min by using 30 crystal cells. Myers [72, 73] studied the use of piezoelectricity for supplying 50 to 100 mW of power for electronic cardiac pacemakers. His demonstration was successful in spite of the problems of mechanical failure he encountered. An interesting experiment was performed by Enger [30, 31]; he used the mechanical energy of a dog's heart to generate electrical energy. The converter used was a lead zirconate titanate crystal.* The same crystal might be capable of producing 25 μJ for each contraction of the human heart. In another experiment, he used the mechanical energy produced by the displacement of the rib cage to power a bimorph crystal which generated 10 μJ/cycle. By fixing the crystal to a spinous process† at one end and to a rib at the other, Enger observed an energy output of 22 μJ/cycle. He later built a piezoelectric pacemaker weighing 3 g which performed successfully when attached to the left ventricle of a dog's heart.

Piezoelectric crystals have the advantage of being inert toward the body and, because of their small size, they can be placed in the heart itself to power a pacemaker. However, because of the high output impedance of piezoelectric generators, the presence of fluids creates a perpetual danger of short circuit. This poses a difficult problem of shielding the crystal from the fluids but not from the necessary strains.

The piezoelectric energy converter developed by Ko [52] operates in the resonant mode, thus permitting the crystal to be sealed in a glass container. Ko's generator was coupled to the mechanical motion through the base mount and a loading weight. It is a 10-cm^3 crystal driven at 80 pulses/min with a motion simulating a dog's heart. The 160 μW generated by this device is enough power for a cardiac pacemaker.

The number of mechanical deformations that a cardiac piezoelectric crystal must sustain has been estimated at about 40 million per year. At this rate, the danger of mechanical failure becomes serious. Furthermore, it is not yet known how great the effect of the reaction of the organism will be on the power output of the crystal.

This field is extremely young and further research and development are needed before piezoelectricity can safely be used clinically.

10.10 Piezoelectric Generator Systems

Natural crystals have been used since 1930 in the generator mode in systems such as record players and microphones. Many factors are of importance in these applications: efficiency, linearity, stability, and frequency response. Piezoelectric PZT materials have, by now, almost completely replaced natural crystals in these applications. This was a welcome change, especially in those devices (e.g., "hi-fi"

---

* This is a $PbZrO_3$-$PbTiO_3$ system [87].

† The median spinelike dorsal process of the neural arch of a vertebra.

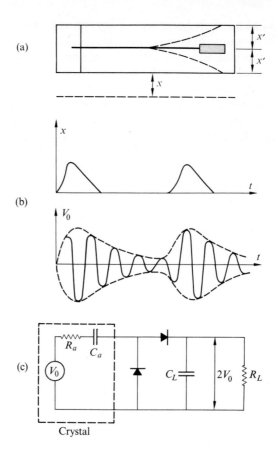

Figure 10.16. Piezoelectric converter system for implants: (a) piezoelectric cell structure, (b) wave form, (c) piezoelectric converter circuit. (From Ko [52].)

record players) where the ruggedness and the high output voltage of the piezoelectric unit is desired.

Because of its high output voltage, PZT-4 has been proposed as a source of high voltage for car ignition systems. In such a system [88], two cylindrical ceramic elements, each 1 cm in diameter and 2 cm in length, are mounted end to end, connected electrically in parallel and mechanically in series. The flat surfaces of the elements are electroded and arranged so that the unit is cycled from $0 \ N/m^2$ to a maximum pressure of $10^8 \ N/m^2$. The compression is supplied by the camshaft of the car engine and tensile stresses are avoided. Under the peak compression, each element contracts by about $10 \ \mu$ when open circuited and $18 \ \mu$ when short circuited. The maximum output voltage obtained is in the 20- to 25-kV range and the power delivered to the load can readily be calculated as a function of the frequency of operation. It is about 1 W for a 60-Hz frequency of operation corresponding to 1800 revolutions/min in a four-cylinder, four-stroke engine. The capacitance of the two elements in parallel is 80 pF, and the output voltage can be considered as a superposition of a dc component of 10 kV and an ac component of 10 kV peak value. The optimum resistance for a matched load has been found to be 30 MΩ for 60 Hz.

Another example of a piezoelectric system is the piezoelectric converter for electronic implants described by Ko [52]. The converter transforms the motions of the body into electrical energy without necessary physical contact between the crystal and the source of motion. The conversion efficiency is increased by operating the crystal in its resonant mode rather than its deflectional mode. This has been achieved by coupling the mechanical motion to the crystal through a loading weight and the base mount. This piezoelectric system is shown in Fig. 10.16(c). The efficiency of the converter could be very large for a properly matched load and a high frequency of vibration. However, because of the low frequency of the body motion, the series capacitance C_0 limits the conversion efficiency to a very low level because ωC_0 is small. The design of the structure is shown in Fig. 10.16(a). The motion is transferred to the piezoelectric crystal which is loaded by a weight to resonate at the frequency of the mechanical driving source. The crystal vibrates at its natural frequency with varying amplitude and generates trains of electrical pulses (Fig. 10.16b). A voltage doubler circuit using one external capacitor is employed for the utilization of the crystal output. The converter has a parallele-pipedic form of size 2 by 5 by 1 cm^3 and of PZT-5a material. Its output power is equal to 160 μW when driven at a mechanical pulse rate of 80 pulses/min and with a dog's heart's motion, corresponding to an output of 4 V and a load of 0.1 MΩ. This power output is sufficient to power some existing telemetry systems as well as a pacemaker [59].

Bendix Corporation [64] (Cincinnati) developed a piezoelectric dosimeter charger under a contract with the Office of Civil Defense. The purpose of this device is to eliminate the need for batteries and furnish mechanical designs for PZT-5 crystals for use with standard V-750 charging pedestals* and lenses gathering ambient light. The instrument is required to charge the electroscope dosimeters used by the Department of Defense. The force is applied to the ceramic units by a spring-loaded screw arrangement. The two PZT-5 crystals are arranged mechanically in series and electrically in parallel. To adjust the low range dosim-eter, a swamping capacitor is provided, and a neon lamp is used for the protection of the dosimeters against overvoltages. A protection against a reversal of polarity is also provided by a special shorting switch. Light from an external source is transmitted by an optical system in the charger for the illumination of the dosimeter scale. This external source of light can be the sun, flashlights, or even candles. The advantages of this system over the regular transistorized charger are many. Among these are ruggedness and simplicity. Furthermore, the unit is light in

---

* Radiological charging contact assemblies.

weight (0.5 kg) and is able to float in water. The components have a long storage life and a good time stability (decrease of 3 % strength in 10 years). Operation is simple and there is almost no need for maintenance. The only serious disadvantage is the effect of humidity on the performance of the device.

10.11 Research and Future Trends

Research in the field of piezoelectric power generation is recent and its potential cannot yet be estimated in all the fields of proposed applications. One factor is clear: there has been progress in the manufacture of new and better piezoelectric materials. The PZT elements are the best piezoelectric crystals now available. The maximum possible energy output is 7.5 J/m^3 for quartz and for PZT. Larger values should be sought and this is possible only for larger values of the maximum working pressure, which is 3.5×10^8 N/m^2 for quartz and 10^8 N/m^2 for PZT-4. Synthetic quartz crystals have been commercially available since 1958, and some of them may weigh as much as 600 g.

To obtain better piezoelectric materials, the piezoelectric theory should be developed on new bases. Work in this direction has started. For instance, Jaffe [46] and his co-workers, in a study of AB compounds crystallizing in a diamond structure, have shown that the piezoelectric constants increase with increasing atomic number of the cation and with decreasing atomic number of the anion.* Much theoretical work has been done in the field of crystal vibrations. Theoretically satisfactory relations have been published in the literature [43, 50] for the coupled modes, the overtones, and the effects of damping. The general transducer theory has been developed by Cady [14], who treated the vibrational state of the transducer as due to the superposition of two systems of progressive waves in opposite directions.

The domains of application of piezoelectricity are numerous: generation and reception of electroacoustic waves, filters, measurements, delay lines, ignition systems, ultrasonic cleaning, nondestructive testing, and applications related to artificial human organs. Piezoelectric generators are high-voltage devices and in this respect are similar to EGD generators. For this reason, Crawford [22] proposed their use in voltage amplification applications.

Finally, an important problem that remains to be investigated thoroughly is the coupling between the input mechanical energy and the crystal. For instance, it has been suggested [88] that electrical energy could be generated by converting the acoustic energy at sonic frequencies in the exhaust of jet engines. Since the power generated for a given stress level is proportional to the frequency, 10 W of power could be delivered to a matched load of 3 MΩ at 600 Hz. The field is promising and deserves the attention of workers in the area of direct energy conversion.

---

*   A *cation* is a positively charged ion and an *anion* is a negatively charged ion.

PROBLEMS

10.1 Tourmaline is one of the best known pyroelectric crystals. Its pyroelectric constant μ and dielectric constant ε_r are given by $\mu = 4 \times 10^{-6}$ C/m² °C and $\varepsilon_r = 7.1$ at 25°C. Find the electric field necessary to create a polarization equal to that obtained by a 1°C increase in temperature. (*Answer:* 74 kV/m.)

10.2 When a strain x_{mn} is applied to a piezoelectric crystal, each component of the polarization $\mathscr{P}_m$ becomes a linear function of all the components x_{mn} through the piezoelectric coefficients h_{mnp}, where m, n, and p are related to the tridimensional physical space such as $m = 1, 2, 3$; $n = 1, 2, 3$; and $p = 1, 2, 3$. Thus, in the most general case, Eq. (10.2) can be written as

$$\mathscr{P}_m = h_{mnp} x_{mn}.$$

(a) How many piezoelectric constants are involved in the above formula? (b) Express $\mathscr{P}_1$, $\mathscr{P}_2$, and $\mathscr{P}_3$ as functions of the individual stresses.

10.3 For quartz, the coefficients h_{nmp} can be written in the matricial form

$$h_{mnp} = \begin{pmatrix} -2.3 & 2.3 & -0.67 & 0 & 0 \\ 0 & 0 & 0 & 0.67 & 4.6 \\ 0 & 0 & 0 & 0 & 0 \end{pmatrix} \times 10^{-12}.$$

All the values are in the MKSC system. What are the values of $\mathscr{P}_1$, $\mathscr{P}_2$, and $\mathscr{P}_3$ when a unidimensional stress is applied on such a crystal? The value of this stress is $x_n = 1$ N/cm².

10.4 Equation (10.9) is a direct deduction of the first principle of thermodynamics. (a) Show the correspondence. (b) Write Eq. (10.9) in the tensorial form of the preceding equations.

10.5 Considering only the piezoelectric effect, find the values of ξ and ζ for a crystal of class 7. Such a crystal is characterized by the following matrices:

$$\varepsilon_{mh} = \begin{pmatrix} 0 & 0 & 0 & 0 & e_{15} & 0 \\ 0 & 0 & 0 & e_{24} & 0 & 0 \\ e_{31} & e_{32} & e_{33} & 0 & 0 & 0 \end{pmatrix},$$

$$\delta_{mn} = \begin{pmatrix} 0 & 0 & 0 & 0 & \delta_{15} & 0 \\ 0 & 0 & 0 & \delta_{24} & 0 & 0 \\ \delta_{31} & \delta_{32} & \delta_{33} & 0 & 0 & 0 \end{pmatrix}.$$

10.6 Prove Eqs. (10.18) to (10.21).

10.7 Calculate the energy stored in a piezoresonator in terms of the parameters shown in Fig. 10.6(b).

10.8 The cut in Fig. 10.17 has been obtained from a piezoelectric crystal. Electrodes are attached at the end faces and an electric current is passed at the proper frequency f_0. Find the resonant frequency of the crystal in terms of its piezoelectric, elastic, and dielectric constants. Neglect all the nonpiezoelectric effects including the thermal expansion of the cut.

10.9 A piezoelectric generator having the parallelepipedic form given in Problem 10.8 is subjected to a compression equal to 1 kg/cm². The generator is made of a PZT-4 material and has the dimensions $S = 0.02$ cm² and $\ell = 1$ cm. (a) Find the open-circuit voltage of the generator, and (b) the generator capacitance. (c) What changes would you introduce in the above characteristics to improve the value of the open-circuit voltage?

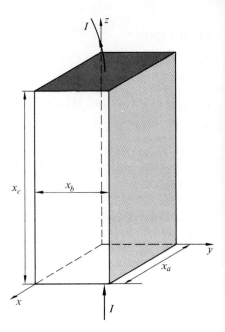

Figure 10.17. A piezoelectric crystal cut.

10.10 Prove Eq. (10.36).

10.11 Find the performance characteristic $V(J)$ of a piezoelectric generator when the crystal is subjected to an acoustic vibration of frequency f_0 and amplitude A_f.

10.12 From the results of Problem 10.11, find the output power and efficiency of the piezoelectric generator in terms of the different coefficients characterizing the crystal and the input vibration.

10.13 Design a piezoelectric generator which would produce 100 mW of electrical power. Take for material an LZ-4a ceramic. The latter has the following characteristics:

$$k_{33} = 0.76,$$
$$\delta_{33} = 300 \times 10^{-12} \text{ C/N},$$
$$\varepsilon_{r33} = 1200 \text{ (free crystal)},$$
$$\delta_{33}^E = 0.147 \times 10^{-10} \text{ m}^2/\text{N},$$
$$\rho = 7.6 \text{ g/cm}^3 = \text{density},$$
$$\text{Mechanical } Q_m = 500,$$
$$\sigma = 0.333 \times 10^{-8} \text{ mho/m at } 100°\text{C}.$$

Find (a) the dimensions of the generator, and (b) the necessary power input. Assume that the input is in the form of a stress in the z-direction.

10.14 In Section 10.9, it was mentioned that Enger built a lead zirconate titanate crystal capable of generating 25 μJ for each contraction of the human heart. Lead zirconate titanate strips were arranged in a bimorphic configuration measuring 3.81 by 1.27 cm. The voltage output was 20 V. Find the mechanical energy produced by the human heart in every contraction.

10.15 To power an artificial heart two natural sources of energy are available: the heat energy produced by the body and the mechanical energy of a given muscle. Study the feasibility of using a crystal which is both pyroelectric and piezoelectric for producing electrical power from both sources of energy.

REFERENCES AND BIBLIOGRAPHY

1. Ballman, A. A., "The Growth and Properties of Piezoelectric Bismuth Germanium Oxide," *J. Crystal Growth*, Vol. 1, No. 1, p. 37, 1967.

2. Baranskii, K. N., "The Excitation of Vibrations of Hypersonic Frequency in Quartz," *Soviet Physics Doklady*, Vol. 114, p. 237, 1957.

3. Bechmann, R., "The Crystal Control of Transmitters," *Wireless Engineer (London)*, Vol. 11, p. 249, 1934.

4. Bechmann, R., "Elastic and Piezoelectric Coefficients of Lithium Sulphate Monohydrate," *Proc. Phil. Soc.*, Vol. 65, Series B, p. 375, 1952.

5. Bechmann, R., "Elastic and Piezoelectric Constants of Alpha-Quartz," *Phys. Rev.*, Vol. 110, p. 1060, 1958.

6. Bedeau, R., *Le Quartz Piézoelectrique et ses Applications dans la Technique des Ondes Hertziennes*, Gauthier-Villars, Paris, 1928.

7. Berlincourt, D. A., and H. Jaffe, "Elastic and Piezoelectric Coefficients of Simple Crystal Barium Titanate," *Phys. Rev.*, Vol. 111, p. 143, 1958.

8. Berlincourt, D. A., C. Cmolik, and H. Jaffe, "Piezoelectric Properties of Polycrystalline Lead Titanate Zirconate Compositions," *Proc. IRE*, Vol. 48, p. 220, 1960.

9. Bernstein, J. L., "The Unit Cell and Space Group of Piezoelectric Bismuth Germanium Oxide," *J. Crystal Growth*, Vol. 1, No. 1, p. 45, 1967.

10. Bogdanov, S. V., B. M. Vul, and R. Y. Razback, "The Shear Piezoelectric Modules of Polarized Barium Titanate," *Kristallografiya*, Vol. 2, p. 115, 1957 (Russian).

11. Bommel, H. E., and K. Dransfield, "Excitation and Attenuation of Hypersonic Waves in Quartz," *Phys. Rev.*, Vol. 117, p. 1245, 1960.

12. Born, M., *Atomtheorie des Festen Zustandes*, B. G. Teubner, Leipzig, Germany, 1923.

13. Cady, W. G., "The Piezoelectric Resonator," *Phys. Rev.*, Vol. 17, p. 531, 1921.

14. Cady, W. G., "The Piezoelectric Resonator and the Effect of Electrode Spacing Upon Frequency," *Physics*, Vol. 7, p. 237, 1936.

15. Cady, W. G., *Piezoelectricity*, McGraw-Hill, New York, 1946.

16. Cady, W. G., *Piezoelectric Phenomena*, AIEE Winter Meeting, New York, Jan. 1947.

17. Cady, W. G., "A Theory of the Crystal Transducer for Plane Waves," *J. Acous. Soc. Am.*, Vol. 21, p. 65, 1949.

18. Cady, W. G., "Crystals and Electricity," *Scientific American*, Vol. 193, No. 12, p. 47, 1949.

19. Cady, W. G., "Piezoelectric Equations of State and Their Application to Thickness Vibration Transducers," *J. Acous. Soc. Am.*, Vol. 22, p. 579, 1950.

20. Cady, W. G., "Composite Piezoelectric Resonator," *Amer. J. Phys.*, Vol. 23, p. 31, 1955.

21. Caspari, M. E., and W. J. Merz, "The Electromechanical Behavior of $BaTiO_3$ Single-Domain Crystals," *Phys. Rev.*, Vol. 80, p. 1082, 1950.

22. Crawford, A. E., "Piezoelectric Ceramic Transformers and Filters," *J. Inst. Radio Eng. (England)*, Vol. 21, No. 4, p. 353, 1961.

23. Crawford, A. E., "Lead Zirconate Titanate Piezoelectric Ceramics," *Brit. J. Appl. Phys.*, Vol. 12, No. 10, p. 529, 1961.

24. Curie, P., *Oeuvres de Pierre Curie*, Gauthier-Villars, Paris, 1908.

25. Dake, H. C., F. L. Fleener, and B. H. Wilson, *Quartz Family Minerals*, Whittlesey House, New York, 1938.

26. Dawson, L. H., "Piezoelectricity of Crystal Quartz," *Phys. Rev.*, Vol. 29, p. 532.

27. De Gramont, A., *Recherches sur le Quartz Piezoelectrique*, Editions de la Rev. d'Optique Theor. et Instr, Paris, 1935.

28. Devonshire, A. F., "Theory of Rochelle Salt," *Phil. Mag.*, Vol. 2, p. 1027, 1957.

29. Eisner, I. I., "On Certain Physical Properties of Rochelle Salt Crystals," *Bull. of Sciences of the USSR, Phys. Sec.*, Vol. 20, No. 2, p. 197, 1956.

30. Enger, C. C., and J. H. Kennedy, *Piezoelectric Power Sources Utilizing the Mechanical Energy of the Human Heart*, 16th Annual Conference on Engineering in Medicine and Biology, Nov. 1963.

31. Enger, C. C., and J. H. Kennedy, "An Improved Bioelectric Generator," *Trans. Am. Soc. Art. Intern. Organs*, Vol. 10, p. 373, 1964.

32. Enger, C. C., and J. H. Kennedy, "A Subminiature Self-Powered Cardiac Pacemaker: Circuit Design," *Trans. Amer. Soc. Art. Intern. Organs*, Vol. 11, p. 148, 1965.

33. Enger, C. C., and F. W. Rhinelander, *A Bioelectric Generator Utilizing the Mechanical Forces of the Rib Cage*, Proceedings of Symposium on Biomedical Engineering, Milwaukee, 1966.

34. Enger, C. C., and M. Klain, *A 3-Gram Self-Powered Pacemaker Implanted on the Surface of the Left Ventricle*, 19th Annual Conference on Engineering in Medicine and Biology, San Francisco, Nov. 1966.

35. Essen, L., "Accurate Measurement of Time," *Phys. Today*, Vol. 3, p. 26, 1960.

36. Fang. P. H., and W. S. Brower, "Temperature Dependence of the Breakdown Field of Barium Titanate," *Phys. Rev.*, Vol. 113, No. 1, p. 456, 1959.

37. Forsbergh, P. W., Jr., "Piezoelectricity, Electrostriction and Ferroelectricity," in *Handbuch der Physik*, Vol. 17, S. Fluggs, Berlin, 1955.

38. Fry, W. J., et al., *Design of Crystal Vibrating Systems for Projectors and Other Applications*, Dover, New York, 1948.

39. Fukuo, H., "Researches in Modes of Vibration of Quartz Crystals Resonators by Means of the Probe Method," *Bull. Tokyo Inst. Tech.*, Ser. A, No. 1, 1955.

40. Gibbs, R. E., and L. C. Tsien, "The Production of Piezoelectricity by Torsion," *Phil. Mag.*, Vol. 22, p. 311, 1936.

41. Hadyl, W. H., K. Harker, and C. F. Onate, "Current Oscillation in Piezoelectric Semiconductors," *J. Appl. Phys.*, Vol. 38, p. 4235, 1967.

42. Hankel, W. G., *Ber. Sächs*, Vol. 33, p. 52, 1881.

43. Hearmon, R. F., "Frequency of Flexural Vibration of Rectangular Orthotropic Plates with Clamped or Supported Edges," *Trans. ASME, J. Appl. Mech.*, Vol. 81 (3), Series E, No. 4, p. 537, 1959.

44. Holman, W. F., "Piezoelectric Excitation of Cane Sugar," *Ann. Physik*, Vol. 29, p. 160, 1909.

45. Jacobsen, E. H., "Piezoelectric Production of Microwave Phonons," *Phys. Rev. Letters*, Vol. 2, p. 249, 1959.

46. Jaffe, B., R. S. Roth, and S. Marzullo, "Properties of Piezoelectric Ceramics in the Solid Solution Series Lead Titanate-Lead Zirconate-Lead Oxide: Tin Oxide and Lead Titanate-Lead Hafnate," *J. Research Nat'l. Bur. Standards*, Vol. 55, 1955.

47. Jaffe, H., "Piezoelectric Ceramics," *J. Am. Ceramic. Soc.*, Vol. 41, p. 494, 1958.

48. Jaffe, H., "Piezoelectricity," *Encyclopedia Britannica*, Vol. 17, p. 9, 1961.

49. Jeffree, J. H., "The Scophony Light Control," *Television and Short-Wave World*, Vol. 9, p. 2605, 1936.

50. Jerrard, R. P., "Vibration of Quartz Crystal-Plates," *Quart. Appl. Math.*, Vol. 18, p. 173, 1961.

51. Kelvin (Lord), "On the Piezoelectric Property of Quartz," *Phil. Mag.*, Vol. 36, p. 331, 1893.

52. Ko, W. H., *Piezoelectric Energy Converter for Electronic Implants*, 19th Annual Conference on Engineering in Medicine and Biology, San Francisco, Nov. 1966.

53. Koga, I., "The Sole Case of the Shear Mode of Thickness for a Thin Oscillating Quartz Plate," *J. Inst. Electr. Eng. Japan.*, Vol. 55, No. 867, p. 822, 1935 (Japanese).

54. Koga, I., et al., "Theory of Plane Elastic Waves in a Piezoelectric Crystalline Material and Determination of Elastic and Piezoelectric Constants of Quartz," *Phys. Rev.*, Vol. 38, p. 573, 1959.

55. Krueger, A. B., and D. A. Berlincourt, "Effects of High Static Stress on the Piezoelectric Properties of Transducer Materials," *J. Acous. Soc. Am.*, Vol. 33, Oct. 1961.

56. Kulcsar, F., "Electromechanical Properties of Lead Titanate Zirconate Ceramics with Lead Partially Replaced by Calcium and Strontium," *J. Am. Ceramic Soc.*, Vol. 42, No. 1, p. 49, 1959.

57. Kulcsar, F., "Electromechanical Properties of Lead Titanate Zirconate Ceramics Modified with Certain 3 or 5 Valent Additions," *J. Am. Ceramic Soc.*, Vol. 42, No. 7, p. 343, 1959.

58. Kurchatov, I. V., *Seignette Electricity*, Moscow, 1933.

59. Kusserow, B. K., and J. F. Clapp, "A Small Ventricule Type Pump for Prolonged Perfusions," *Trans. Am. Soc. Art. Intern. Organs*, Vol. 10, 1964.

60. Langevin, A., "On the Vibration of the Piezoelectric Modulus of Quartz with Temperature," *J. Physique (France)*, Vol. 7, p. 95, 1936.

61. Langevin, L., French Patent 505,703, issued Aug. 5, 1920.

62. Langevin, P., and J. Solomon, "On the Laws of the Liberation of Electricity by Torsion of Piezoelectric Bodies," *Compt. Rend. (France)*, Vol. 200, p. 1257, 1935.

63. Le Corre, Y., "Les Coefficients de Couplage et les Rendements Electromécaniques des Materiaux Piézoelectriques," *J. Physique Radium*, Vol. 18, p. 51, 1957.

64. Lehmann, R., and J. Stirnkob, *Piezoelectric Charger CD V-751X*, Final Report Contract No. OCD-05-6283-AD412337, July 1963.

65. Lindgren, N., "The Artificial Heart—Exemplar of Medical Engineering Enterprise," *IEEE Spectrum*, Vol. 2, No. 9, p. 67, 1965.

66. Lippmann, G., "Principle of the Conservation of Energy," *Ann. Chimie Physique*, Vol. 24, Series 5, p. 145, 1881 (French).

67. Loehr, M., et al., "The Piezoelectric Heart," *Trans. Am. Soc. Art. Intern. Organs*, Vol. 10, 1966.

68. Mason, W. P., *"Electromechanical Transducers and Wave Filters*, Van Nostrand, Princeton, N.J., 1942.

69. Mason, W. P., *Piezoelectric Crystals and Their Application to Ultrasonics*, Van Nostrand, Princeton, N.J., 1950.

70. Mason, W. P., and H. Jaffe, "Methods of Measuring Piezoelectric Elastic, and Dielectric Coefficients of Crystals and Ceramics," *Proc. IRE*, Vol. 42, p. 921, 1954.

71. Mason, W. P., *Physical Acoustics and the Properties of Solids*, Van Nostrand, Princeton, N.J., 1958.

72. Myers, G. H., et al., "Biologically Energized Cardiac Pacemakers," *IEEE Trans.*, Vol. BME-10, April 1963.

73. Myers, G. H., et al., "Biologically Energized Cardiac Pacemakers," *Am. J. Med. Electronics*, Vol. 3, p. 233, 1964.

74. Nye, J. F., *Physical Properties of Crystals*, Clarendon Press, Oxford, England, 1957.

75. Parsonnet, V., et al., "Cardiac Pacemaker Using Biological Energy Sources," *Trans. Am. Soc. Art. Intern. Organs*, Vol. 9, p. 174, 1963.

76. Parsonnet, V., et al., "A Self Perpetuating Biologically Energized Cardiac Pacemaker," *Surg. Forum*, Vol. 14, 1963.

77. Parsonnet, V., et al., "The Potentiality of the Use of Biological Energy as a Power Source for Implantable Pacemakers," *Ann. N.Y. Acad. Sci.*, Vol. 3, June 1964.

78. Pockels, F., *Uber Den Einfluss des Elektrostatischen Feldes auf das Optische Vertialten Piezoelektrischer Krystalle*, Dieterish'sche Verlagsberchhandlung, Gottingen, Germany, 1894.

79. Raillard, H., *Development of an Implantable Cardiac Pacemaker*, Proceedings of the International Solid-State Circuits Conference, 1962.

80. Reynolds, W. L., *Utilization of Bioelectric Potentials*, Quarterly Report, Ames Research Center, NASA, Feb. 1964.

81. Richardson, E. G., *Ultrasonic Physics*, Elsevier, Houston, Texas, 1952.

82. Richardson, E. G., "Industrial Applications of Ultrasonics," *Brit. J. Appl. Phys.*, Vol. 6, p. 413, 1955.

83. Roy, O., and R. Wehnert, "Keeping the Heart Alive with Biological Battery," *Electronics*, Vol. 39, No. 6, p. 105, March 21, 1966.

84. Rzhanov, A. V., "Piezo-Effect on Barium Titanate," *J. Exper. Theo. Phys. USSR*, Vol. 19, p. 502, 1949.

85. Scheibe, A., *Piezoelektrizität des Quarzes*, Theodor Steinkopf, Dresden, Germany, 1938.

86. Schmidt, G., "Piezoelectricity and Electrostriction of Rochelle Salt," *Zs. Physik*, Vol. 161, p. 597, 1961 (German).

87. Shirane, G., and K. Suzuki, "Crystal Structure of Pb (Zr-Ti)O$_3$," *J. Phys. Soc. Jap.*, Vol. 7, p. 333, 1952.

88. Spring, K. H., "Miscellaneous Conversion Methods," in *Direct Energy Conversion*, K. H. Spring (ed.), Academic Press, New York, 1965.

89. Stewart, T. L., and E. S. Stewart, "Hypersonic Resonance of Quartz at 3500 Mc," *J. Acous. Soc. Am.*, Vol. 33, p. 538, 1961.

90. Tamaru, T., "Determination of the Piezoelectric Constants of Tartaric Acid Crystals," *Phys. Zs.*, Vol. 6, p. 379, 1905.

91. Tseng, C. C., and R. M. White, "Propagation of Piezoelectric and Elastic Crystal Waves on the Basal Plane of Hexagonal Piezoelectric Crystals," *J. Appl. Phys.*, Vol. 38, No. 10, p. 4274, 1967.

92. Van der Veen, A. L., "Symmetry of Diamond," *Zs. Kristallographie*, Vol. 51, p. 545, 1913.

93. Vigoureux, P., *Quartz Resonators and Oscillators*, His Majesty's Stationary Office, London, 1931.

94. Vigoureux, P., and C. F. Booth, *Quartz Vibrators and Their Applications*, Her Majesty's Stationary Office, London, 1950.

95. Voigt, W., *Lehrbuch der Kristallphysik*, B. G. Teubner, Leipzig, Germany, 1910.

96. Von Hippel, A., "Piezoelectricity, Ferroelectricity and Crystal Structure," *Zs. Physik*, Vol. 133, p. 158, 1962 (German).

97. Walker, A. C., "Growing Piezoelectric Crystals," *J. Franklin Institute*, Vol. 250, p. 481, 1950.

98. Warner, A. W., "High Frequency Crystal Units for Primary Frequency Standards," *Proc. IRE*, Vol. 42, p. 1452, 1954.

99. Watanabe, Y., "Piezoelectric Resonator in High Frequency Oscillation Circuits," *Proc. IRE*, Vol. 18, p. 695, 1930.

100. Welkowietz, W., "Ultrasonics in Medicine and Dentistry," *Proc. IRE*, Vol. 45, p. 1059, 1957.

101. White, D. L., "High Quartz Crystals at Low Temperatures," *J. Appl. Phys.*, Vol. 29, p. 856, 1958.

102. Yamada, T., N. Niizeki, and H. Toyoda, "Piezoelectric and Elastic Properties of Lithium Niobate Single Crystals," *Japan. J. Appl. Phys.*, Vol. 6, No. 2, p. 151, 1967.

11 FERROELECTRIC POWER GENERATION

11.1 Introduction

Ferroelectric power generation is a method which uses the change with temperature of the value of the permittivity of a dielectric material to convert thermal energy into electricity. A dielectric is a nonconducting material such as is used between the plates of capacitors to increase their capacitance.

Some classes of piezoelectric crystals are spontaneously polarized. However, surface charges in the materials belonging to these classes become noticeable only when the polarization is changed rapidly. This can be done by a change in the temperature of the crystal. Such materials are said to be *pyroelectric* and belong to the following 10 classes: 1, 3, 4, 7, 10, 14, 16, 19, 23, and 26. In one class (class 11), the spontaneous polarization may be reversed in direction under the effect of an external electric field. These materials are known as ferroelectric.

It was J. Curie* who first made a distinction between the effects of uniform and nonuniform heating. Quantitative measurements of the pyroelectric effect had been started by several workers shortly before World War I. Among these, one notes Hayashi's [34] dissertation at Gottingen in 1912, Röntgen's [87] investigation on the pyroelectric crystals (1914), and, especially, Ackerman [1], who performed the most complete investigation, under the direction of Voigt, in studying the dependence of pyroelectricity on temperature. It was only in the late 1950's that work was started to use ferroelectricity for power generation. S. R. Hoh [35] had directly converted heat into electricity by alternately heating and cooling a barium titanate capacitor at International Telephone and Telegraph Laboratories. Interest in this new device has grown steadily since then.

The ferroelectric converter produces alternating currents at high voltages. Its efficiency is less than Carnot's limit since it operates on a temperature cycle. This device is relatively inexpensive and its specific weight compares favorably with the best of the other direct energy converters. Another advantage of the

---

* Ref. 24 of Chapter 10.

ferroelectric converter is the fact that it can convert solar energy into electricity with a higher power output per unit weight than can the photovoltaic converters. The field is, however, in too early a stage of development to ascertain the importance of this method in converting solar electromagnetic energy.

11.2 Basic Dielectric Concepts

Consider a *dipole* made up—by definition—of a pair of equal and opposite point charges, $+q$ and $-q$, separated by a distance d. The absolute potential V produced by this dipole at a distance r is

$$V = \frac{\not{p} \cdot \mathbf{r}}{4\pi\varepsilon_0 r^3},\tag{11.1}$$

where $\mathbf{r}$ is much larger than $\mathbf{d}$ and where $\not{p}$ is the dipole moment,

$$\not{p} = q\mathbf{d}.\tag{11.2}$$

A dielectric material can be considered to be a material made up of a large number of equal positive and negative point charges, each positive charge occupying the same point as a negative charge so that the material is electrically neutral. Under the effect of an externally applied electric field, the positive charges move in the direction of the electric field and the negative charges move in the opposite direction, thus creating a large number of dipoles in the dielectric material. The polarization $\mathscr{P}$ of the material is then defined as the dipole moment $\not{p}$ per unit volume,

$$\mathscr{P} = d\not{p}/d\mathscr{V}.\tag{11.3}$$

The effect of this polarization is to increase the electric flux density $\mathbf{D}$, so that

$$\mathbf{D} = \varepsilon_0\mathbf{E} + \mathscr{P}.\tag{11.4}$$

In linear dielectric materials, $\mathscr{P}$ is directly proportional to the electric field. If the material is isotropic and homogeneous, it is said to be a "class A dielectric," for which

$$\mathbf{D} = \varepsilon_0(1 + \chi)\mathbf{E},\tag{11.5}$$

where χ is the electric susceptibility. The dielectric coefficient of the material is defined as $\varepsilon_r = 1 + \chi$ and Eq. (11.5) becomes $\mathbf{D} = \varepsilon_0\varepsilon_r\mathbf{E}$. Gauss' law describes the nature of the charges and fields at the interface between a conductor and a dielectric,

$$\oint \mathbf{D} \cdot d\mathbf{S} = q,$$

or at the boundary and if $\mathbf{D}$ and $d\mathbf{S}$ are in the same direction,

$$D \, dS = q_s \, dS,$$

leading to $|D| = q_s$, where q_s is the surface charge density at the interface. If, now, a dielectric slab is inserted between the conducting plates to form a parallel plate capacitor of area S and thickness d, the electric field will be

$$E = q_s/\varepsilon_0\varepsilon_r. \tag{11.6}$$

The potential difference between the capacitor plates is $V = Ed$. The capacitance C_F of the capacitor is defined by

$$C_F = q/V = q_s S/V, \tag{11.7}$$

and taking into consideration Eqs. (11.6) and (11.7),

$$C_F = \varepsilon_0\varepsilon_r S/d. \tag{11.8}$$

The dipoles created in the material by the externally applied field establish a depolarizing field $\mathbf{E}$ opposed to the applied field. A depolarizing factor m_ℓ is then defined by the ratio

$$m_\ell = \varepsilon_0 E_{\mathrm{dep}}/\mathscr{P}.$$

11.3 Ferroelectric Hysteresis and the Curie Temperature

Hysteresis. The usefulness of a ferroelectric material depends on the fact that the polarization $\mathscr{P}$ is not directly proportional to the applied electric field. If such a material is placed in a field-free region and then subjected to a steadily increasing electric field, the polarization will increase following the path OAB shown in Fig. 11.1. The curve OAB is not linear but has an inflection point near A. As the electric field is further increased, the curve becomes linear in the region BC. Now, if the electric field is reduced toward zero, the polarization will follow a new path CBD and, as the electric field becomes negative and equal to the maximum value reached in the positive region, the curve CBD will be continued to form $CBDH'C'$. By reversing once more the electric field, the curve $C'D'HBC$ will be traced, and the cycle $C'D'HCDH'C'$ can be repeated.

Thus, note the existence of two values of the polarization for each value of E, depending on whether the applied field is increasing or decreasing. Such a phenomenon is called *hysteresis*. The value of the polarization for $E = 0$ is called *remnant polarization* $\pm\mathscr{P}_r$ (points D and D'). The linear part of the curves (BC and $B'C'$) if extended would cross the polarization axis at $\pm\mathscr{P}_s$, called the *spontaneous polarization*. The value of the electric field making the polarization vanish is $\pm E_c$, the *coercive field*.

The nonlinear curve OAB can be explained in terms of changes in crystallite domains under the effect of an externally applied electric field. Figure 11.2(a) is a plane representation of the polycrystalline structure of a ferroelectric material composed of individual crystal sections called *crystallites*. Each crystallite has a spontaneous polarization but, since the different crystallites are randomly arranged, the resultant polarization is equal to zero (point O in Fig. 11.1). Each crystallite can actually be divided into domains, as shown in Fig. 11.2(b). Under an applied electric field, the domains parallel to it decrease until eventually the entire

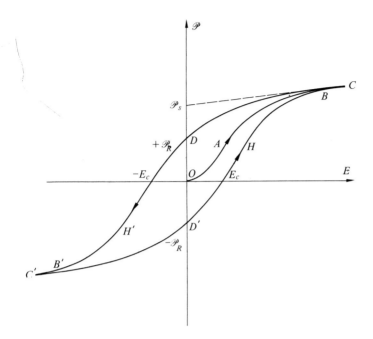

Figure 11.1. The hysteresis loop of a ferroelectric material. Curve OAB is often called the *virgin curve*.

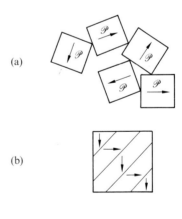

Figure 11.2. Crystallites and changes in crystallite domains under the effect of an applied electric field on a ferroelectric material: (a) crystallites, (b) crystallite domains, (c) crystallite domains under the effect of an applied electric field.

crystallite becomes a regular dielectric (region BC of the curve). The hysteresis effect as well as the existence of a spontaneous polarization are often explained in terms of the crystalline structure of the ferroelectric material.

The Curie temperature. In a solid, each dipole is subject to the field of all the others. Hence, the total field acting on a dipole can be considered as the sum of all the fields; this sum is called the *local field,*

$$E_{\text{loc}} = E + E_{\text{dep}} + E_{bd}, \tag{11.9}$$

where E_{dep} is the depolarizing field and E_{bd} is the field due to the bound surface charges in the vicinity of the dipole. From Eq. (11.4),

$$E_{\text{dep}} = -\mathcal{P}/\varepsilon_0, \tag{11.10}$$

and E_{bd} can readily be calculated by considering a sphere centered around the dipole, the radius of which tends toward zero. It is found that

$$E_{bd} = \mathcal{P}/3\varepsilon_0.$$

In Eq. (11.4), the electric flux must be continuous across the surface between the dielectric and free space. Thus, calling E_i the field inside the dielectric, one has

$$D = D_i = \varepsilon_0 E = \varepsilon_0 E_i + \mathcal{P}, \tag{11.11}$$

or

$$E_i = E - \mathcal{P}/\varepsilon_0. \tag{11.12}$$

Replacing E, E_{dep}, and E_{bd} by their values in Eq. (11.9) yields

$$E_{\text{loc}} = E_i + \mathcal{P}/3\varepsilon_0. \tag{11.13}$$

This expression is known as the *Lorentz form of the local field.*

When an electric field is applied to a crystal composed of a large number of identical atoms, it will slightly displace the nuclei of the atoms in the direction of the electric field and the center of the atomic electron clouds in the opposite direction, thus creating dipoles of separation d at the sites of the atoms. By equating the force on the nucleus to the force due to the applied field, the separation d is found to be

$$d = \frac{4\pi\varepsilon_0 r_a^3}{n_z e} E_{\text{loc}},$$

where n_z is the atomic number, e is the electronic charge, and r_a is the radius of the atom. From Eqs. (11.2) and (11.3), the total induced polarization is

$$\mathcal{P} = n\varkappa E_{\text{loc}}, \tag{11.14}$$

where n is the number of atoms per unit volume placed in the field, and $\varkappa = 4\pi\varepsilon_0 r_a^3$. Introducing Eq. (11.13) into Eq. (11.14) leads to

$$\mathcal{P} = \frac{n\varkappa E_i}{1 - n\varkappa/3\varepsilon_0}, \tag{11.15}$$

and the susceptibility is

$$\chi = \frac{n\varkappa}{\varepsilon_0(1 - n\varkappa/3\varepsilon_0)} .$$ (11.16)

When

$$\frac{n\varkappa}{3\varepsilon_0} = 1,$$

χ becomes infinite. This is known as the *Mossotti catastrophe*. By assuming that the energy $\mathscr{E}$ associated with an atomic dipole follows a Maxwellian distribution function, $\varkappa$ can be calculated, leading to the following value of the polarization [77]:

$$\mathscr{P} = n\left(\varkappa_e + \varkappa_i + \frac{\not\!p_p^2}{3kT}\right)E_i.$$ (11.17)

This expression is known as the *Debye formula* in which $\varkappa_e$ and $\varkappa_i$ are respectively the electronic and ionic polarizabilities and $\not\!p_p$ is the permanent dipole moment. The polarizabilities $\varkappa_e$ and $\varkappa_i$ depend on the electronic structure of the atoms and are independent of temperature. Comparing Eq. (11.17) to Eq. (11.14), one notes that

$$\varkappa = \left(\varkappa_e + \varkappa_i + \frac{\not\!p_p^2}{3kT}\right).$$ (11.18)

The Curie temperature is the critical temperature corresponding to the Mossotti catastrophe. For negligible $\varkappa_e$ and $\varkappa_i$, its value becomes

$$T_c = \frac{n\not\!p_p^2}{9\varepsilon_0 k},$$

and Eq. (11.15) becomes, in the most general case,

$$\chi = \frac{\mathscr{C}_F T_c}{T - T_c} .$$ (11.19)

This is known as the Curie-Weiss law, where 3 has been replaced by $\mathscr{C}_F$, the Curie constant, which depends on the type of material.

Above the Curie temperature, ferroelectric crystals are no longer polar. The dielectric constant shows a very high peak at the Curie point and then decreases rapidly, as can be seen from Eq. (11.19).

Ferroelectric materials useful for power generation should have a large polarization, often equal to the dielectric displacement. The dc permittivity can thus be obtained from the relation $\varepsilon_0\varepsilon_r = \mathscr{P}/E$. It is equal to the slope of the curve $\mathscr{P}(E)$ shown in Fig. 11.1. As can be seen, ε_r is a function of the applied field, and it becomes constant only at temperatures well above the Curie temperature. The initial dielectric constant of a ferroelectric crystal is defined by the slope of the virgin curve (OAB) at point O.

The similarity between this ferroelectric hysteresis and the ferromagnetic one is apparent, and the similar ferromagnetic effect can be used for power generation (Chapter 12).

11.4 Time Constant

It was seen in the last section that a good ferroelectric converter should have a large permittivity. However, in many cases dc measurements of the hysteresis loops are impossible because of the low insulation resistances of the considered ferroelectric compositions, especially at high temperatures. Therefore, the conductivity σ of the material is as important as the permittivity for conversion applications, and the ratio ε/σ becomes a measure of the quality of a ferroelectric converter. Note that this ratio is equal to RC_F, since

$$RC_F = \frac{d}{\sigma S} \cdot \frac{\varepsilon S}{d}.$$

The product RC_F is called the *time constant*. It is the time necessary for the ferroelectric capacitor to discharge itself to $1/e$ of its initial voltage. The capacitor voltage V decreases exponentially in time, according to

$$V = V_B \exp(-t/RC_F), \tag{11.20}$$

where V_B is the voltage of the battery used for charging.

For a ferroelectric material, the product RC_F varies with the electric field applied to the capacitor, and Eq. (11.20) is not necessarily valid. However, it has been demonstrated [37] that the actual relation differs little from the exponential function, and the above relation has been most often used for measuring RC_F.

Also, RC_F is a function of temperature and it usually decreases with a temperature increase. Additives to the host material greatly influence the value of RC_F. This subject will be treated in more detail below.

11.5 Ferroelectric Devices [2–5, 12, 48–53]

Here, two ferroelectric applications will be reported: the second-harmonic dielectric amplifier and the ferroelectric memory system. Both applications have been studied by Fotland and Mayer [30]* on an Air Force contract.

Interest in highly reliable lightweight amplifiers working at high temperatures triggered a great interest in solid state materials [59]. The nonlinearity of the ferroelectric materials can be used for amplifying the second harmonic of an input voltage. The amplifier works over the audio frequency range and uses a 100-kHz ac power supply.

The $C_F(V)$ characteristic of a ferroelectric material can be idealized by the following linear function:

$$C_F = C_0(1 - \mathscr{Y}V). \tag{11.21}$$

When the ferroelectric capacitor is connected to a sinusoidal voltage source having an output

$$V = \sqrt{2} V_{\text{rms}} \cos(\omega t), \tag{11.22}$$

---

* Horizons Incorporated, Cleveland, Ohio.

and a dc bias V_{dc} such as $V_{dc} > \sqrt{2}\, V_{rms}$, it generates a current I given by

$$I = d(C_F V)/dt.$$

Thus, taking into consideration Eqs. (11.21) and (11.22), one obtains

$$I = 2\frac{dC_F}{dV}\, \omega V_{rms}^2 \sin (2\omega t) - 2C_0 \omega V \sin (\omega t).$$

For $V_{dc} \leqslant \sqrt{2}\, V$, the ratio of the second-harmonic current I to the fundamental I_i becomes

$$\frac{I_2}{I_i} = \frac{V_{dc}}{V_0} \cdot \frac{dC_F}{dV}. \qquad (11.23)$$

The term $(1/C_0)(dC_F/dV)$ is often called the *extrinsic nonlinearity factor* $\mathscr{Y}$. Since it is a function of the thickness of the material, an *intrinsic nonlinearity factor* $\mathscr{Y}_i$ is defined as

$$\mathscr{Y}_i = \frac{d}{C_0} \cdot \frac{dC_F}{dV}.$$

Two basic second-harmonic amplifier circuits are shown in Fig. 11.3. Circuit (a) is a low-Q_m bridge in which the third harmonic cancels the fundamental at the load, leaving only the second harmonic. Circuit (b) consists of a ferroelectric capacitance and an inductance resonating at the second harmonic of the power

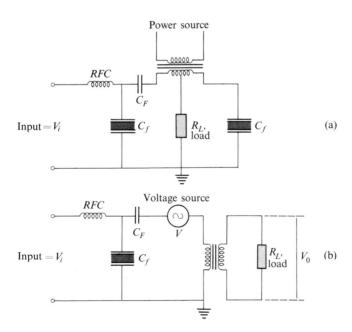

Figure 11.3. Second-harmonic amplifier circuit: (a) Q_m bridge, (b) lumped LC_F circuit.

supply frequency. The power gain of the nonresonant amplifier of Fig. 11.3 is given by

$$G = I_2 R_i R_L / V_i^2,$$

where R_i is the mid-frequency input impedance equal to

$$R_i = \frac{1}{2\pi\,\Delta f C_F}. \tag{11.24}$$

It has been determined experimentally that for the optimum power output

$$R_L = 1/4\omega C_F. \tag{11.25}$$

Thus, taking account of Eqs. (11.23), (11.24), and (11.25), the gain-bandwidth product becomes

$$G\,\Delta f = \frac{I_1^2 \mathscr{Y}}{8\pi C_F^2 \omega} = \tfrac{1}{4}\mathscr{T}f,$$

where $\mathscr{T}$ is the material parameter of fundamental importance for amplifier operation. It is equal to

$$\mathscr{T} = (V\mathscr{Y})^2 \cong \left(\frac{I_1 \mathscr{Y}}{\omega C_F}\right)^2.$$

The principal limitation on this device arises from the hysteresis losses, which become very large at high fields and high frequencies. These losses may raise the crystal temperature above the Curie point, leading to the destruction of the crystal nonlinearity. Experiments with $BaTiO_3$ crystals led to a maximum $G\,\Delta f$ of about 10^5. Nevertheless, ferroelectric amplifiers have great prospects with the development of ferroelectric materials having high Curie points and low hysteresis losses. The same loss and Curie temperature limitations apply to the development of ferroelectric frequency multipliers.

A ferroelectric memory [81–84] consists of a thin ferroelectric plate having columns of electrodes on one side and rows of electrodes on the other. Information is stored by pulsing an appropriate row and is read by pulsing an appropriate column, thus selecting the desired ferroelectric area. When the application of a positive pulse polarizes a ferroelectric element (binary 1), a negative pulse will reverse the direction of the polarization (binary 0). This reversal of polarization results in a large current pulse through the element. The application of a negative pulse on a negatively polarized ferroelectric will result in a small current pulse charging the ferroelectric capacitor. These two pulses are usually called *switching* and *nonswitching transients*, respectively.

In order to utilize a ferroelectric material in a coincidence selection system, its hysteresis loop must be almost rectangular, thus limiting the reduction of the remanent polarization by the disturbing pulses. Single crystals of $BaTiO_3$ have such a property. From the hysteresis loop of Fig. 11.1, it can be seen that the hysteresis effect results in a reversal in $d\mathscr{P}/dE$ at $E = 0$ (points D and D'). This

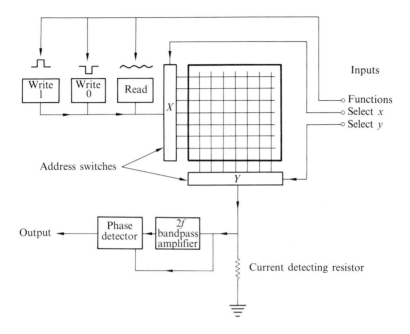

Figure 11.4. Ferroelectric storage system. The read pulser consists of an ac generator of frequency f. (From Fotland and Mayer [30].)

represents a change in the sign of dC/dE at $E = 0$, and a method of determining this sign makes digital information storage feasible. This is done by applying a small sinusoidal voltage across the ferroelectric capacitor and observing the phase of the resulting second-harmonic current. A storage system is shown in Fig. 11.4 in which information is read by applying a negative pulse. In this system, the "read" and "write" generators consist of pulsers. The binary bit stored in any position in the matrix is determined by a second-harmonic bandpass amplifier and phase detector.

From Fortland and Mayer's [30] experiments, it was concluded that certain ceramics are capable of storing information and being interrogated without suffering any degradation in the process. The experiments were performed with sodium-zinc niobate. The successful operation of a complete memory system is possible, although a great deal of research is needed to improve the ceramic technologies, to determine the maximum density of the matrices, and to develop proper selection and address equipment for this type of memory.

11.6 Ferroelectric Generators

Operating characteristics. By placing a slab of ferroelectric material between the two plates of a capacitor, as shown in Fig. 11.5, heat can be transformed directly

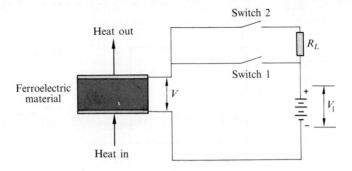

Figure 11.5. Basic ferroelectric converter. This figure corresponds to line *bc* or line *da* of Fig. 11.6.

into electricity. The capacitance of the condenser drops when heat is added because of the drop in ε_r. Consequently the voltage rises and the capacitor is made to discharge through a load, delivering electrical power. The cycle is completed by cooling the capacitor. The different steps of the cycle shown in the charge voltage diagram of Fig. 11.6 can be summarized as follows:

Line ab: switch 1 closed, switch 2 open. The condenser is charged by the battery from q_1 at voltage V_0 to q_2 at voltage V_2.

Line bc: switch 1 open, switch 2 open. Heat is added and the capacitance of the condenser decreases at constant charge q_2 and the voltage increases from V_1 to V_2.

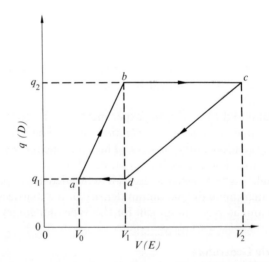

Figure 11.6. Ferroelectric converter cycle. The converted electrical energy is given by the area obtained from the difference of the areas of triangles *bcd* and *abd*.

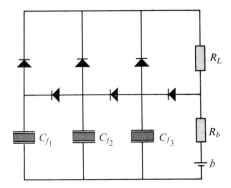

Figure 11.7. DC ferroelectric generator. The battery b serves to initially charge the ferroelectric cells.

Line cd: switch 1 open, switch 2 closed. The condenser discharges through the load and the battery and its charge is reduced to the initial value q_1 at voltage V_1.

Line da: switch 1 open, switch 2 open. The heat is rejected at a lower temperature and the capacitance increases at constant charge q_1, decreasing the voltage to the initial value V_0.

The cycle can then be repeated. The energy delivered by the battery is

$$\mathscr{E}_1 = \tfrac{1}{2}(V_1 - V_0)(q_2 - q_1),$$

and the energy delivered to the load and the battery after each cycle is

$$\mathscr{E}_2 = \tfrac{1}{2}(V_2 - V_1)(q_2 - q_1). \tag{11.26}$$

Therefore, the net thermal heat energy converted to electricity is $\mathscr{E} = \mathscr{E}_2 - \mathscr{E}_1$ or

$$\mathscr{E} = \tfrac{1}{2}(q_2 - q_1)(V_2 + V_0 - 2V_1).$$

In practical circuits, switches 1 and 2 of Fig. 11.5 are replaced by diodes and the load itself can be a storage battery.

There are three important types of ferroelectric energy converters: dc output generators, high-voltage generators, and ac output generators.

DC generators. An example of a dc output generator is shown in Fig. 11.7. This is a multiple capacitor converter in which the battery b serves only to prime (initially charge) the ferroelectric capacitors and to overcome possible leakages in them. To prevent output current from the battery, a large resistance R_b is placed in series with it. The ferroelectric cells C_{f1}, C_{f2}, and C_{f3} are temperature cycled in such a way that at any given instant the cells are at different temperatures. The total charge can then be recirculated between the cells, leading to a dc output at the load.

The heat energy can be taken directly from solar radiation as shown by the example reported by Hoh [35] and illustrated in Fig. 11.8. The converter can be water-borne (or balloon-borne), with several lenses on the surface of the buoy

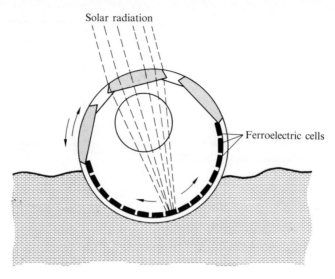

Figure 11.8. Ferroelectric energy converter buoy. The waves swing the ferroelectric system, thus creating the required temperature cycling. (After Hoh [35].)

to concentrate solar energy. The ferroelectric cells are placed in the focal planes of the lenses and they are heated alternately as the waves swing and rotate the buoy. The buoy can carry the electronic equipment powered by the ferroelectric generator, such as radio or sonar transmitters. An arrangement should be made to keep the lens in the direction of the solar radiation.

High-voltage generators. As stated in the introduction, ferroelectric generators are good sources of high voltage. If the capacitance of a ferroelectric cell is C_c at the Curie temperature T_c, its electrical charge is

$$q = C_c V_c, \tag{11.27}$$

where V_c is the charging voltage. If the capacitor is cooled to a temperature T_a, its capacitance decreases to the value C_a, where

$$C_a = C_c \varepsilon_{ra}/\varepsilon_{rc}, \tag{11.28}$$

as can be deduced from Eq. (11.8). Since the charge remains the same,

$$q = C_a V_a. \tag{11.29}$$

When Eqs. (11.27), (11.28), and (11.29) are taken into consideration, the output voltage is

$$V_a = V_c \varepsilon_{rc}/\varepsilon_{ra}. \tag{11.30}$$

The ratio $\varepsilon_{rc}/\varepsilon_{ra}$ can be as large as six for some ferroelectric materials. Further research may lead to much higher ratios.

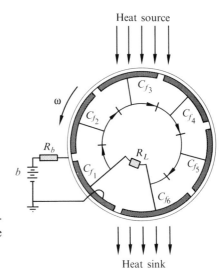

Heat source

Heat sink

Figure 11.9. High-voltage ferroelectric converter. The battery b serves to initially charge the ferroelectric cells. (After Hoh [35].)

A much higher output voltage can be obtained by cascading several ferroelectric cells in such a way that the output voltage of each stage will serve to charge the following stage. This is shown in Fig. 11.9. In this example, four ferroelectric cells are attached to a rotating cylinder (or sphere) which works as a common electrode. The cylinder is then heated to a temperature T_c on one side and cooled to a temperature T_a on the other side, the sources and sink of heat being stationary. The initial charges and leakage losses are supplied by the battery b. The total output of n cascaded ferroelectric cells can be found to be

$$V_{an} = V_c(\varepsilon_{rc}/\varepsilon_{ra})^n. \tag{11.31}$$

Here again solar radiation can be used as source of heat.

AC generators. An obvious disadvantage of dc ferroelectric generators is the necessity of temperature cycling. This cycling is particularly suited for ac power generation. An example is shown in Fig. 11.10(a). In this device, two large ferroelectric cells are arranged on opposite sides of a rotating cylinder in front of a stationary heat source such as solar radiation. The other side of the heat source can be cooled through radiation in the case of space applications or by other means (conduction or convection) for terrestrial applications. The initial charge is supplied by a capacitor. The energy converted into electricity is derived from the temperature cycling of the ferroelectric cells and not from the kinetic energy of rotation of the cylinder.

Figure 11.10(b) shows the configuration of a three-phase generator. In this configuration, the charge of a ferroelectric cell at the Curie temperature is $q_c = C_c V_c$, whereas at the lower temperature T_a, $q_a = C_a V_c$. The difference between the two charges is $\Delta q = V_c(C_c - C_a)$. This charge is forced alternately

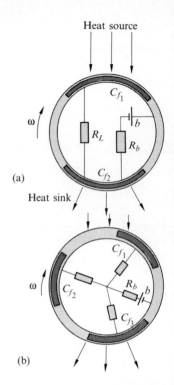

Figure 11.10. AC energy converters: (a) Single-phase generator; (b) three-phase generator.

through the load twice during every temperature cycle. The output current is $I = 2\omega \Delta q$, or $I = 4\pi f V_c(C_c - C_a)$, where f is the frequency of the temperature cycle. Under open-circuit conditions, the voltage can be obtained from Eq. (11.30).

11.7 Ferroelectric Materials

There are four families of ferroelectric materials: (1) the tartrate family, of which Rochelle salt is the most important representative, (2) the KDP family, (3) the GASH family, and (4) the oxygen-octahedra family, of which the perovskite sub-group is of the greatest importance for power generation.

Rochelle salt has been mentioned in Chapter 10. It is ferroelectric only between its two Curie points, i.e., for $-18°C < T < 23°C$. The ferroelectric crystal has a monoclinic symmetry and its spontaneous polarization occurs along the A-axis of the original orthorhombic crystal. The behavior of ε_r is shown in Fig. 11.11, curve A, as reported by Habluetzel [32]. Kurchatov [54] has shown that after substitution of 12 percent of the potassium ions by NH_4^{--}, Rochelle salt shows a single Curie point at a low temperature.

Two other ferroelectric materials are of the tartrate family: LAT (LiNH$_4$-C$_4$H$_4$O$_6$·H$_2$O) and LTT (LiTeC$_4$H$_4$O$_6$·H$_2$O). Their Curie points are at low temperatures and therefore they are not very useful for power generation. LAT

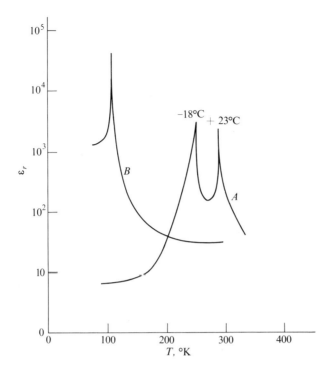

Figure 11.11. The dielectric constant as a function of temperature for KDP and Rochelle salt. Curve A: Rochelle salt (from Habluetzel [32]); curve B: KDP (from Busch [15]).

is ferroelectric only for temperatures below $-173°C$, its Curie temperature, whereas LTT has one of the lowest Curie points ($10°K$).

Potassium dihydrogen phosphate (KDP) has also been mentioned in Chapter 10. Its Curie temperature is about $-150°C$, above which the crystal is piezoelectric, while at temperatures below $-150°C$ it is both piezoelectric and ferroelectric. The dielectric behavior of the crystal with temperature is shown in Fig. 11.11, curve B, as reported by Busch [15], and the spontaneous polarization is shown in Fig. 11.12, curve A.

The GASH (guanadine aluminum sulfate hexahydrate) family is a relative newcomer to the ferroelectric group. GASH itself is of great interest, since it is ferroelectric over a very wide temperature range. Its dielectric constant at room temperature is, however, very small (around 6 along the polar axis). The Curie point of the material is between 200 and 300°C but because the crystal starts losing its water of crystallization at such high temperatures it has not been possible to determine it exactly.

Other materials of the GASH family are obtained by replacing the aluminum cation of GASH by gallium or cerium (or both) and the sulfate anion by selenia

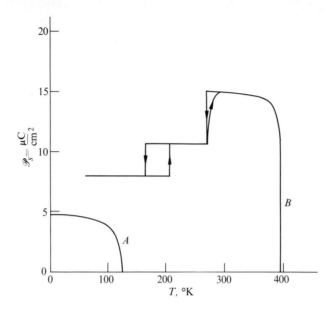

Figure 11.12. Spontaneous polarization of barium titanate and KDP. Curve *A*: KDP (after Von Arx and Bantle [96]); curve *B*: barium titanate single crystal (after Merz [63]).

(SeO_4). The new ferroelectric materials are consequently isomorphs of GASH and have similar properties.

The best known subgroup of the oxygen-octahedra family is the perovskite-type subgroup, represented by barium titanate, $BaTiO_3$. The general chemical composition common to all the perovskite group is XYO_3, where X is mono- or divalent, Y is a small highly polarizable ion surrounded by an octahedron of oxygen atoms, and adjacent octahedra are closely bonded together by sharing the X corners of the cubic crystal. Some oxygen-octahedra materials are tungsten trioxide (WO_3), cadmium niobate ($Cd_2Nb_2O_7$), lead niobate ($PbNb_2O_6$), and the ilmenite group ($FeTiO_3$). Other perovskite-type crystals which have some ferroelectric properties are $PbTiO_3$, $LiTaO_3$, and $KNbO_3$. The latter is very similar in behavior to $BaTiO_3$.

Barium titanate. At the present state of the art, barium titanate ($BaTiO_3$), of the perovskite subgroup, is the best known and the most useful ferroelectric material to be considered for power generation. Barium titanate exhibits three different ferroelectric phases as shown in Fig. 11.12, curve *B*, representing the spontaneous polarization as a function of temperature. The Curie temperature is near 120°C, above which the dielectric constant follows the Curie-Weiss law and the structure becomes cubic. In the temperature range 5°C < T < 120°C, the crystal has a tetragonal symmetry and the spontaneous polarization is along the tetragonal axis. The symmetry becomes orthorhombic in the temperature range $-80°C < T < 5°C$. Below $-80°C$, the symmetry is rhombohedral.

In 1946 [97] $BaTiO_3$ was discovered to be ferroelectric and since then a great deal of experimental and theoretical work has been performed, leading to its use in many applications. The crystal has stable chemical properties, high mechanical strength, and the best dielectric properties now available.

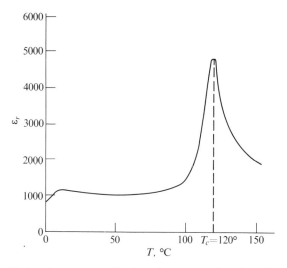

Figure 11.13. Dielectric constant of barium titanate as a function of temperature. The Curie point is at 120°C. (After Roberts [85].)

The properties of polycrystalline barium titanate samples differ slightly from those of the single crystal. The values reported for the dielectric constant vary from one investigator to another. These values are functions of the frequency of measurement as well as of the impurities added to the material. The experimental result of Fig. 11.13 has been reported by Roberts [85] for a frequency of 400 kHz and with no bias.

Measurement of the time constant of the pure $BaTiO_3$ crystal shows that it is not practical for power generation in its pure form. The value of RC_F is only 10 sec for the commercially pure grade (CP). This value is multiplied by a factor of 100 when special care is given to the preparation of the crystal. For instance, for the high-purity grade of small particle size, RC_F equals 840 sec. The greatest influence on the value of RC_F occurs through the addition of impurities. As an example, Fig. 11.14 shows the variation of the time constant as a function of the addition of indium oxide for $(Ba_{0.77} \cdot Sr_{0.23})TiO_3$. It can be seen that the time constant of the material increases from a value not much larger than 10 sec for zero percentage of indium oxide to a maximum of 6000 sec for 0.04% In_2O_3 and then decreases very rapidly for higher percentages.

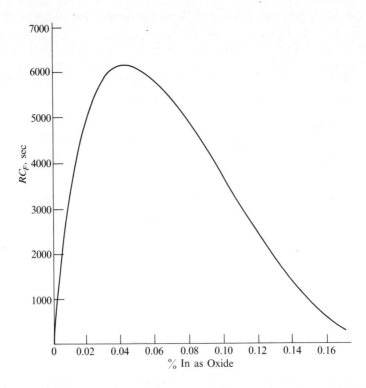

Figure 11.14. Experimental variation of the time constant of $(Ba_{0.77}, Sr_{0.23})TiO_3$ with indium oxide (In_2O_3) doping concentration. (From Jain and Ravindran [44].)

Additives such as $Cr_2O_3 \cdot BaO$ and various stannates improve both the permittivity and the resistivity of the material, leading to optimum values of the product RC_F at a certain level of doping. For instance, the time constant equals 12,000 sec (3 hr 20 min) for a pressed barium titanate sample fired 1 hr at 1450°C in air and with $(0.1Cr_2O_3 + 1.0CaSnO_3)$ additive.

The time constant is also a function of temperature and it decreases very rapidly for increasing temperatures. As an example, the variation of RC_F with temperature is shown for various additives in Fig. 11.15. This is a very serious problem and it is due to the rapid degradation of the dielectric strength, the insulation resistance, and the operating life at high temperatures. To eliminate this disadvantage, introduction of high valence cations or low valence anions has been suggested. The cations may be uranium in the form of oxides and the anions may be fluorine. These compounds can be introduced in a gaseous form, but a better approach is to introduce them in the form of solid fluorides. For example, UF_4 partly decomposes during the firing of the barium titanate ceramic, thus liberating some fluorine. The use of uranium tetrafluoride has proved successful as shown in Fig. 11.15.

Polarization is another important parameter of power conversion. Figure 11.12 shows that a value of 15 μC/cm^2 is obtained for single crystal barium titanate. At a field of 0.15 kv/m, values of 21 μC/cm^2 were measured for barium titanate samples doped with fluorine. Fluorine doping also increases the polarization of barium titanate. Husimi [39] reported values as high as 44 μC/cm^2 and Blood, et al. [11], 150 μC/cm^2 for special ceramic samples probably having a surface layer with space charges.

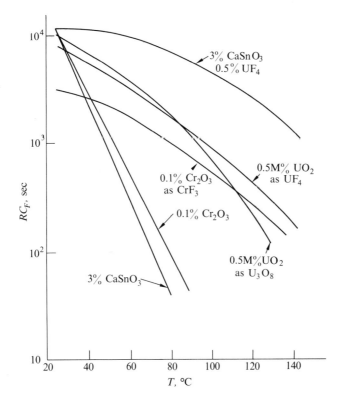

Figure 11.15. Experimental time constants of barium titanate with several additives. (Adapted from Hoh [37].)

Polarization may be increased by a deficiency in oxygen or an excess of BaO. Lead can also have a very important effect on the polarization, and a value of 100 μC/cm^2 has been reported by Burfoot [14] for lead titanate. The latter has a very high Curie point (490°C). The Curie point can be chosen at the desired value by using solid solutions of barium titanate with compounds of higher Curie points, such as $PbTiO_3$, or of lower Curie points, such as $CaTiO_3$, $SrTiO_3$, and $SeSnO_2$, among others.

11.8 Power Output, Losses, and Efficiency

Power output. The energy output of a ferroelectric generator is given by Eq. (11.26). The voltage output V_2 can be related to the electric field E_2 by the relation $V_2 = E_2 d$. This is the peak voltage of the pulse-shaped output. The peak current of the discharge can then be determined by the relation $I_L = V_2/R_L$, where R_L is the resistance of the load. The peak power output is

$$\hat{P}_{out} = I_L V_2 = d^2 E_2^2/R_L.$$

The average output current becomes, for $q_1 = 0$ (see Fig. 11.5), $I = qf$, where f is the frequency of the temperature cycling. Since

$$q = \mathscr{P}S, \tag{11.32}$$

where $\mathscr{P}$ is the polarization, the average current is $I = \mathscr{P}Sf$, where S is the surface area of the ferroelectric cell. The average power output can be obtained by multiplying the energy output by the frequency of the temperature cycling $P = \mathscr{E}_2 f$. Taking into consideration Eqs. (11.26) and (11.32), $P = \frac{1}{2}S\mathscr{P}\,\Delta Vf$, and, replacing ΔV by its value from Eq. (11.30) yields

$$P = \tfrac{1}{2}S\mathscr{P}V_1 f\left(\frac{\varepsilon_{r1}}{\varepsilon_{r2}} - 1\right).$$

It is often preferable to have ε_{r1} at the Curie point and ε_{r2} at a lower temperature. It is, however, possible to operate between the Curie point and a higher temperature, but this will lead to a smaller value of the power output. In the case where temperature 1 is the Curie point,

$$P = \tfrac{1}{2}\mathscr{V}\mathscr{P}E_c f\left(\frac{\varepsilon_{rc}}{\varepsilon_{r2}} - 1\right), \tag{11.33}$$

where $\mathscr{V}$ is the volume of the ferroelectric cell.

Losses. The most important source of loss is the heat required to bring the uncharged ferroelectric cell to the Curie point. This amount of heat ΔP_ℓ is itself equal to the sum of the specific heating of the ferroelectric cell, the heat of transition $\Delta P_{\ell t}$ necessary to overcome the spontaneous polarization of the ferroelectric material, and the heat $\Delta P_{\ell e}$ required by the electrodes. Thus

$$\Delta P_\ell = Mc_v f\,\Delta T\mathscr{V} + \Delta P_{\ell t} + \Delta P_{\ell e}, \tag{11.34}$$

where c_v is the heat capacity per unit mass at constant volume and M is the total mass of the ferroelectric material. Note that $\Delta P_{\ell t}$ is not required if the ferroelectric cell operates between the Curie point and a higher temperature, and $\Delta P_{\ell e}$ can often be neglected for small converter masses. The only important source of loss is therefore given by the first term on the right-hand side of Eq. (11.34).

It may also be necessary to take into consideration the loss of energy by the charging battery. Furthermore, the temperature cycling may necessitate some kinetic energy input to rotate the overall structure of the ferroelectric generator. This energy input is considered as a source of loss.

Efficiency. The efficiency of a ferroelectric cell is given by the ratio

$$\eta = P/(P + \Delta P_\ell).$$

Taking into account Eq. (11.33) and the first term of Eq. (11.34), this ratio becomes

$$\eta = \frac{\mathscr{P}E_c(\varepsilon_{rc}/\varepsilon_{r2} - 1)}{\mathscr{P}E_c(\varepsilon_{rc}/\varepsilon_{r2} - 1) + 2Mc_vT_2(\Delta T/T_2)}. \tag{11.35}$$

In this equation, both the ratio $\varepsilon_{rc}/\varepsilon_{r2}$ and c_v are functions of the temperature difference ΔT. Note that because of the temperature cycling, this efficiency should be less than the Carnot efficiency $\eta_c = \Delta T/T_2$. Equation (11.35) can also be written as

$$\eta = \left[1 + \frac{2Mc_v\,\Delta T}{E_c^2\varepsilon_0\varepsilon_{rc}(\varepsilon_{rc}/\varepsilon_{r2} - 1)}\right]^{-1}. \tag{11.36}$$

When the dielectric permittivity follows the Curie Weiss law,

$$\varepsilon_r = \varepsilon_{r1} + \mathscr{C}_F T_c/\Delta T,$$

where ε_{r1} is a constant of the material which is not necessarily unity, Eq. (11.36) becomes

$$\eta = \frac{1}{1 + 2Mc_v\varepsilon_0\mathscr{C}_F T_c/D_c^2}. \tag{11.37}$$

This is the simplified equation of efficiency where $D_c = \varepsilon_0\varepsilon_{rc}E_c$. To determine a more accurate expression, the thermodynamics of the cycle should be taken into consideration. This calculation has been done by Clingman and Moore [19], who found that

$$\eta = \frac{\dfrac{1}{M}\displaystyle\int_0^{D_c} \dfrac{D}{\varepsilon_0}\left[\dfrac{1}{\varepsilon_r(D, T_2)} - \dfrac{1}{\varepsilon_r(D, T_c)}\right] dD}{\displaystyle\int_{T_c}^{T_2} c_v(D_c, T)\, dT + T_2\displaystyle\int_0^{D_c} \dfrac{D}{\varepsilon_0}\cdot\dfrac{\partial}{\partial T}\left[\dfrac{1}{\varepsilon_r(D, T_2)}\right] dD}.$$

For many ferroelectric materials E is not directly proportional to the electric displacement D, but is given by

$$E = a_0 D + b_0 D^3,$$

where

$$a_0 = 1/\varepsilon_0\varepsilon_{r1}$$

and

$$b_0 = \frac{16}{27}\cdot\frac{1}{\varepsilon_0^2 E_0^2}\left(\frac{\Delta T}{\mathscr{C}_F T_c}\right)^3.$$

The result of this case leads to a higher value for the efficiency than the result

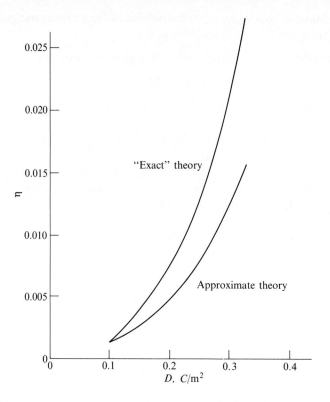

Figure 11.16. Energy conversion efficiency as a function of electric displacement: $T_c = 125°C$, $T_h = 155°C$, and $\eta_c = 7\%$. (From Clingman and Moore [19].)

given by Eq. (11.37). This "exact" result is shown in Fig. 11.16 as reported by Clingman and Moore [19], and compared to the approximate result of Eq. (11.37). The "exact" efficiency is also a function of the temperature difference as shown in Fig. 11.17. The "exact" value of the efficiency is given by

$$\eta = \frac{(\Delta T/T_c) + \frac{1}{2}D_c^2 \varepsilon_0 \mathscr{E} \, \Delta b_0}{1 + 8\pi M c_v \, \Delta T \varepsilon_0 \mathscr{E}/D_c^2 + \frac{1}{2}D_c^2 \varepsilon_0 \mathscr{E}_F[T_c b_0'(T_c) + \Delta b_0]}, \qquad (11.38)$$

where b_0' is a constant.

11.9 Specific Power Output

The specific power output is the electric output of a ferroelectric cell per unit mass. If ρ is the mass density of the ferroelectric material, then, taking into consideration Eq. (11.33), one has

$$\frac{P}{M} = \frac{1}{2}\mathscr{P}E_c f\left(\frac{\varepsilon_{rc}/\varepsilon_{r2} - 1}{\rho}\right).$$

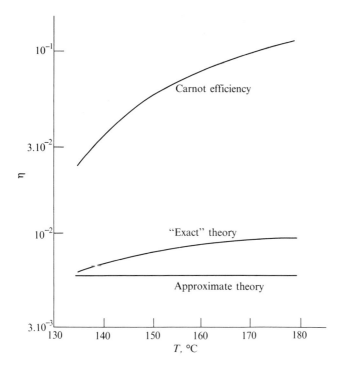

Figure 11.17. Energy conversion efficiency as a function of the heat source temperature: $T_c = 125°C$ and $D = 0.2\ C/m^2$. (From Clingman and Moore [19].)

When the heat energy is obtained from the solar radiation, as is the case in satellite applications, the power input can be written as

$$P_{in} = S\alpha_\lambda \Phi_s \cos(\theta),$$

where α_λ is the *absorptivity* of the ferroelectric cell surface, Φ_s is the solar irradiation, and θ is the angle between the normal to the cell plate and the incident radiation. In the case of a rotating cylinder, each plate is exposed to the solar radiation twice every cycle. In this case, $\theta = \omega t$, and

$$\mathscr{E}_{in} = \int_0^{(1/2)f} P_{in}\ dt = \frac{S\alpha_\lambda \Phi_s}{\pi f}. \tag{11.39}$$

This absorbed heat causes a temperature rise ΔT in the ferroelectric cell such that $\mathscr{E}_{in} = Mc_v\ \Delta T$, or

$$\Delta Tf = \frac{\alpha_\lambda \Phi_s}{\rho\pi\ dc_v}. \tag{11.40}$$

The power output is $\mathscr{E}_{out} = \eta\mathscr{E}_{in}$, and the power per unit mass becomes

$$\frac{P}{M} = \frac{\eta\alpha_\lambda\Phi_s}{\rho\,d\pi} = \eta c_v f\,\Delta T. \tag{11.41}$$

From Eqs. (11.40) and (11.41), note that both ΔTf and P/M are inversely proportional to the thickness of the ferroelectric material. This means that the specific output is largest for the thinnest cell. For instance, for 5% efficiency with $\Phi_s = 1.4\ \text{kW/m}^2$, $\rho = 5500\ \text{kg/m}^3$, and $d = 1\ \mu$, the specific weight becomes equal to 3.65 kW/kg. This is a very satisfactory value for a solar converter. Therefore, one of the qualities of ferroelectric power generators is that in spite of their low efficiency, the specific power output can be made large if special care is given to the construction. For the same data given above, the output power density will be 20 W/m^2 and the product ΔTf will be 138.5°K/sec, leading to a temperature variation of, say, 13.85°K at 10 Hz.

11.10 Ferroelectric Crystal Growth and Ceramic Preparation

Examples of preparation have been reported by Fotland and Mayer [30]. The crystals were prepared by dissolving the ferroelectric material in fused salt and then cooling the solution slowly. The form of the crystals and their characteristics depend greatly upon the cooling rate and the impurities introduced into the melt. The cooling rate is approximately 40°C/hr and the soaking temperatures* were varied from 1050 to 1200°C for different experiments.

The crystals were grown in a capped crucible of platinum, and the fused salt (flux) was KF, NaF, or NaCl. In all the experiments, 12.5 g of ferroelectric raw material was used along with 36 g of flux and 0.016 g of Fe_2O_3. The function of the Fe_2O_3 is to compensate for the electron acceptor levels which are reduced during the crystal growth. If care is not taken to stabilize the temperature, the resulting crystals may be a large number of very small cubes. This can be explained by the formation of a great number of nucleation centers during the cooling of the solution. By stabilizing the temperature of the crucibles with the aid of thermal mass and temperature control systems, Fotland and Mayer obtained large crystal plates about 1 cm thick. With this method, crystals of several materials were obtained: barium titanate, $Na_{0.9}Cd_{0.1}NbO_3$, $NaNbO_3$, and mixtures of $BaTiO_3$ and $PbTiO_3$.

The preparational procedure of a ceramic is determined by the goals sought in the finished ferroelectric ceramic. The general procedure can be described as follows: the raw material is powdered, weighed, and put into jars. It is then tumbled for about $\frac{1}{2}$ hr to initiate mixing. The contents of the jars are then ground in a mortar for 15 min and passed through a micropulverizer. The fine powder, in the form of slugs (2.54 cm in diameter and 1.27 cm in height) is pressed under a 360-kg load and the slugs are then calcined at about 980°C, broken and ground,

---

* Initial temperature in the crucible.

and calcined again at 1100°C. After this second calcination the slugs are broken again and ground in a rubber lined ball mill using barium titanate balls. The powder obtained is mixed with a binder (1.5% carboax and 1.5% polyvinyl alcohol in a water solution) and dried in an oven at about 250°C. The dry powder is compressed under loads in the 340- to 550-kg range to thicknesses in the 2- to 95-mil range. The resulting pellets are preburned for 30 min at 815°C to remove the binder. The temperature is then raised to the vitrification firing temperature, about 1260°C, for 1 hr. The firing oven is a zirconium oxide crucible. The characteristics of the finished ceramic depend on the exact values of temperatures and times mentioned above. The firing temperature may vary from 1090 to 1260°C, the firing time may vary from 30 min to 6 hr, and the pressure may vary from 750 to 60,000 psi. Batches of $Na_{0.9}Zn_{0.1} \cdot NbO_3$ were prepared by Fotland and Mayer by this method. Additives were chosen to improve the squareness ratio of the ferroelectric material to be used in memory applications.

11.11 Spontaneous Polarization of Barium Titanate

By considering the displacement of the Ti^{4+} ion in $BaTiO_3$ along the x-direction only, quantitative predictions of the spontaneous polarization can be made. The total polarization associated with the stable positions $x = \pm r_0$ is equal to the sum of the spontaneous polarization and the polarization induced by the local field $\mathscr{P} = \mathscr{P}_{sp} + \mathscr{P}_{ind}$. The spontaneous polarization is given by

$$\mathscr{P}_p = 4er_0 \tag{11.42}$$

for a cell of unit volume. The local field E induces a force on the titanium ion equal to

$$F = 4eE_{loc}. \tag{11.43}$$

This force should be balanced by a force of reaction which tends to bring the ion back to the center of the crystal. Assume that this force is due to an energy distribution given by

$$\mathscr{E} = \mathscr{E}_{max} \cos(\pi x/r_0).$$

The restoring force F_{res} is then

$$F_{res} = -\frac{\partial \mathscr{E}}{\partial x} = \frac{\mathscr{E}_{max}\pi}{r_0} \sin\left(\frac{\pi x}{r_0}\right). \tag{11.44}$$

Of interest is the value of the force in the neighborhood of the stable positions, such as $x = +d$ with d very small. In this case, Eq. (11.44) becomes

$$F_{res} = \mathscr{E}_{max}\pi^2 d/r_0^2.$$

This force should be equal to the force given by Eq. (11.43). Thus the value of d can be obtained readily,

$$d = \left(\frac{2r_0}{\pi}\right)^2 \frac{eE}{\mathscr{E}_{max}}.$$

The induced polarization per unit volume is given by $\mathscr{P}_{\text{ind}} = -ed$, or

$$\mathscr{P}_{\text{ind}} = -\left(\frac{2r_0}{\pi}\right)^2 \frac{e^2 E_{\text{loc}}}{\mathscr{E}_{\text{max}}},$$

and taking into consideration Eq. (11.42), the total polarization becomes

$$\mathscr{P} = 4er_0\left(1 - \frac{r_0}{\pi^2} \cdot \frac{eE_{\text{loc}}}{\mathscr{E}_{\text{max}}}\right).$$

By first applying an electric field in the positive direction on the barium titanate, and then in the reverse direction as shown in Fig. 11.1, the titanium ion will jump from the position $x = r_0$ to the position $x = -r_0$ and back. The resulting energy change will be equal to $2\mathscr{E}_{\text{max}}$ and the corresponding local field becomes

$$E_{\text{loc}} = \tfrac{1}{2}\mathscr{E}_{\text{max}}/er_0. \tag{11.45}$$

When the titanium ion is at $x = r_0$, $E_{\text{loc}} = 0$ and $\mathscr{P} = 4er_0$. By increasing the local field in the negative direction, there is an inflection point due to the variation in energy. If the field continues to be increased until the polarization is cancelled for the value given by Eq. (11.45), the titanium ion will suddenly jump to the position $x = +r_0$. This corresponds to a double hysteresis loop characteristic of barium titanates, which has also been experimentally observed [64].

11.12 Ferroelectric Generator Systems

One of the ferroelectric systems proposed by Hoh [35] is an ac circuit consisting of two ferroelectric panels connected through the load. An initial charge is supplied between the load and the ground. The panels are then temperature cycled in such a way that when one panel is heated the other is cooled. Thus a charge Δq is forced back and forth through the load, resulting in an alternating current. To make up for the dielectric losses, a feedback made up of a transformer-diode combination is provided. In Hoh's experiments [35], 12 converter panels were mounted on a rotating cylinder and heat was supplied by a radiant heat source. The dimensions of the panels were 3.8 by 5 cm and the output current was rectified by a commu-tator. To ensure contacts between the panel of highest temperature and that of lowest temperature, the angular settings of the commutator springs were varied. The electrodes on the ferroelectric cells were made up of a silver paint, and the outer electrodes were covered by a special substance to increase heat absorption and emission. The cylinder had a rotational speed varying between 10 and 30 rpm. For an initial input of 90 V, the dc output voltage was 50 V and the output current was 20 μA for $\omega = 10$ rpm. This means that the output power was about 1 mW.

The high-voltage converter of Fig. 11.9 can be made to use both the radiation of the sun and the effect of the wind. The wind will rotate the configuration, thus providing the necessary temperature cycling.

The application of ferroelectric generators in space involves many difficulties, the most important being related to the rejection of heat by radiation. Since the heat is absorbed by the converter twice every cycle and radiated continuously, as in a blackbody,

$$P_{out} = Se_{\partial}\mathscr{B}_1 T^4, \tag{11.46}$$

where e_{∂} is the emissivity and $\mathscr{B}_1$ is the Stephan-Boltzmann constant,

$$\mathscr{B}_1 = 5.67 \times 10^{-8} \text{ W/m}^{2 \cdot \circ}\text{K}^4.$$

The radiation is assumed to be in the neighborhood of absolute zero temperature. For rotating flat plates, Eq. (11.46) should be equal to the input power obtained from Eq. (11.39),

$$\frac{S\alpha_\lambda \Phi_s}{2\pi} = Se_{\partial}\mathscr{B}_1 T^4.$$

Thus,

$$T = \left(\frac{\alpha_\lambda \Phi_s}{2\pi e_{\partial}\mathscr{B}_1}\right)^{1/4}.$$

In outer space $\Phi_s = 1.4 \text{ kW/m}^2$ and $\alpha_\lambda = e_{\partial}$; thus $T = 250^\circ\text{K}$. This temperature of -23°C is far below the Curie points of the doped barium titanate series. Two solutions to the problem are proposed: use of optical concentration and use of ceramics of low Curie points (or both).

11.13 Future Trends

In spite of the pessimistic predictions of the early workers, it seems that ferroelectric power generation has attractive qualities in many applications. However, a great amount of research is needed in both the theoretical and the experimental fields before the potential of this method can be ascertained.

To obtain higher efficiencies, higher temperature differences are required, as shown by Fig. 11.17. From Eq. (11.40), it can be seen that this high temperature difference corresponds to a low output frequency. To avoid this, the frequency can be multiplied to the desired level by the use of a multiple panel system.

It was mentioned above (Eq. 11.41) that the highest specific output power is obtained for the lowest cell thickness. This means that development of techniques for the production of thin ferroelectric films is required. Many techniques were proposed, such as chemical precipitation, vacuum evaporation, flame spraying, electrophoresis,* and sputtering. However, these methods have failed to yield films of good ferroelectric qualities for power generation. Surely, more theoretical understanding of the ferroelectric behavior of thin films is needed to direct the experimental programs for obtaining thin films which have high polarization and high dielectric strength and which can withstand thermal shocks without damage.

The experimental ferroelectric generators have low efficiencies, but values exceeding those of the best direct energy converters are predicted. The quality

---

*   Electrophoresis is the movement of suspended particles under the effect of an applied electric field [81, Chapter 12].

of a ferroelectric converter lies in its use at low temperatures. If adequate ferro-electric materials are obtained, this method promises specific power output, radiation resistance, and cost that are better than photovoltaic cells for the conversion of solar energy. For cascaded ferroelectric cells, extremely high voltages can be obtained. The highest voltage possible will be limited only by the dielectric strength of the ferroelectric material.

The serious disadvantage of ferroelectric power generators is the need for temperature cycling. Although this cycling may enhance the rejection of heat by radiation in space applications, it is clear that this is a serious handicap to further development and research. Another problem is the "aging" of the ferroelectric material, related to the deterioration of its qualities with time. It seems, nevertheless, that ferroelectric generators can be of great interest in such applications where lightweight, unattended generators are required and where solar energy can be used in conjunction with the wind for temperature cycling.

PROBLEMS

11.1 (a) Show that the orbits followed by N electrons and N ions entering a unit volume in the presence of uniform magnetic and electric fields are cycloids. The magnetic field $\mathbf{B}$ is perpendicular to the electric field $\mathbf{E}$ and the initial velocity of the charge carriers is zero. (b) Show that the result of such a motion leads to the establishment of a polarization

$$\mathscr{P} = N(M_e + M_i)\mathbf{E}/B^2,$$

where M_e is the mass of the electrons and M_i is the mass of the ions.

11.2 Consider a monoatomic gas of n atoms per unit volume in the presence of an electric field $\mathbf{E}$. Since the distances between the molecules of the gas are much larger than the size of the molecules, interactions between molecules become negligible. By equating the force exerted by the electric field on each molecule to the Coulomb restoring force, show that a polarization $\mathscr{P}$ is established and is equal to

$$\mathscr{P} = 4\pi\varepsilon_0 n r_a^3 \mathbf{E},$$

where r_a is the radius of each atom.

11.3 From Eq. (11.14) and Problem 11.2, find the value of the electronic polarizability $\varkappa_e$. This polarizability can be measured and values of some noble elements are given in the table below. Find the radii of these elements in Angstroms. Note that the values of r_a found are different from the values given in the table. This discrepancy is due to the several effects which were neglected.

Gas	$\varkappa_e$, m^3	r_a, Å
Helium	0.22×10^{-40}	0.95
Neon	0.43×10^{-40}	1.15
Argon	1.80×10^{-40}	1.40
Krypton	2.70×10^{-40}	1.60
Xenon	4.40×10^{-40}	1.75

11.4 In polyatomic molecules, the dipole moment of the entire molecule is equal to the vector sum of the individual moments. Consider n molecules per unit volume having each a dipole moment μ. In the presence of an electric field, the distribution of the moments ceases to be uniform. It can be obtained from Boltzmann's distribution law which states that, the probability for μ lying between θ and $\theta + d\theta$ is proportional to $\exp(-\mathcal{E}/kT)\,d\Omega$, where θ is the angle between $\mathbf{r}$ and $\mathbf{E}$, $d\Omega$ is the element of solid angle corresponding to $d\theta$, and $\mathcal{E}$ is the potential energy of the dipole. (a) Find the value of $\mathcal{E}$, and (b) show that the orientational polarizability per molecule is $\varkappa_0 = \mu_p^2/3kT$. This is Eq. (11.18).

11.5 From Fig. 11.1 draw the $C_F(V)$ characteristic of a ferroelectric material. Under which conditions can Eq. (11.21) be considered as satisfactory?

11.6 By writing the balance of forces on a dielectric atom, study the effect of a high-frequency electric field. (a) Write the equation of motion of the electrons. (b) Show that the polarizability becomes complex and calculate its value by considering both Hook's law and the damping effect.

11.7 The wind blowing at a speed of 30 km/hr powers a wheel of 5-m radius in the presence of solar irradiation. The wheel is coupled to a ferroelectric generator the elements of which face on one side the sun and on the other side are in contact with a water sink. The material is Rochelle salt with $T_c = 23°C$ and $\varepsilon_{rc} = 1400$. The temperature of the water is $T_a = 10°C$ at which $\varepsilon_{ra} = 105$. The capacitance of the generator at the Curie point is $C_c = 1\ \mu F$. Find (a) the output voltage and (b) the output current of this generator.

11.8 Calculate the power output in a load of 100 Ω and the efficiency of the generator of Problem 11.7. What type of material would be ideal for this type of application (i.e., solar-wind energy utilization)? Assuming that the heat input of the sun is 0.8 kW/m^2 and that the ambient temperature is 23°C, what part of this heat would be necessary to reach 23°C and maintain it, or is it necessary to add heat from another source?

11.9 Prove Eq. (11.31).

11.10 To dramatize the fact that the Clausius-Mossotti theory is an oversimplification, (a) calculate the Curie temperature of water, knowing that $\mu_p = 6.0 \times 10^{-30}$ C·m and $\varkappa = 2.5 \times 10^{-34}$ C·m^2/V. (b) Following the Clausius-Mossotti theory, what would be the state of water at the ambient temperatures after the occurrence of a spontaneous polarization? [*Answers:* (a) $T = 1350°K$; (b) solid state.] From this, it can be concluded that the Mossotti catastrophe does not occur in practice and is a result of the oversimplified theory. The most stable arrangement in a simple cubic lattice is that in which the dipoles are alternately arranged; a material having such a property is called an *antiferroelectric* material.

11.11 Assume that you can fabricate at will a ferroelectric material to be used as an energy converter. What should be the characteristics of such an ideal material?

11.12 Prove Eq. (11.38).

11.13 (a) For which material has the "exact" curve of Fig. 11.16 been drawn? (b) Can you deduce the different parameters given in Eq. (11.38) from this curve?

11.14 (a) Prove the values given in the examples treated in Section 11.9. (b) What would be the power output and the specific output if all the parameters are the same but for a solar flux equal to 0.7 kW/m^2?

11.15 Design a ferroelectric generator producing 30 W of ac electrical power. The heat energy is produced by a hot flow of water. (a) Choose the material, (b) calculate the dimensions of the generator, and (c) calculate its efficiency. The generator should work under normal conditions of temperature and pressure (25°C, 1 atm).

REFERENCES AND BIBLIOGRAPHY

1. Ackerman, W., "Dependence of Pyroelectricity on Temperature," *Ann. Physik*, Vol. 46, p. 197, 1916.

2. Alonso, R., and T. Conley, *Ferroelectric Applications to Digital Computers*, U.S. Aberdeen Proving Ground Ballistics Research Laboratory, PB 120239, AD 83112, Dec. 1955.

3. Anderson, J. R., "Ferroelectric Storage Elements for Digital Computers and Switching Systems," *Electrical Engineering*, Vol. 71, No. 10, p. 916, 1952.

4. Anderson, J. R., "New Type of Ferroelectric Shift Register," *IRE Trans. Prof. Group. Electronic Computer*, Vol. EC5, p. 184, 1956.

5. Anderson, J. R., *Ferroelectric Storage Array*, U.S. Patent 2,905,928, 1959.

6. Ballman, A. A., and H. Brown, "The Growth and Properties of Strontium Barium Metaniobate $Sr_{1-x}Re_xNb_2O_6$, a Tungsten Bronze Ferroelectric," *J. Crystal Growth*, Vol. 1, p. 311, Dec. 1967.

7. Beadle, C. W., and J. W. Dally, "Experimental Methods for Investigating Strain Wave Propagation and Associated Charge Release in Ferroelectric Ceramics," *Experimental Mechanics*, Vol. 4, p. 70, March 1964.

8. Berlincourt, D. A., "Recent Developments in Ferroelectric Transducer Materials," *IRE Trans. on Ultrasonic Eng. PGUE-4*, p. 53, Aug. 1956.

9. Birks, J. B. (ed.), *Progress in Dielectrics*, Wiley, New York, 1960.

10. Bline, R., and S. Svetina, "Cluster Approximations for Order Disorder-Type Hydrogen-Bonded Ferroelectric," *Phys. Rev.*, Vol. 147, p. 423, 1966.

11. Blood, H. L., S. Levine, and N. H. Roberts, "Anomalous Polarization in Undiluted Ceramic $BaTiO_2$," *J. Appl. Phys.*, Vol. 27, No. 6, p. 660, 1956.

12. Buck, D. A., *Ferroelectrics for Digital Information Storage and Switching*, MIT Digital Compt. Lab. Report R-212, 1952.

13. Buessem, W. R., and P. A. Marshall, *Crystal Chemistry of Ceramic Dielectrics*, Linden Labs. Inc., State College Pa., Dec. 1962.

14. Burfoot, J. C., "Ferroelectrics," *Wireless World*, Vol. 65, p. 326, 1959.

15. Busch, G., "Neue Seignette-Elektrika," *Helvetica Physica Acta*, Vol. 11, p. 269, 1938.

16. Callaby, D. R., and E. Fatuzzo, "Oscillations of Ferroelectric Bodies," *J. Appl. Phys.*, Vol. 35, No. 8, p. 2443, 1964.

17. Ceve, P., M. Modesto, and M. Schara, "Growth of the Ferroelectric Crystals, $NaH_3(SeO_3)_2$ and $NaD_3(SeO_3)_2$," *J. Crystal Growth*, Vol. 1, p. 160, Aug. 1967.

18. Childress, J. D., "Application of a Ferroelectric Material in an Energy Conversion Device," *J. Appl. Phys.*, Vol. 33, No. 5, p. 1793, 1962.

19. Clingman, W. H., and R. G. Moore, "Application of Ferroelectricity to Energy Conversion," *J. Appl. Phys.*, Vol. 32, No. 4, p. 675, 1961.

20. Cochran, W., "Crystal Stability and the Theory of Ferroelectricity," *Phys. Rev. Letters*, Vol. 3, p. 412, 1959.

21. De Bretteville, A. P., Jr., "Antiferroelectric $PbZrO_3$ and Ferroelectric $BaTiO_3$ Phenomena," *Ceramic Age*, Vol. 61, pp. 18, 92, 1953.

22. Devonshire, A. F., "Theory of Barium Titanate I," *Phil. Mag.*, Vol. 40, p. 1040, 1949.

23. Devonshire, A. F., "Theory of Barium Titanate II," *Phil. Mag.*, Vol. 42, p. 1065, 1951.

24. Devonshire, A. F., "Review of Ferroelectricity and Antiferroelectricity in the Perovskites," *Rep. Brit. Electr. Res. Assoc.*, Rep. L/T 298, 1953.

25. Devonshire, A. F., "Theory of Ferroelectrics," *Advances in Physics*, Vol. 3, p. 85, 1954.

26. Dungan, R. H., et al., "Lattice Constants of Dielectric Properties of Barium Titanate-Barium Stannate-Strontium Titanate Bodies," *J. Am. Ceramic Soc.*, Vol. 35, p. 318, 1952.

27. Fang, P. H., and W. S. Brower, "Temperature Dependence of the Breakdown Field of Barium Titanate," *Phys. Rev.*, Vol. 113, No. 1, p. 456, 1959.

28. Fang, P. H., and I. E. Stegun, "Ferroelectric Switching and the Sievert Integral," *J. Appl. Phys.*, Vol. 34, No. 2, p. 284, 1963.

29. Fotland, R. A., "Ferroelectrics as Solid-State Devices," *Elec. Manufacturing*, March 1958.

30. Fotland, R. A., and E. F. Mayer, "Ferroelectric Devices," WADC-58347; Astia AS204093, 1958.

31. Gerson, R., P. C. Chou, and W. J. James, "Ferroelectric Properties of $PbZrO_3$-$BiFeO_3$ Solid Solutions," *J. Appl. Phys.*, Vol. 38, No. 1, p. 55, 1967.

32. Habluetzel, J., "Sehweres Seignettesaltz Dielektrische Untersuchungen an $UNaC_4$-$H_2D_2O_6 \cdot 4H_2O$-Kristallen," *Helvetica Physica Acta*, Vol. 12, p. 489, 1939.

33. Hamilton, D. L., "Existence and Origin of a Polarization Threshold Field in Bismuth Titanate," *J. Appl. Phys.*, Vol. 38, No. 1, p. 10, 1967.

34. Hayashi, F., *Observation on Pyroelectricity*, Doctorate Dissertation, Gottingen University, Germany, 1912.

35. Hoh, S. R., *Ferroelectric Energy Converters*, Solid State Research Conference, Cornell University, Ithaca, N.Y., June 8, 1959.

36. Hoh, S. R., "Ferroelectric Energy Converters," *Elec. Comm.*, Vol. 37, p. 23, 1961.

37. Hoh, S. R., "Conversion of Thermal to Electrical Energy with Ferroelectric Materials," *Proc. IEEE*, Vol. 51, No. 5, p. 838, 1963.

38. Hoh, S. R., and F. E. Pirigyi, "Barium Titanates with Improved Insulation Resistance and Time Constant," *J. Am. Ceramic Soc.*, Vol. 46, p. 816, 1963.

39. Husimi, K., "Ultra-Low-Velocity Component of Spontaneous Polarization in $BaTiO_3$ Single Crystal," *J. Appl. Phys.*, Vol. 29, No. 9, p. 1379, 1958.

40. Ichikawa, M., and T. Mitsui, "Ferroelectricity in $H \cdot NH_4(ClCH_2COO)_2$," *Phys. Rev.*, Vol. 152, No. 12, p. 495, 1966.

41. Jackson, W., "The Structure, Electrical Properties and Potential Applications of the Barium Titanate Class of Ceramic Materials," *Proc. Inst. Elect. Engrs.*, Pt. III, Vol. 97, p. 285, 1950.

42. Jaffe, B., "Antiferroelectric Ceramics with Field-Enforced Transitions, New Non-Linear Circuit Element," *Proc. IRE*, Vol. 49, p. 1264, 1961.

43. Jaffe, B., "Properties of Ferroelectric Ceramics in the Lead-Titanate-Zirconate System," *Proc. IRE*, Vol. 109, p. 351, 1962.

44. Jain, G. C., and K. Ravindran, "Doping of Ferroelectric Solutions of $(Ba, SrTiO_3)$

with In_2O_3 to Improve the RC Time Constant," *Adv. Energy Conversion*, Vol. 6, No. 4, p. 233, 1966.

45. Jaynes, E. T., *Ferroelectricity*, Princeton University Press, Princeton, N.J., 1953.

46. Jona, F., and G. Shirane, *Ferroelectric Crystals*, Pergamon Press, New York, 1962.

47. Känzig, W., *Ferroelectrics and Antiferroelectrics Solid State Physics*, Vol. 4, Academic Press, New York, 1962.

48. Kaufman, A. B., "Obtaining Nondestructive Read-Out with Ferroelectric Memories," *Electronics*, Vol. 34, p. 47, Aug. 25, 1961.

49. Kaufman, A. B., *Ceramic Memories for Ordnance Fuzing*, Proceedings of Timers for Ordnance Symposium, H. Diamond Labs., Vol. 3, p. 27, Nov. 1966.

50. Kaufman, A. B., "Ceramic Memory; Design and Use in Programmable Timers," *Electromechanical Design*, p. 34, April 1967.

51. Kaufman, A. B., "Memories Shot from Guns," *Electronics*, Vol. 41, p. 98, Feb. 5, 1968.

52. Kell, W. L., et al., "Use of Ferroelectric Ceramics for Vibration Analysis," *J. Sci. Instruments*, Vol. 34, p. 271, 1957.

53. Kovit, B., "The Ferroelectric Power Source," *Space Aeronautics*, Vol. 3, p. 15, Dec. 1959.

54. Kurchatov, I. V., *Le Champ Moleculaire dans les Dielectriques (le Sel de Seignette)*, Hermann, Paris, 1936.

55. Kwok, P. C., and P. B. Miller, "Free Energy of Displacive Ferroelectrics," *Phys. Rev.*, Vol. 151, p. 387, 1966.

56. Land, C. E., R. H. Plumlee, and D. G. Schueler, "Ferroelectrics: Switching Properties, Logic and Memory Devices," *Electromechanical Design*, p. 46, Oct. 1966.

57. Lane, A. L., "Barium Titanate Admittance-Temperature Characteristics," *J. Acoust. Soc. Am.*, Vol. 25, p. 873, 1953.

58. Linde, R. K., "Depolarization of Ferroelectrics at High Strain Rates," *J. Appl. Phys.*, Vol. 38, No. 11, p. 4839, 1967.

59. Mason, W. P., and R. F. Wick, "Ferroelectronics and the Dielectric Amplifier," *Proc. IRE*, Vol. 42, p. 1606, 1954.

60. Mason, W. P., "Piezoelectric and Ferroelectric Devices," Part 17 of *Molecular Science and Molecular Engineering*, A. R. Von Hippel (ed.), Wiley, New York, 1959.

61. McIrvine, E. C., "Comment on the Ferroelectric Polarization Field Effect," *Phys. Rev.*, Vol. 148, p. 528, 1966.

62. Megaw, H. D., *Ferroelectricity in Crystals*, Methuen, London, 1957.

63. Merz, W. J., "The Electric and Optical Behavior of $BaTiO_3$ Single Domain Crystals," *Phys. Rev.*, Vol. 76, No. 8, p. 1221, 1949.

64. Merz, W. J., "Double Hysteresis Loop of $BaTiO_3$ at the Curie Point," *Phys. Rev.*, Vol. 91, No. 3, p. 513, 1953.

65. Merz, W. J., and J. R. Anderson, "Ferroelectric Storage Devices," *Bell Labs. Rec.*, Vol. 33, p. 335, 1955.

66. Miller, B., "Conversion to High Voltage Via Ferroelectrics in Works," *Electronic News*, p. 7, Sept. 7, 1959.

67. Miller, N. C., and P. A. Casabella, "Na^{23} Quadrupole Interaction in Ferroelectric Rochelle Salt," *Phys. Rev.*, Vol. 152, No. 12, p. 228, 1966.

68. Mitsui, T., "Theory of the Ferroelectric Effect in Rochelle Salt," *Phys. Rev.*, Vol. 111, p. 1259, 1958.

69. Muirhead, J. C., and W. J. Fenrick, *Studies on Shock Wave Pressure-Time Gauges*, Suffield Tech. Note No. 139, March 1964.

70. Muirhead, J. C., and W. J. Fenrick, "On the Anomalous Behavior of Blast Pressure Gauges with Ferroelectric Ceramic Elements," *J. Sci. Instruments*, Vol. 41, p. 483, 1964.

71. Muirhead, J. C., "On Charge Release in Ferroelectric Ceramics," *Proc. IEEE*, Vol. 53, No. 3, p. 327, 1965.

72. Müser, H. E., and H. Flunkert, "Upper Curie Temperature and Domain Structure in Different Fields of Seignette Salt Crystal," *Zs. Phys.*, Vol. 150, p. 21, 1958 (German).

73. Müser, H. E., "Measurement of the Dielectric Non-Linearity of Rochelle Salt," *Zs. Angew. Phys.*, Vol. 12, p. 300, 1960 (German).

74. Nelson, F. W., *Ferromagnetic and Ferroelectric One-Shot Explosive Electric Transducers*, Sandia Corp. Tech. Memo 230-B-56-51, Nov. 1956.

75. Nettleton, R. E., "Lattice-Dynamical Theory of Switching in Barium Titanate Single Crystals," *J. Appl. Phys.*, Vol. 38, No. 7, p. 2775, 1967.

76. Nettleton, R. E., "Renormalisation and the First-Order Ferroelectric Transition in Perovskites," *J. Appl. Phys.*, Vol. 38, No. 7, p. 2766, 1967.

77. Nussbaum, A., *Electromagnetic and Quantum Properties of Materials*, Prentice-Hall, Englewood Cliffs, N.J., 1966, Chap. 6.

78. Peck, D. B., *Ferroelectric Transducer*, U.S. Patent 2,782,600, Feb. 1959.

79. Penney, G. W., et al., "Dielectric Amplifiers," *Commun. and Electronics*, Vol. 5, p. 68, 1953.

80. Perry, C. H., and E. F. Young, "Infrared Studies of Some Perovskite Fluorides," *J. Appl. Phys.*, Vol. 38, No. 11, p. 4616, 1967.

81. Pulvari, C. F., "An Electrostatically Induced Permanent Memory," *J. Appl. Phys.*, Vol. 22, p. 1039, 1951.

82. Pulvari, C. F., *Snapping Dipoles of Ferroelectrics as a Memory Element for Digital Computers*, Catholic University of America, School of Engineering and Architecture, Progress Report 4, AD35404, Jan. 1953.

83. Pulvari, C. F., "Ferroelectrics and their Memory Applications," *IRE Trans. on Component Parts*, Vol. CP-3, p. 3, March 1956.

84. Pulvari, C. F., "Research on Barium Titanate and Other Ferroelectric Materials for Use as Information Storage Media," Catholic University of America, Electrical Engineering Department, Report WADC Tech. Report 58-657, AD209530, Sept. 1958.

85. Roberts, S., "Dielectric and Piezoelectric Properties of Barium Titanate," *Phys. Rev.*, Vol. 71, p. 890, 1947.

86. Roberts, S., "Polarization of Ions of Perovskite-Type Crystals," *Phys. Rev.*, Vol. 81, No. 3, p. 865, 1951.

87. Röntgen, W. C., "Pyro- and Piezoelectric Investigations," *Ann. Physik*, Vol. 45, p. 737, 1914.

88. Sachse, H., *Ferroelectrika*, Springer-Verlag, Berlin, 1956.

89. Saito, Y., and S. Yamanaka, "On the Anomaly in Residual Polarization of $BaTiO_3$ Ceramics," *Trans. AIEE*, Vol. 78, p. 70, March 1959.

90. Samara, G. A., "Pressure and Temperature Dependences of the Dielectric Properties of the Perovskites $BaTiO_3$ and $SrTiO_3$," *Phys. Rev.*, Vol. 151, p. 378, 1966.

91. Sawaguchi, E., and M. Charters, *Aging of Ferroelectrics*, RADC-TR-59-128, July 1959.

92. Schubring, N. W., et al., "Polarization Reversal in Ferroelectric KNO_2," *J. Appl. Phys.*, Vol. 38, No. 4, p. 1671, 1967.

93. Shibata, T., et al., "Electrostatic Transformer Type Particle Accelerator Using Ceramic $BaTiO_3$—Ferrostac," *J. Phys. Soc. Japan*, Vol. 14, p. 227, 1959.

94. Shirane, G., G. Jona, and R. Pepinski, "Some Aspects of Ferroelectricity," *Proc. IRE*, Vol. 43, No. 12, p. 1738, 1955.

95. Triebwasser, S., "Behavior of Ferroelectric $KNbO_3$ in the Vicinity of the Cubic-Tetragonal Transition," *Phys. Rev.*, Vol. 101, p. 993, 1956.

96. Von Arx, A., and W. Bantle, "Polarisation und Spezifische Wärme von KH_2PO_4," *Helvetica Physica Acta*, Vol. 16, p. 211, 1943.

97. Von Hippel, A., et al., "High Dielectric Constant Ceramics," *Ind. Eng. Chem.*, Vol. 30, p. 1097, 1946.

98. Von Hippel, A., "Dielectrics Made to Order," *Electronics*, Vol. 24, p. 126, 1951.

99. Von Hippel, A., "Dielectric Materials and Applications," MIT Press, Cambridge, Mass., and Wiley, New York, 1954.

100. Warner, E., *Ferroelectric Devices*, Horizons, Inc., Final Report on Improved Electronic Components, WADC Tech. Report 56-362, AD97124, PD121974, Dec. 1956.

101. Wingrove, E., et al., "Hysteresis Loops in Dielectric Amplifiers," *Trans. Am. Inst. Elect. Engrs.*, Pt. I, Commun. and Electronics, Vol. 75, p. 283, 1956.

12 MISCELLANEOUS
METHODS OF POWER GENERATION

A. FERROMAGNETIC POWER GENERATION

12.1 Ferromagnetic Theory

Ferromagnetic materials are substances in which magnetic effects are strong and the permeability is a function of both the history of the specimen and the applied magnetic field. The phenomenon is similar to ferroelectricity, where the permeability μ plays the role of permittivity and where the field is magnetic instead of electric. By definition,

$$\mu = \mathbf{B}/\mathbf{H}, \qquad (12.1)$$

where $\mathbf{B}$ is the applied magnetic field and $\mathbf{H}$ is the resulting magnetic field intensity. In a ferromagnetic material, permeability changes with temperature in the vicinity of the Curie point, thus allowing the conversion of thermal energy into electrical energy by the use of a temperature cycle.

Edison [6] in 1887, was the first to patent a thermomagnetic power generator. His device used the magnetic field induced by a varying voltage around a ferromagnetic core which was subjected to a thermal cycle to produce an altered flux linkage. However, in spite of Giauque and MacDougall's [8] work on gadolinium sulfate $[Gd_2(SO_4)_3]$ at low temperatures, a thorough theoretical and experimental investigation did not start until the late 1940's with Brillouin and Iskenderian's [4] work.

In the presence of a magnetic medium, the total magnetic field present can be considered to be the sum of the field $\mathbf{B}_i$ induced by electric currents and the field $\mathbf{B}_m$ produced by the magnetic medium, $\mathbf{B} = \mathbf{B}_i + \mathbf{B}_m$.

The magnetic field $\mathbf{B}$ produced by a current in a loop of radius $\mathbf{r}_0$ at a distance $\mathbf{r}$ from its center is $\mathbf{B} = \nabla \times \mathbf{A}$ when $\mathbf{r} \gg \mathbf{r}_0$. Here $\mathbf{A}$ is the vector potential defined by

$$\mathbf{A} = \frac{\mu_0}{4\pi r^3} \, m \times \mathbf{r},$$

with m being the magnetic dipole moment, $\mathit{m} = IS\mathbf{k}$, where S is the surface area of the loop, I is the current in the loop, $\mathbf{k}$ is the unit vector in the direction perpendicular to the plane of the loop, and μ_0 is the permeability of free space. The magnetic polarization or magnetization $\mathscr{M}$ is defined to be the magnetic dipole per unit volume, $\mathscr{M} = d\mathit{m}/d\mathscr{V}$. Consider a coil wound around a toroidal magnetic material. The induced magnetic field can be obtained from Maxwell's equation,

$$\nabla \times \mathbf{B} = \mu_0 \mathbf{J} + \mu_0\, \partial \mathbf{D}/\partial t + \mu_0 \mathbf{J}_M, \tag{12.2}$$

where $\mathbf{J}$ is the conduction current density, $\partial \mathbf{D}/\partial t$ is the displacement current density, and $\mathbf{J}_M$ is the conduction volume current due to the magnetization, $\mathbf{J}_M = \nabla \times \mathscr{M}$. Equation (12.2) can then be written

$$\nabla \times \mathbf{H} = \mathbf{J} + \partial \mathbf{D}/\partial t,$$

with $\mathbf{B} = \mu(\mathbf{H} + \mathscr{M})$, or, by putting $\mathscr{M} = \chi_M \mathbf{H}$ and $\mu_0 = 1 + \chi_M$, then $\mathbf{B} = \mu_0 \mu_r \mathbf{H}$, where χ_M is the magnetic susceptibility and μ_r is the magnetic coefficient. For $\mu_r < 1$, the material is said to be *diamagnetic*, and for $\mu_r > 1$, it is called *paramagnetic*.

A paramagnetic material can be considered to be an ensemble of magnetic domains [13] wherein the individual molecules are aligned with each other and are considered as elementary magnets. It is a characteristic of a ferromagnetic material that these magnets are aligned spontaneously within the domain region, thus explaining the occurrence of hysteresis in the magnetization-magnetic field intensity characteristic.

Figure 12.1 shows the $\mathscr{M}(H)$ characteristic for an unmagnetized ferromagnetic material to which a steadily increasing H-field is applied. Initially, a small field produces a large increase in $\mathscr{M}$ due to the wall motion (path OR). If the applied field is now decreased, path RO will be retraced; the region OR is reversible. After point R, the rotation of the domains becomes important, leading to an irreversibility of the curve and a slower increase in magnetization (path RA), until complete saturation is reached (path ABC). Similarly to the ferroelectric hysteresis curve, the points at which the curve crosses the axes are known as the *coercive field intensity* $\pm H_c$ and the *remnant magnetization* $\pm \mathscr{M}_R$. It is not necessary to reach saturation to reverse the direction of increase of the applied field; consequently H_c and $\mathscr{M}_R$ depend on the maximum value attained by the field. The maximum values (corresponding to saturation) are known as *coercivity* and *retentivity*.

The saturation corresponds to an equilibrium situation between the magnetizing effect of the applied field and the demagnetizing effect of thermal agitation. These two effects cancel each other at the Curie temperature, above which the material is no longer ferromagnetic. The Curie temperature is given here again by the Curie-Weiss law,

$$\chi_M = \frac{\mathscr{C}_M T_c}{T - T_c}, \tag{12.3}$$

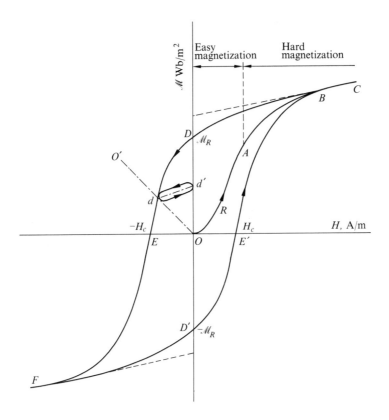

Figure 12.1. The hysteresis curve of a ferromagnetic generator. Cycle $FE'CEF$ is the hysteresis curve of the permanent magnet and cycle OO' results from the temperature cycling of the ferromagnetic gap material.

where $\mathscr{C}_M$ is a constant given by

$$\mathscr{C}_M = \mu_0 N m^2 / k T_c,$$

where k is Boltzmann's constant, N is the number of spins per unit volume, and m is the magnetic moment. The number of spins is equal to the number of spins parallel to the magnetic field and those antiparallel to it. By using the well-known Maxwell-Boltzmann probability distribution law for the spins, the value of the magnetization as a function of the temperature can be found. It is given by

$$\frac{\mathscr{M}}{\mathscr{M}_s} = \tanh\left(\frac{\mathscr{M}}{\mathscr{M}_s} \cdot \frac{T_c}{T}\right). \tag{12.4}$$

This function is shown in Fig. 12.2. It shows an intense magnetization below the Curie point even in the absence of an applied magnetic field.

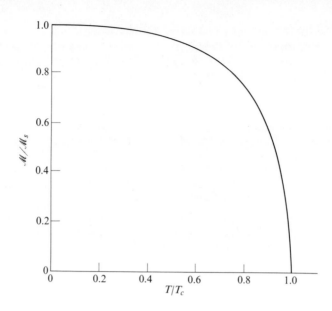

Figure 12.2. Spontaneous magnetization as a function of temperature.

12.2 Ferromagnetic Materials

The most important magnetic materials are iron, cobalt, and nickel. Alloys using these materials in different proportions as well as other materials such as silicon and molybdenum are also employed in many applications.

At present, iron is by far the most used ferromagnetic material. Alloys using iron as the primary constituent are found in the magnetically soft materials used in relays, transformers, and motors as well as in magnets. The Curie temperature of iron is 770°C, and the maximum relative permeability of the commercial product is 5000. This value reaches 350,000 for a polycrystalline material. Figure 12.3 shows the permeability μ_r versus the magnetic field **B** for various iron specimens. Hysteresis loops for different maximum fields are represented in Fig. 12.4 for the laboratory material of maximum permeability (curve A in Fig. 12.3). From Fig. 12.3, note that the method of preparation of the material has a great influence on the ferromagnetic properties of the finished metal. The saturation induction corresponding to the asymptotic value in the virgin curve decreases with temperature from a maximum value at 0°K to zero at the Curie point. The time and temperature of the annealing greatly affect the properties of ordinary iron. For instance in the 800 to 900°C temperature range, the cooling rate should be about 5°C/min to obtain high permeability. Aging of the ferromagnetic properties is seriously affected by impurities such as carbon and nitrogen.

Cobalt is used in many ferromagnetic alloys such as alnicos [3] (aluminum, iron, nickel, and up to 24% cobalt), KS steel, New Honda steel, Co-W steel, Co-Cr steel, and Permalloy for the hard materials, and Permendur and Permivar for soft materials. The proportion of cobalt in these alloys ranges from about 8% for Permivar to more than 50% for Permendur. The Curie temperature of pure cobalt varies between 1121 and 1145°C. Cobalt's permeability is rather low; it varies linearly with the magnetic field, increasing from an initial value of 68 (at $H = 0$).

Nickel is used in high-permeability alloys such as the Permalloys, Numetal, and Hypernik, which contain from 40 to 80% Ni with the remainder mainly iron. The permanent magnets of the alnico type contain from 14 to 28% nickel together with iron, aluminum, and cobalt. The Curie point of cobalt is about 358°C, and the maximum relative permeability reaches about 500 for the best specimens.

Silicon is used with iron to increase its permeability and its resistivity, and to decrease the hysteresis losses as well as the effect of aging. The largest permeability reached was obtained by Williams [23] in a single crystal iron-silicon alloy containing 3.8% silicon; the value was 1.4×10^6. The commercial grain oriented iron-silicon alloy has 3% silicon and a permeability of 0.5×10^6.

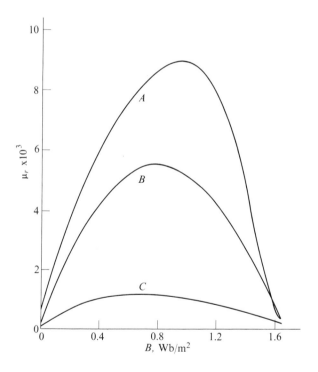

Figure 12.3. Permeability curves of pure iron. Curve A: laboratory material annealed for 1 hour at 925°C and cooled slowly with the furnace; curve B: typical commercial sheet; curve C: cold worked metal with 64% reduction of area. (From Bozorth [3].)

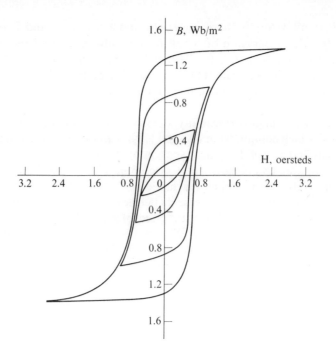

Figure 12.4. Hysteresis loops (curve A, Fig. 12.3) of iron. (Adapted from Bozorth [3, Fig. 3.5, p. 57].)

Iron-nickel alloys have some very special mechanical and thermal characteristics. Invar (36% nickel) has low thermal expansion, which can be made to match that of glass or iron by changing the proportion of nickel. These alloys are also characterized by their extremely low temperature coefficient. The Curie point can be made to vary from a low level of 0°C for about 25% nickel to a maximum of 610°C for 65% nickel [3]. The maximum permeability of the alloy is enhanced by rapid cooling. The most important commercial binary alloys* are 78 Permalloy, 45 Permalloy, Hipernik, compressed Permalloy powder, Isoperm, and Thermoperm. The addition of molybdenum and chromium increases the electrical resistivity of iron-nickel alloys as well as the relative permeability, the latter reaching a maximum of 0.9×10^6 for some Supermalloy specimens. Modifications of Permalloy using copper were introduced for some desired non-magnetic properties (Numetal, Isoperm). Other iron-nickel alloys contain cobalt, manganese, aluminum, silicon, titanium, vanadium, tantalum, or tungsten.

Iron-cobalt alloys are characterized by their high saturation induction of 2.43 Tesla,† which is higher than that of iron at room temperature. The addition

---

* Alloys composed of two different constituent metals.

† 1 Tesla $= 1$ Wb/m^2.

of vanadium leads to an alloy which can be worked at low temperatures and which has an extremely high resistivity (35 $\mu\Omega$·cm for 4% Va) [3, Fig. 6.11]. High resistivity results in lower eddy current losses. A detailed study of the different materials has been given by Bozorth [3].

12.3 Ferromagnetic Generator Analysis

An analysis of a thermomagnetic generator has been made by Elliott [7]. He used a soft ferromagnetic material to fill the gap of a permanent magnet, as shown in Fig. 12.5. An output coil is wound around the magnet. The ferromagnetic gap material is then temperature cycled near its Curie point, thus causing a periodic change in the value of the magnetic flux through the circuit.

Considering the hysteresis curve of Fig. 12.1 as the hysteresis curve of the permanent magnet, note that when the magnetic circuit is homogeneous and closed and when H is removed, the magnet state will be represented by point D on the demagnetizing curve. When the ferromagnetic gap material is heated above its Curie point, μ_r will tend toward unity, thus operating the circuit and bringing the magnet to a certain point d which is the intersection of the demagnetizing curve with the air gap line (OO') given by the relation

$$\mathcal{M}_m/H_m = -\mu_0 \ell_m/\ell_g,$$

where m is a subscript related to the permanent magnet and ℓ_g is the length of the gap. Upon cooling the ferromagnetic "gap" to its Curie point, its permeability rises rapidly and the system moves to point d' for H going to zero. The slope of the line dd' is given by $\mu_0\mu_r$ for the permanent magnet. Further cycling will cause the system to oscillate between points d and d' along a loop created by the hysteresis losses. In order to have a specific output as large as possible, the distance dd'

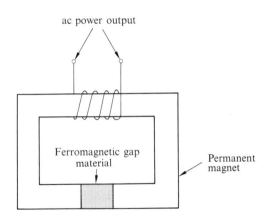

Figure 12.5. Schematic thermomagnetic generator. The ferromagnetic gap material is temperature cycled near its Curie point.

should be made large. This can be obtained with a permanent magnet having large H_c and large μ_r.

When the cycling temperature applied to the gap material is varied from T to $T + dT$ and the induction from B to $B + dB$ $(B = \mathcal{M})$, the heat absorbed by the material is

$$d\mathcal{Q} = C(T, B)\, dT + f_m\, dB, \tag{12.5}$$

where C is the heat capacitance at T and B, and f_m is a "heat of magnetization" coefficient. For mathematical simplicity, the permeability μ is assumed independent of the field H, and Eq. (12.5) becomes [4].

$$d\mathcal{Q} = C(T, B)\, dT + TH\, dH\, d\mu/dT. \tag{12.6}$$

Maximum efficiency could be reached by using an ideal Carnot cycle consisting of two adiabatic paths in the $H(T)$ characteristic. For an adiabatic transformation $(d\mathcal{Q} = 0)$, Eq. (12.6) shows that an increase dT in temperature is accompanied by a rise dH in the field intensity since $d\mu/dT$ is a negative quantity. As an approximation, Brillouin and Iskenderian assumed that the permeability varies linearly with temperature and that the ferromagnetic gap material is nearly saturated throughout the heat cycle; thus

$$B(T, H) = B(T, O) + \mu_0 \mu_\delta H,$$

where μ_δ is a constant value equal to some few units.

When the coil current was properly matched with the load, Brillouin and Iskenderian [4] found that the optimum power output for a sinusoidal temperature cycling could be determined:

$$\mathscr{E}_{max} = \tfrac{1}{2}\omega \mathscr{V} \mu_0 \mu_\delta H_0^2, \tag{12.7}$$

where the transient effects have been neglected. If the saturation flux density of the gap material is larger than that of the permanent magnet at point d', Eq. (12.7) becomes

$$\mathscr{E}_{max} = \tfrac{1}{8}\mu_0 \omega \mathscr{V} \mu_r H_c^2,$$

where $\mu_r H_c^2$ represents the figure of merit of the permanent magnet material. This value should be as large as possible. Brillouin and Iskenderian found, for cycling between the Curie point and a slightly lower temperature, that

$$\eta = 0.55\eta_c, \tag{12.8}$$

where η_c is the Carnot efficiency. This figure seems too optimistic, since most of the causes of loss have been ignored.

A reasonable value of the Carnot efficiency is no larger than 1%. This makes the probable efficiency of a thermomagnetic generator extremely small: $\tfrac{1}{2}\%$ for Eq. (12.8) [4], and no larger than $10^{-4}\%$ for Elliott's experimental work [7].

In Elliott's experiments, the ferromagnetic gap used was gadolinium chosen for its low Curie point of 16°C, thus allowing the recovery of low-grade energy.* The material used for the permanent magnet was of the alnico type. Gadolinium is an expensive rare earth having a specific power output of 7 W/kg, comparing poorly with the other direct energy converters.

Another approach to thermomagnetic power generation has been attempted by Strauss [17]. He used the ferromagnetic gap as the working material by wrapping the output coil around the gap rather than around the permanent magnet. He predicted, however, very low efficiencies of about $10^{-5}\%$.

Work reported in the literature can be considered as a proof of the possibility of generating electricity by using the ferromagnetic effect. How practical such a generator will be cannot be ascertained at the present state of the art.

B. NERNST POWER GENERATION

12.4 Nernst and Ettinghausen Effects (Fig. 12.6)

When a current of electrons flows through some types of materials in the presence of a magnetic field, the electrons are deflected by the field. This creates an electric field perpendicular to both the initial direction of the current and the magnetic field, thus establishing a difference of potential between points A and B of Fig. 12.6. In this way, a Hall voltage is created in a manner similar to that discussed in

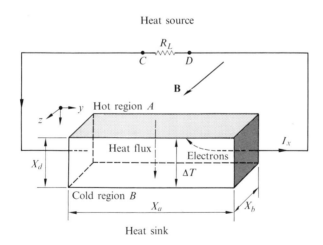

Figure 12.6. Nernst-Ettinghausen effect in a slab.

---

*   *Low-grade energy* is the energy lost to the cold sink by a heat engine and which can be partly recovered by a tailer (see Section 1.7).

Chapter 4. If there is no external load between A and B, electrons will accumulate on surface A. Consider that the heat carriers are also the charge carriers; since only the slower ("colder") electrons will be affected by the field, a temperature gradient will be established between regions A and B, B being the hotter region. The creation of such a temperature gradient in the presence of crossed electric current and magnetic field is known as the *Ettinghausen effect* [35]. In the case where holes are the charge carriers, an electric field will be established in the opposite direction to that created by the electrons flowing in the opposite direction of the holes. The temperature gradient will be, however, in the same direction for both types of carriers [48].

The Ettinghausen temperature gradient in turn creates a voltage between regions A and B equal to the difference between isothermal and adiabatic Hall voltages. In an intrinsic semiconductor, both holes and electrons are present. If this material is placed in a magnetic field, there will be no net Hall voltage if the two types of charge carriers have the same mobilities and carrier densities. Carriers of opposite signs will be deflected by the magnetic field in the same direction (toward the A-region, for instance). Therefore, there will be a transport of kinetic energy from side A to side B, with electron-hole recombination at side A and generation of electrons and holes at side B. These effects will heat region A and cool region B, thus greatly enhancing the Ettinghausen temperature gradient. This effect was first suggested as a refrigerating method by O'Brien and Wallace [47].

If now a temperature gradient is established between regions A and B in a direction perpendicular to the magnetic field, the reverse of the Ettinghausen effect will take place: an emf will be created in a direction perpendicular to both the heat flow and the magnetic field. This is known as the *Nernst effect*. It was Bridgeman [30] who first proposed the use of this effect to generate electrical energy directly from heat. However, it was only in the early 1960's that an analysis of the Nernst generator was presented by Harman and Honig [37], and by Angrist [27].

Ettinghausen and Nernst effects are two of the four effects known under the broad name of *galvano-thermomagnetic* phenomena (not to be confused with the thermomagnetic generation, which used ferromagnetic materials, described in an earlier section). The two others are *Hall* and *Righi-Leduc effects*. The galvano-thermomagnetic phenomena are related to the thermoelectric effects as can be seen in the very comprehensive book by Harman and Honig [43].

Nernst and Ettinghausen discovered the Nernst effect in 1886. A year later Ettinghausen discovered the opposite phenomenon, known by his name. The Hall effect was discovered in 1879 by Hall at Johns Hopkins University. He found that an emf is created in the same direction as a current flowing perpendicularly to a strong magnetic field. In 1888 and 1889 Righi [48] and Leduc [46] simultaneously discovered that in some materials and in the presence of crossed magnetic field and thermal flow, a temperature gradient is created in a direction perpendicular to both the magnetic field and the heat flow.

The Ettinghausen coefficient C_E is defined as

$$C_E = \frac{\Delta T_E x_b}{BI},$$

where x_b is the dimension of the generator in the direction of the magnetic field B, I is the electron current, and ΔT_E is the Ettinghausen temperature difference. O'Brien and Wallace [47] showed that the passage of the current I resulted in a maximum temperature difference

$$\Delta T_{E\max} = \tfrac{1}{2} C_E^2 B^2 \kappa \sigma,$$

where κ and σ are the thermal and electrical conductivities, respectively; or

$$\Delta T_{E\max} = \tfrac{1}{2} Z_E T_c^2,$$

with

$$Z_E = C_E^2 B^2 \kappa \sigma / T_c^2.$$

In a manner similar to that seen in thermoelectricity, Z_E is taken as the figure of merit for Ettinghausen cooling.

When a temperature gradient is applied between regions A and B, the produced Nernst electric field E_y can be expressed as

$$E_y = C_N B \, dT/dx, \tag{12.9}$$

where C_N is the Nernst coefficient. A Nernst figure of merit is also defined as

$$Z_N = C_N^2 B^2 \sigma / \kappa. \tag{12.10}$$

Delves [33] showed that the relation between Z_N and Z_E is given by

$$\frac{Z_E}{Z_N} = \frac{1}{1 - Z_N T}. \tag{12.11}$$

For small values of $Z_N T$, this relation is approximately unity, leading to the relation found by Bridgeman [30],

$$C_E/C_N = T/\kappa \qquad \text{(for } Z_N T \ll 1\text{)}. \tag{12.12}$$

From Eq. (12.11), note that the product $Z_N T$ is limited:

$$Z_N T \leqslant 1. \tag{12.13}$$

12.5 Nernst Materials

Nernst's figure of merit for several materials is shown in Fig. 12.7. They are semiconductors of the intrinsic or the extrinsic type. For an extrinsic nondegenerate semiconductor, the electrical conductivity is a function of the mobility ℓ, $\sigma = ne\ell$. The thermal conductivity κ is the sum of the conductivity κ_L due to the vibration of the crystal and the conductivity due to the motion of the electrons. It is found that

$$\kappa = \kappa_L + 2\sigma T k^2/e^2.$$

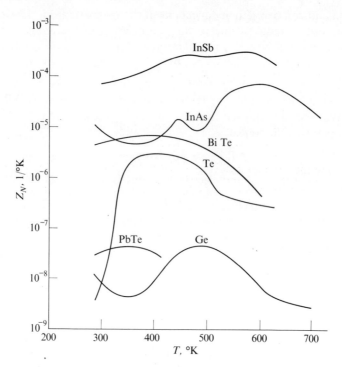

Figure 12.7. Figure of merit Z_N of several Nernst materials as a function of temperature for $B = 1$ Wb/m^2. Note the figure of merit of InSb. (From Angrist [27].)

For low magnetic fields, Nernst's coefficient becomes

$$C_N = \frac{3\pi}{16} \cdot \frac{k}{e} \ell, \qquad (12.14)$$

and, finally, the figure of merit can be written as

$$Z_N T = \frac{\mathscr{C}_1 \exp(\mathscr{E}_w/kT)\ell^2 B^2 (kT)^{5/2}}{\kappa_L + \mathscr{C}_2 \exp(\mathscr{E}_w/kT)\ell(kT)^{5/2}}, \qquad (12.15)$$

where $\mathscr{C}_1 = 3.66 \times 10^{78}$ and $\mathscr{C}_2 = 9.51 \times 10^{101}$ in the MKSC system. For good performance, large mobilities and magnetic fields are needed.

For high magnetic fields, Eq. (12.14) becomes

$$C_N = \frac{3\pi}{16} \cdot \frac{k}{e} \cdot \frac{\ell B}{1 + \ell^2 B^2} \cdot \frac{1}{B}.$$

For small magnetic fields, this relation is obviously reduced to Eq. (12.14). The maximum value of Z_N is obtained for $\ell B = 1$, leading to very small values of $Z_N T$. By doping the semiconductor, C_N can be made much larger than in the extrinsic case and can be made independent of the magnetic field. However, the performance is often reduced due to the contribution of the ambipolar* effect on the thermal conductivity. Furthermore, the anisotropy of the semiconductor crystal has sometimes a good influence on the Nernst figure of merit.

Colwell [32] and his co-workers found that if the Hall coefficient is made negligible over a wide temperature range, the Nernst figure of merit could be increased. This requires the use of intrinsic semiconductors having equal hole and electron densities and mobilities, or semimetals such as pyrolitic graphite, for which a figure of merit $Z_N T = 0.045$ was measured at 400°C and 34 kG (kilogauss). However, this is not the largest figure possible for this material. Increase of the temperature and the applied magnetic field could lead to a value increased to about three times that quoted above.

The semiconductors which were considered for Nernst power generation are silicon (Si), germanium (Ge), tellurium (Te), lead telluride (PbTe), bismuth titanate (Bi_2Ti_3), indium arsenide (InAs), and indium antimonide (InSb). Silicon has an extremely low Nernst figure of merit, about 0.78×10^{-8}. Due to its low electrical conductivity and high thermal conductivity, germanium is also a rather poor Nernst material ($Z_N T \cong 0.5 \times 10^{-4}$). Tellurium seems to be a better material than germanium as can be seen Fig. 12.7. The two materials of highest Nernst performance are indium antimonide and indium arsenide. Both materials have very large electron mobilities (6.5 and 2.3 m²/V·sec, respectively), which explain the high value of their Nernst coefficients.

12.6 Nernst Generator Analysis

Consider the generator shown in Fig. 12.6. This device works between T_h and T_c with the flux of heat in the x-direction. The steady state is assumed and the Righi-Leduc effect is neglected. The parameters σ, κ, and C_N are assumed to be independent of temperature. The applied magnetic field in the z-direction is constant and uniform.

The Nernst output voltage V_y of such a generator can be obtained from Eq. (12.9) by multiplying the electric field by the distance x_a. After integration,

$$V_y = C_N B(T_h - T_c)x_a/x_d, \tag{12.16}$$

and since the internal resistance of the material is

$$R_i = x_a/(\sigma x_b x_d),$$

Eq. (12.16) becomes

$$V_y = \sigma R_i C_N B(T_h - T_c)x_b.$$

With the assumption that the resistance of the load is $R_L = m_L R_i$, the output

---

*   Defined in Section 3.8.

current becomes

$$I = \frac{\sigma C_N B(T_h - T_c)x_b}{m_L + 1},$$ (12.17)

and the output power is $R_L I^2$, or

$$P_{out} = \frac{1}{(m_L + 1)^2} \sigma C_N^2 B^2 (T_h - T_c)^2 \frac{x_a x_b}{x_d}.$$ (12.18)

At the hot side of the generator, the heat input per unit time is

$$\mathcal{Q}_{in} = \mathcal{Q}_c - \mathcal{Q}_E - \mathcal{Q}_j,$$ (12.19)

where $\mathcal{Q}_c$ is the heat transferred by conduction, $\mathcal{Q}_E$ is the heat rejected by the Ettinghausen effect, and $\mathcal{Q}_j$ is the Joulean loss. These heats are respectively equal to

$$\mathcal{Q}_c = \kappa \, \Delta T x_a x_b / x_d,$$

$$\mathcal{Q}_E = \kappa B C_E I x_a / x_d,$$

and

$$\mathcal{Q}_j = \frac{1}{2} \cdot \frac{x_a}{x_b x_d} \cdot \frac{I^2}{\sigma},$$

since half the heating passes to the hot region and the other half to the cold region. When Eqs. (12.12) and (12.17) are taken into consideration, the total heat input becomes

$$\mathcal{Q}_{in} = \sigma \kappa \, \nabla T \left[1 - \frac{Z_N T_c}{(m_L + 1)} - \frac{1}{2} \frac{Z_N \, \Delta T}{(m_L + 1)^2} \right] \frac{x_a x_b}{x_d},$$

where Z_N is the figure of merit given by Eq. (12.10). The efficiency of the generator is equal to the ratio of the power output to the input:

$$\eta = \frac{m_L Z_N \, \Delta T}{(m_L + 1)^2 - (m_L + 1)Z_N T_c - \frac{1}{2} Z_N \, \Delta T}.$$

The maximum efficiency is obtained in a manner similar to that for the thermo-electric device by setting $d\eta/dm_L = 0$. For this value it is found that

$$m_L = (1 - Z_N \bar{T})^{1/2},$$

where $\bar{T} = \frac{1}{2}(T_h + T_c)$, and the maximum efficiency becomes

$$\eta_{max} = \frac{1 - (1 - Z_N \bar{T})^{1/2}}{1 + (T_c/T_h)(1 - Z\bar{T})^{1/2}} \eta_c,$$ (12.20)

where η_c is the ideal Carnot efficiency. Calculations based on Eqs. (12.19) and (12.20) are shown in Fig. 12.8, where the thermal efficiency and the power output density are shown as functions of the temperature and magnetic field. Both increase with increase in the temperature and magnetic field.

The performance of the generator can be improved by staging a certain amount of individual legs. For an N-stage cascaded generator, the total efficiency is given

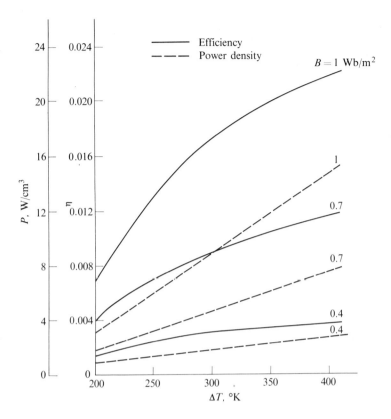

Figure 12.8. Thermal efficiency and power density for a single leg InSb Nernst generator at different values of the applied magnetic field: $x_a/x_d = 0.25$, $x_b = 5$ mm.

by a relation identical to that found in thermoelectricity:

$$\eta_t = 1 - \prod_{i=1}^{N}(1 - \eta_i). \tag{12.21}$$

Note here that staging will increase the current output or the voltage (or both) and, consequently, the power output, but there will be hardly any gain in efficiency. For infinite staging, Eq. (12.21) becomes

$$\eta_t = 1 - \exp\left[\int_{T_c}^{T_h} \eta(T)\, dT/T\right].$$

One of the most important losses of a Nernst generator is the power required to produce the magnetic field. On the other hand, it seems highly improbable that fields exceeding 10^5 oersteds can be obtained and, consequently, only materials having hole and electron mobilities exceeding 1 m²/V·sec can be of much interest.

However, the mobility decreases with increasing temperature and with decreasing energy gap; the first effect is a limitation to the performance of the Nernst generator, whereas the second can be considered as an advantage, since materials of low energy gap are usually required (the opposite of thermoelectric power generation). Consequently, the prospects of Nernst power generation seem good at low temperatures (e.g., between room temperature and cryogenic temperatures). In such a case, thermal efficiencies exceeding 5% and specific power outputs larger than 10 W/kg might be obtained in the near future.

C. THERMOPHOTOVOLTAIC POWER GENERATION

12.7 Thermophotovoltaic Effect

In most of the direct energy converters using a heat source for the input energy, the heating of the electrons (i.e., the exchange of energy between the heat source and the electrons) is often coupled with the heating of the material lattice. This is the case in thermoelectric and thermionic power generation. In both cases, the heating of the material creates losses and raises severe technological problems at higher temperatures.

In the photovoltaic effect, light generates electron-hole pairs with a relaxation time for energy transfer to the lattice of about 10^{-6} to 10^{-3} sec. This figure is of the order of the minority carrier lifetime and at least six orders of magnitude lower than the relaxation time of conduction. Therefore, electrons can be energized in a photovoltaic cell by photons without heating the material lattice. This is precisely the basic idea behind the thermophotovoltaic (TPV) effect. In this effect, heat is converted through incandescence into light, which in turn is converted into electricity by the known photovoltaic effect.

This method of power conversion was considered in France [54] and Germany [65, 69] in the late 1950's and first proposed in this country by Aigrin during a lecture series at the Massachusetts Institute of Technology in 1961. Since then work has been carried out by the Energy Conversion Group at MIT [72, 73, 74, 77, 78], at Illinois Institute of Technology [70, 71], and at General Motors [53, 57].

The advantage of the TPV converters as compared to the photovoltaic converter is that the spectral characteristics as well as the power levels of the input radiation may be controlled to yield optimum conversion efficiencies. The power level can be controlled by use of concentrators for the solar energy, or by some other method in cases where the heat source is nuclear or radioisotopic. The spectrum can be controlled by variation of the temperature and the emissivity of the heat source or by the use of thin film filters.

One of the most important sources of loss in the photovoltaic cell is the loss by reflection. In a TPV converter, this loss can be returned to the light source without difficulty by using special geometric arrangements. Another possibility

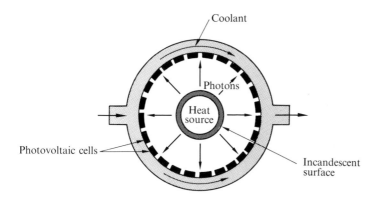

Figure 12.9. Coaxial thermophotovoltaic energy converter.

is the placing of antireflecting coatings at the surface of the TPV converter facing the input radiation. The best proposition, however, is the use of a coaxial configuration (as shown in Fig. 12.9) with the photovoltaic cells surrounding the incandescent source. In this arrangement, the reflected radiation is either returned to the emitter or accepted by another cell.

12.8 Thermophotovoltaic Materials

The two materials in widest use are silicon and germanium. Silicon is, however, of less importance in TPV power generation, since the advantage of an energy gap matching the solar spectrum (which is the main characteristic of silicon pn junctions) is no longer required. Germanium remains the best element available. However, for power generators using radioisotopic sources, the radiation damage to the cell becomes an important problem which should be dealt with in any material consideration.

Germanium cells are of two types: the alloyed junction cell type built by Delco Radio Division [61] and the diffused cell type built by the Allison Division of General Motors [61]. The alloyed cells are in a more advanced state of development, although diffused elements hold great promise. Figure 12.10 shows relative spectral responses for alloyed and diffused germanium cells as a function of wavelength. Allison diffused junction cells have a junction depth of approximately 10 μ and their maximum response is at 1.35- and 1.52-μ wavelength for the two cases shown in this figure. Alloyed junction cells maximize at about 1.60- and 1.68-μ wavelength for two different cases and break sharply in the 1.3- to 1.5-μ wavelength range. In all these cases, the junctions are located at 100 to 500 μ within the cell. The absorption coefficient of germanium is reported in Chapter 8; its reflectance $\mathscr{A}_r$ is equal to about 0.35 in the 0.8- to 1.5-μ range of λ and increases to a higher maximum at smaller and larger wavelengths ($\mathscr{A}_r = 0.42$ for $\lambda = 0.6 \mu$ and $\mathscr{A}_r = 0.58 \mu$ for $\lambda = 2.6 \mu$) for a pn diffused germanium photocell. In the

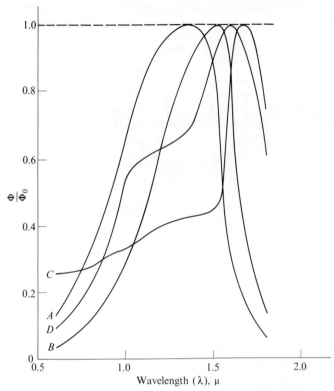

Figure 12.10. Relative spectral responses of germanium photocells. Curve A: diffused germanium photocell [61]; curve B: diffused germanium photocell [61]; curve C: Ruth and Moyer cell [66]; curve D: alloyed germanium photocell [61].

useful spectral region (0.8 to 1.8 μ) the total reflectance of an alloyed cell is similar to that of a n-on-p diffused cell

In the presence of a radioisotopic source of energy, the problem of gamma radiation should be considered and eventually solved. The effect of this radiation, as seen before, is a change of majority carrier concentration and, of greater importance, a degradation of the minority carrier lifetime. By considering the number of electrons removed by a photon at a certain energy, it would be possible to predict the decrease in carrier concentration and lifetime. The problem of constructing photovoltaic cells which have a good resistance to radiation has already been considered (Chapter 8). From different studies, it seems that such cells should have the following characteristics:

1. A shallow junction, thus limiting the effect of lifetime degradation.
2. Incorporated diffusion also limiting the degradation of the cell lifetime.
3. The substrate should be p-type, since p-type germanium is less affected by radiation damage than the n-type.

12.9 Radiant Exchange Systems and Radiant Sources

A radiant exchange system controls the spectrum of the radiation issued by the heat source and incident on the TPV cell system. It consists of a multilayer, interference-type optical filter placed on a glass substrate. The interference-type filter consists of alternating layers of thin metal and dielectric films. With this system, the transmittance and reflectance of the composite coating can be controlled by properly selecting the number, thickness, and indices of refraction of the layers. Gritton and Bourke [61] report that a reflectance of 0.98 can be obtained for the 1.6- to 3.5-μ wavelength range with a filter transmittance of 0.9 in the 1.25- to 1.6-μ region, and three diminishing transmittance peaks for lower wavelengths. The filter is essentially a bandpass designed for the germanium spectral response. The substrate material is also subject to the hazard of irradiation damage, and consideration should be given to the choice of a material having good resistance to this damage.

The radiant source should be chosen so that it will match the spectral response of the photovoltaic cell. For germanium TPV cells, for instance, this response should be in the 0.8- to 1.8-μ wavelength range. A spectrally selective emitter can be obtained by the use of a controlled mixture of rare earth oxides such as ceria (CeO_2) and thoria (ThO_2). There is no published work for the useful wavelength range and most of the research that has been done was directed toward obtaining emission bands in the visible region. To obtain high source emission in the 0.8- to 1.8-μ range, the temperature of the heat source should be relatively high. For a resulting maximum spectral intensity in the 1.3- to 1.6-μ region, and a temperature of an emitting gray body in the 1800 to 2200°K range, 60 to 75% of the total emitted energy will lie above 1.6 μ. Gritton and Bourke [61] investigated two types of "radiators": a tungsten sphere and a coated source emitting blackbody radiation. They found that for the tungsten sphere, the spectral emittance decreases with increasing wavelength, resulting in a smaller percentage of the total energy occurring above the cutoff wavelength of the cell. The advantage of a near-blackbody source is that a large percentage of the emitted radiation returned to the source is absorbed and radiated again. It has identical spectral emittance and absorptance at a given wavelength and temperature. Therefore, the blackbody source will limit the amount of loss due to multiple reflections between the emitter and the cells.

The several phosphors experimented with by Watts [70, 71] and his co-workers have emission band peaks at too high wavelength levels. These phosphors are radium, sulfur zincate (ZnS:Cu), and cadmium zincate (ZnCdS:Cu), among others.

12.10 Thermophotovoltaic Generator Analysis

As seen in Chapter 8, the series resistance of a photovoltaic cell creates a serious limitation to the overall efficiency of the cell. In a photovoltaic cell this series resistance is determined by the thickness of the diffused layer receiving the input

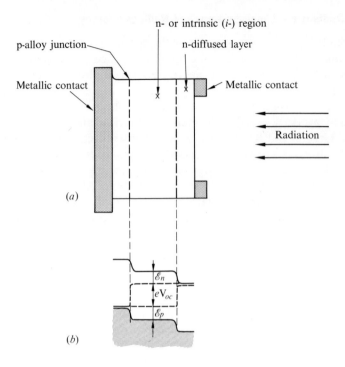

n- or intrinsic (*i*-) region

p-alloy junction

n-diffused layer

Metallic contact

Metallic contact

Radiation

(*a*)

$\mathscr{E}_n$

eV_{oc}

$\mathscr{E}_p$

(*b*)

Figure 12.11. The PIN thermophotovoltaic junction: (a) the PIN junction, (b) its corresponding energy band diagram.

radiation. In a TPV cell, the depth of the junction is several orders of magnitude greater than that of the photovoltaic cell. The collection efficiency is determined by the dimensions of the blank upon which the junction is formed. Therefore, to decrease the series resistance of the cell, a high conductivity diffused layer may be placed facing the radiation as shown in Fig. 12.11(a). This figure illustrates a PIN thermophotovoltaic cell, the energy band diagram of which is shown in Fig. 12.11(b).

If one assumes an equilibrium in the distribution of the conduction and valence bands, the total energy gap is

$$\mathscr{E}_g = \mathscr{E}_n + eV_{oc} + \mathscr{E}_p, \tag{12.22}$$

with $\mathscr{E}_n$ and $\mathscr{E}_p$ being the quasi-Fermi* levels for electrons and holes, respectively, whereas V_{oc} is the open circuit voltage. Electron and hole concentrations n and p are given, by definition of the quasi-Fermi level, by

$$p = N_v \exp\left(-\mathscr{E}_p/kT_c\right) \tag{12.23}$$

---

* Partial potential of an *assembly* of electrons or of holes.

and

$$n = N_c \exp\left(-\mathscr{E}_n/kT_c\right), \tag{12.24}$$

where T_c is the lattice temperature (also the temperature of the heat sink), N_c is the effective density of states in the conduction band, and N_v is the effective density of states in the valence band. The intrinsic electron concentration is given by

$$n_i^2 = N_c N_v \exp\left(-\mathscr{E}_g/kT_c\right). \tag{12.25}$$

By taking into consideration Eqs. (12.23), (12.24), and (12.22), the product np can be written

$$np = N_c N_v \exp\left[(eV_{oc} - \mathscr{E}_g)/kT\right], \tag{12.26}$$

or, introducing Eq. (12.25) into Eq. (12.26), as

$$np = n_i^2 \exp\left(eV_{oc}/kT_c\right). \tag{12.27}$$

Therefore, the open-circuit voltage becomes

$$V_{oc} = \frac{kT_c}{e} \ln\left(\frac{np}{n_i^2}\right).$$

In the TPV cell, electron-hole pairs are generated by the absorption of photons. If radiative recombination* and thermal recombination and generation phenomena are considered, optimum efficiency will be obtained when the incident radiation spectrum perfectly matches the recombination spectrum of the cell material. Neglecting the traps,† the recombination light spectrum for an intrinsic semiconductor is characterized by a frequency

$$v = \mathscr{E}_g/h, \tag{12.28}$$

and a bandwidth proportional to the ratio $kT_c/\mathscr{E}_g$.

With the assumption that the recombination radiation is returned to the source of heat, the overall efficiency can be written as

$$\eta = \frac{VI}{hv(g_r - r_r)},$$

where g_r and r_r are the radiative generation and recombination rates, respectively, and V and I are the voltage and the current at the load. If the thermal recombination r_t and generation g_t rates are considered, the condition of the steady state becomes

$$(g_r - r_r) + (g_t - r_t) = I/e, \tag{12.29}$$

where g_t is essentially a function of the lattice temperature, whereas r_t is a function

---

* A recombination process is said to be *radiative* when the energy liberated by the recombination goes to a photon.

† A *trap* is an empty quantum state slightly below the conduction band. It is created by an impurity or a negative ion vacancy in the semiconductor crystal.

of np which increases with the increasing radiation. Under the effect of radiation, the net thermal recombination process will be given by $r_n = -(g_t - r_t)$. The overall efficiency can now be expressed as a function of the current and the thermal recombination rates; it is found that

$$\eta = \frac{\eta_0}{1 + er_n/I},$$

(12.30)

where $\eta = eV/h\nu$ is known as the *quantum efficiency*. It can be seen from Eq. (12.30) that when the number of electrons and holes removed by the output current is much larger than the number of electrons and holes lost by thermal recombination, the overall efficiency approaches the quantum efficiency.

Equation (12.29) can be written in terms of N_c, N_v, and np if the probabilities of available state of photon generation n_g, hole recombination p_r, and electron recombination n_r are introduced. Thus

$$n_g N_c N_v - (p_r + n_r)np = I/e,$$

and when Eqs. (12.25) and (12.27) are introduced, the open-circuit voltage is,

$$V = \frac{1}{e}\left\{ \mathscr{E}_g - kT_c \ln\left[\frac{p_r + n_r}{n_g(1 - I/eg_r)} \right] \right\}.$$

This voltage tends toward $\mathscr{E}_g/e$ for T_c approaching zero. When Eq. (12.28) is taken into account, the value of the quantum efficiency becomes

$$\eta_0 = 1 - \frac{kT_c}{h\nu} \ln\left[\frac{p_r + n_r}{n_g(1 - I/eg_r)} \right].$$

The probability coefficients can be expressed in terms of the temperature T_h of the source of heat:

$$n_g = \frac{n_{g1}}{\exp(h\nu/kT_h) - 1},$$

$$p_r = p_{ro} + \frac{p_{r1}}{\exp(h\nu/kT_h) - 1},$$

where n_{g1}, p_{ro}, and p_{r1} are functions of the frequency ν of the radiation but are independent of temperature. Now, the quantum efficiency becomes

$$\eta_0 = 1 - \frac{T_c}{T_h} - \frac{kT_c}{h\nu} \ln\left\{ \frac{\left(\frac{p_r + n_r}{n_{g1}}\right)\left[1 - \exp\left(-\frac{h\nu}{kT_h}\right) + \frac{p_{c1}}{n_{g1}}\exp\left(-\frac{h\nu}{kT_h}\right)\right]}{1 - I/eg_r} \right\}.$$

(12.31)

At thermal equilibrium, it is found that $n_{g1} = p_{ro} = p_{r1}$ and that $n_r = 0$ [77]. In this case Eq. (12.31) becomes

$$\eta_0 = \eta_c - \frac{kT_c}{h\nu} \ln\left(\frac{1}{1 - I/eg_r} \right),$$

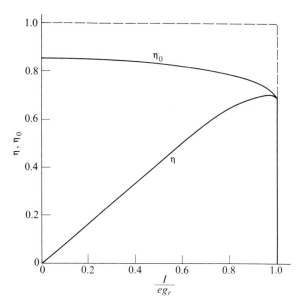

Figure 12.12. The quantum efficiency η_0 and the overall efficiency η versus load current for a TPV converter. (From White, Wedlock, and Blair [77].)

where η_c is Carnot's efficiency. The radiative recombination is usually negligible as compared to the thermal recombination. In this case, Eq. (12.29) yields

$$1 + er_n/I = eg_r/I,$$

and the overall efficiency becomes

$$\eta = \eta_0 I/eg_r.$$

This relation is shown in Fig. 12.12, along with the quantum efficiency η_0 as a function of the ratio I/eg_r.

The maximum theoretical efficiency of the TPV converter is therefore slightly lower than the Carnot efficiency. The actual efficiency will certainly be much lower than the values suggested in Fig. 12.12 and will be affected by various effects such as surface recombination, finite width of the incandescent spectrum, and the resistivity of the cell. White [77] and his co-workers predicted a 35% efficiency from heat source to electrical power with present technological and scientific methods.

D. OTHER METHODS OF DIRECT ENERGY CONVERSION

12.11 Solid State Converters

Several solid state effects were suggested as a basis for power generation, some of them mentioned in previous chapters. These effects are pyroelectricity, the Hall effect in semiconductors, the Righi-Leduc effect, and the different effects related to thermoelectricity, as well as the change of state or phase in materials. A large number of methods have been proposed, but only four of them will be treated here in some detail: magnetothermoelectric conversion, photoelectromagnetic conversion, superconducting conversion, and magnetostrictive conversion.

Magnetothermoelectric conversion. When a magnetic field is applied to a thermoelectric semiconductor it affects its figure of merit Z in such a way that it may be useful for power generation. A transverse magnetic field decreases the electrical and thermal conductivities of a nondegenerate* semiconductor. The Seebeck coefficient is increased with a lattice scattering in the acoustic mode in some extrinsic semiconductors such as bismuth-antimony alloy. In this material the presence of an optimum magnetic field of 10^4 oersteds led to maximum values of Z and a value of ZT equal to 1.3 at 300°K, as shown in Fig. 12.13. The effect of a transverse magnetic field is not always advantageous, and it was found that the

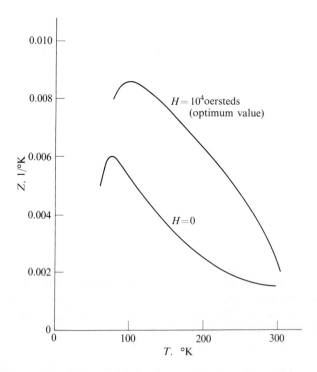

Figure 12.13. The effect of the magnetic field on the thermoelectric figure of merit of bismuth antimonide (BiSb). (After Wolfe and Smith [105].)

---

*   When a semiconductor is highly doped, the impurity atoms become so close to each other that their energy levels become broadened into a band. This semiconductor is said to be *degenerate*.

Seebeck coefficient (as well as the figure of merit) is decreased in an extrinsic semiconductor with impurity and optical mode scatterings.* The effect of a magnetic field on the Seebeck coefficient is rather complicated, since Nernst-Ettinghausen phenomena are not always negligible in this case.

Photoelectromagnetic conversion. The photoelectromagnetic effect was first investigated by Kikoin and Noskov [90] in a cuprous oxide (Cu_2O) material. They found that when a plate of cuprous oxide at the temperature of liquid air is placed in a tangential magnetic field and then illuminated by an axial white or black light, a difference of potential is created in a direction perpendicular to both the magnetic field and the radiation. The direction of the potential difference was found to change without altering its magnitude by reversing the direction of the magnetic field or that of the radiation. The difference of potential was a function of the magnetic field strength, reaching 5 V in some cases with a field strength of 2500 G. This effect was found later in germanium. Under the effect of the energetic light, electrons tend to diffuse more rapidly than those in the opposite dark region (the Dember effect), and under the influence of the magnetic field, a Hall space charge separation is created, leading to the observed potential difference. When a load is connected to the open-circuit cell, electric power can be generated. By assuming that the recombination velocity is zero at the illuminated face and infinity at the dark face, the maximum efficiency is given by

$$\eta_{\max} = \mathscr{A} H^2 \mathscr{b}_e \mathscr{b}_h, \tag{12.32}$$

where $\mathscr{b}_e$ and $\mathscr{b}_h$ are the electron and hole mobilities and $\mathscr{A}$ is a constant.

Superconducting conversion. This method directly converts heat energy into electricity by using the phase transition in a superconductor. The operation of the system is similar to that of the ferroelectric power generator and a heat cycle is needed. The superconducting energy converter consists of a cylindrical core of superconducting material and a pickup coil wound around the superconductor. The overall system is placed in a magnetic field produced by an external permanent magnet. A temperature cycle periodically changes the state of the core from normal to superconducting. When the core is normal, it accepts the external magnetic field, but when it becomes superconducting, the external field is rejected due to the Meissner effect. With each change in the core flux, a current is produced in the pickup coil, thus creating an alternating current which can yield electrical energy to an external load. The thermodynamic analysis of this system has been done by Chester [82], who estimated the efficiency of the idealized cycle at about 44% for a niobium core at 8°K.

Magnetostrictive conversion. The magnetostriction effect can be used to convert mechanical energy directly into electricity. In this respect it is very similar to the piezoelectric effect mentioned previously. Magnetostriction effects involve the

---

*   Scattering between photons and molecules generating space charge carriers.

change of the dimensions of a material under the influence of a magnetic field. The material exhibits positive magnetostriction when it stretches in the direction of the magnetic field (as in Permalloy) or negative magnetostriction when it shortens in that direction (as in nickel). The change in length of the specimen is due to a change of state from demagnetized to saturated. There are many proposed generator (transducer) arrangements, in all of which input energy is supplied by straining the core by pressure waves, creating a change in the core flux which induces a voltage in a driving coil.

12.12 Liquid State Converters

Diverse forces in materials in the liquid state are used for directly generating electric power. Almost all the phenomena occurring in solid state materials can be combined with hydrodynamic, temperature, or pressure gradient forces to convert mechanical or thermal energy into electricity. The following four important cases will be considered: electron convection, electrokinetic conversion, ferrohydrodynamic conversion, and photogalvanic conversion.

Electron convection. A converter based on this principle is shown in Fig. 12.14. It consists of a truncated cone enclosing a pool of alkaline liquid metal. The liquid is heated and electrons as well as neutral atoms are emitted (or actually boiled

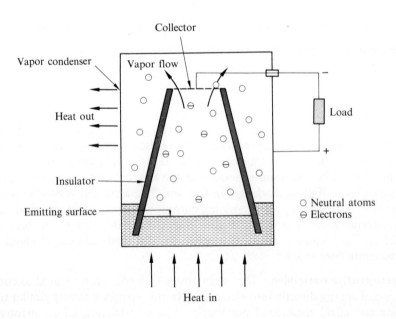

Figure 12.14. Schematic diagram of an electron convection converter. (After Hoh and Harries [88].)

off). The neutral atoms act in a manner similar to EGD by transporting the electrons toward a collector. Thus a potential difference is established between the emitter surface and the collector, creating an electric field in the direction opposite to the flow of electrons. This is a high-voltage generator, the output of which is independent of the energy of the emitted electrons. Low work function liquid materials can also be used, allowing the use of lower emitter temperatures. Experiments have been conducted with this generator by Hoh and Harris [88], who obtained an output voltage of 18 V and currents up to 100 μA for a cesium liquid emitter at a temperature as low as 540°C. The efficiency of such a system is given by

$$\eta = 1 - \ell E/v,$$

where ℓ is the mobility of the electrons, v is the neutral gas velocity, and E is the electric field.

Electrokinetic conversion. When certain types of fluids are confined in a fine capillary tube and subjected to the effect of a pressure gradient, a flow is induced and a difference of potential is created between the ends of the capillary tube in the direction of the pressure gradient. This effect is known as the *electrokinetic effect* and was first discovered by Helmholtz [87] in 1879. The reverse effect leads to so-called electrokinetic pumping, in which a difference of potential creates a pressure gradient. In the electrokinetic converter, mechanical energy is converted directly into electricity. For a fluid electrolyte, this can be explained by faster absorption by the capillary walls of ions than of electrons. In liquid metals, the polarization is due in part to a diffuse electron scattering in the fluid near the walls. The theory of fluids such as hydrocarbons, liquid hydrogen, and mercury is more complicated. Osterle [95] studied an electrokinetic generator and found the maximum efficiency of the device to be

$$\eta_{max} = \tfrac{1}{2}(\Delta p_r)^2 S_p^2 \sigma/\ell,$$

where S_p is the so-called streaming potential, σ is the conductivity, Δp_r is the pressure gradient, and ℓ is the thickness of the capillary tube. The streaming potential is given by

$$S_p = \frac{\varepsilon(-V_T)}{v\sigma},$$

where v is the fluid viscosity, ε is the permittivity, and V_T is the voltage at the surface of the tube (which is negative for a positive charge in the field).

Ferrohydrodynamic conversion. A ferromagnetic fluid is a liquid in which there are strongly polarizable and extremely stable dispersions of magnetic media. When a magnetic field is applied to such a fluid, an appreciable body force is created. The latter is a function of temperature and can be the basis of an ac ferrohydrodynamic heat engine. The engine consists of a duct in which the ferromagnetic fluid is drawn by a coaxial solenoid and heated to a temperature above the Curie temperature, greatly reducing its magnetization. A force is created opposed to the initial flow of the fluid, and electric current can be induced directly in the solenoid coils or in an independent coil. The system becomes similar to an MHD generator, since the ferromagnetic fluid is cycled between a high and a low temperature. The efficiency of such a generator has been estimated by Rosensweig [100] and his co-workers to be

$$\eta = \eta_c \left[1 + \frac{P}{M_f} \cdot \frac{1}{Z}\left(\frac{d_t}{H}\right) \right]^{-1},$$

where η_c is the Carnot efficiency, P is the generated power, M_f is the fluid mass, d is the diameter of the tube, H is the magnetic field intensity, and Z is the figure of merit given by

$$Z = \left(\frac{\Pi_m \Psi}{c_0}\right)^2 \left(\frac{\kappa T_c}{\rho}\right),$$

where Π_m is the pyromagnetic coefficient of the fluid, Ψ is the volumetric loading ratio, c_0 is the volumetric specific heat, and ρ is the density.

Photogalvanic conversion. This method uses the effect of solar radiation in photochemical reactions to generate electricity in a manner similar to fuel cells, thus avoiding the Carnot cycle. Most of the chemical reactions respond only to ultraviolet and visible light. Thus more than half the solar energy is of no use in this method of power generation. However, the advantage of this method is that the products of reactions can be stored for other use. In spite of the fact that the fundamentals of photochemistry are well understood, much work is needed to develop a power converter based on this principle. Two examples of photogalvanic reactions will be given without further detail:

$$\text{Thionine } (C_{12}H_9N_3S) + 2Fe^{++} \underset{\text{dark}}{\overset{\text{light}}{\rightleftharpoons}} \text{leukothionine} + 2Fe^{3+}$$

and

$$Fe^{++} + HgCl_2 \underset{\text{dark}}{\overset{\text{light}}{\rightleftharpoons}} Fe^{3+} + HgCl + Cl^-.$$

12.13 Particle Collecting Converters

In this category are devices that directly collect the products of fission to create a difference of potential between the source of the products (emitter) and the collector. There are three types: α-, β-, and γ-collecting devices. Other types of converters in this category are related to thermionic conversion. In such devices, electrons are emitted by a strong electric field, by secondary emission, or by light (as in the case of photoemissive converters).

Alpha conversion. As in the device described by Plummer and Anno [97], an α-particle energy converter consists of a cylindrical emitter of high-energy α-particles coaxial to a collecting surface. The emitted electrons are turned back to the emitter by a negatively biased grid. In such a system, the kinetic energy of the emitted α-particles is converted into electricity when an external load is connected between the anode and the cathode.

Beta conversion. A source of high-energy electrons such as krypton-85 can be made to emit toward a concentric collecting electrode. Krypton-85 has a half-life of about 10.4 years, thus making the krypton battery a reliable source of high-voltage electrical energy capable of long period operation. Such units are commercially available for charging batteries or other applications. Their output is about 1 mW at 1 kV.

Gamma conversion. A unit based on this effect consists of layers of high and low atomic number materials separated by a dielectric or vacuum. Gamma rays interact with the high atomic number material and electrons are emitted through the Compton effect.* These electrons are collected by the low atomic number material anode and, if an external load is established between anode and cathode, electric energy can be produced. This system has been proposed for SNAP-2, developing 3 kW of electric power [98].

Photoemissive conversion. In this case, electrons are emitted from a photosensitive surface by the action of light. An anode close to the photosensitive cathode collects the electrons. In spite of the similarity with thermionic emission, there is no difference of temperature in this case and therefore the device escapes Carnot's limitation on efficiency. The efficiency is

$$\eta = I_0 V_c / \Phi_s \mathscr{E}_s,$$

where I is the saturated photoelectric current density, Φ_s is the number of photons per unit area per unit time in the solar spectrum, $\mathscr{E}_s$ is their average energy, and V_c is the contact potential difference.

12.14 Other Converters

Many methods used in measuring pressure, force, temperature gradients, and radiation use direct energy conversion principles and may evolve into direct energy converters. Other phenomena of less importance can be found in the literature, such as the production of a potential difference up to 200 V in freezing water. The contact potential between two electrochemically dissimilar electrodes can also be used for direct energy conversion.

Two other sources of energy deserve some attention; these are the geothermal energy and the thermal and magnetohydrodynamic energies of the oceans. Geothermal energy stems from the heat existing inside the earth, at a distance from the surface which in some cases is short enough to trigger economic interest. The energy of the oceans is unlimited and can be divided into three categories: (1) such oceanic "rivers" as the Gulf Stream, which alone generates 182 trillion kWh

---

*   Collision between photons and electrons.

per year, (2) the temperature gradients between the surface (of the hot regions) and the deep water, (3) the tides already produce a considerable amount of energy in France [84] and are actively considered for the production of energy in many countries: US and Canada (Passamaquoddy Bay), USSR [80] (Kislaja Bay near Mourmansk), and England (Severn Bay).

It is clear that this field is wide open to research, and new methods which no one has yet thought of may develop into important ways of producing energy. This is just the beginning of development of this field. Consider the reaction when an electron and its antimatter equivalent, the positron, meet; they will mutually annihilate each other in a burst of energy. This energy will be tapped someday, and will allow an interstellar exploration by man. Consider gravitational energy. Man has not yet the slightest idea of how to control it. The field is wide open for creative minds.

PROBLEMS

12.1 In Eq. (12.1), μ is a scalar only if $\mathbf{B}$ and $\mathbf{H}$ are in the same direction. What does μ become when $\mathbf{B}$ is no longer in the same direction as $\mathbf{H}$? In what type of materials can this happen?

12.2 A coil is wound around a toroidal iron ring of radius r_0 and area πr^2. The ring has a very narrow air gap of width d. Assuming that the toroidal iron has a uniform magnetization $\mathcal{M}$, draw graphs showing the variation of μ, H, B, $\mathcal{M}$, and VH, along the center line of the ring with and without the gap.

12.3 Prove Eq. (12.4).

12.4 Using Fig. 12.4, draw the ferromagnetic cycle of a generator working under the best conditions possible. Use gadolinium as gap material and deduce the values of ℓ_g and ℓ_m which you think are the best.

12.5 Calculate the efficiency of a Nernst generator for maximum power output.

12.6 The following data have been deduced from experimental results [32]:

$$C_N = 58.8 \times 10^{-6} \text{ V/Tesla·°K},$$
$$B = 3.5 \text{ Wb/m}^2,$$
$$T = 130°C,$$
$$\kappa = 110 \text{ W/m·°K},$$

and

$$\sigma = 2.2 \times 10^{-4} \text{ mho/m}.$$

The material is a pyrolytic graphite. Calculate (a) the electronic mobility, (b) the Nernst figure of merit at the given temperature and magnetic field, and (c) compare with the materials of Fig. 12.7.

12.7 (a) Find the value of κ_L for the material of Problem 12.6. (b) Can the Ettinghausen effect be considered as negligible?

12.8 Design a Nernst generator operating between 0 and 100°C in the presence of a 4-Tesla magnetic field. The generator is desired to produce an electrical output of 2 W. Use indium antimonide, which has the following characteristics: $\sigma = 7.5 \times 10^4$ mho/m and

$\kappa = 11.5\,\text{W/m·°K}$; Z_N is given in Fig. 12.7. Find (a) the dimensions of the device, and (b) its maximum efficiency.

12.9 A blackbody source at 2000°K irradiates a TPV converter made of germanium ($eV_g = 0.67$ eV). The TPV converter is maintained at a temperature of 30°C and the ratio of the output current to the recombination current is 80%. Find the overall efficiency of the TPV converter.

12.10 For what temperatures has Fig. 12.12 been drawn?

12.11 Wedlock [72] predicted an efficiency of 35% for a TPV cell powered by a tungsten source at about 2400°K. The cell was a PIN germanium. Justify this prediction.

12.12 Prove Eq. (12.32).

12.13 (a) By writing the equation for motion of electrons in an electron convection converter, show that in the steady state the drift velocity of the electrons is

$$v_d = -eE/v,$$

where v is the viscosity coefficient. (b) Find the expressions for the output current, voltage, and power.

12.14 Water has the following properties [95]: $\varepsilon = 7 \times 10^{-10}$ C/V·m, $v = 10^{-3}$ N·sec/m^2, $\sigma = 10^{-4}$ mho/m, and $V_T = -0.2$ V. (a) Calculate the value of the streaming potential for an electrokinetic generator using water as working fluid. (b) The maximum efficiency of such a generator is 0.4% at 1 atm pressure difference and fluid flow rate of 1.25 l/sec. Find the dimensions of the generator which has the form of a cylindrical pipe.

12.15 An effect similar to the ferrohydromagnetic effect mentioned in Section 12.12 is *electrophoresis*, the movement of suspended particles through a fluid under the effect of an applied electric field. Leiby and Fields [91] proposed this effect for direct energy conversion. Study this type of generator and its feasibility. (See Ref. 81 on electrophoresis.)

REFERENCES AND BIBLIOGRAPHY

A. Ferromagnetic Power Generation

1. Belker, R., and W. Döring, *Ferromagnetismus*, Springer-Verlag, Berlin, 1939.

2. Bitter, F., *Introduction to Ferromagnetism*, McGraw-Hill, New York, 1937.

3. Bozorth, R. M., *Ferromagnetism*, Van Nostrand, Princeton, N.J., 1961.

4. Brillouin, L., and H. P. Iskenderian, "Thermomagnetic Generator," *Elect. Comm.*, Vol. 25, p. 300, 1948.

5. Callaby, D. R., and F. Fatuzzo, "Oscillations of Ferroelectric Bodies," *J. Appl. Phys.*, Vol. 35, No. 8, p. 2443, 1964.

6. Edison, T., British Patent No. 16,709, 1887.

7. Elliott, J. F., "Thermomagnetic Generator," *J. Appl. Phys.*, Vol. 30, No. 11, p. 1774, 1959.

8. Giauque, W. F., and D. P. MacDougall, "The Production of Temperatures below 1°K by Adiabatic Demagnetization of Gadolinium Sulfate," *J. Am. Chem. Soc.*, Vol. 57, No. 6, p. 1175, 1935.

9. Glass, M. S., "Principles of Design of Magnetic Devices for Altitude Control of Satellites," *Bell System Tech. J.*, Vol. 46, No. 4, p. 93, 1967.

10. Judeinstein, A., "Measurement of Magnetic Properties of Thin Films," *Elect. Comm.*, Vol. 41, p. 65, 1966.

11. Judeinstein, A., and J. H. Tyszka, "Magnetic Thin Film Memory Having Low Access Currents," *Elect. Comm.*, Vol. 42, p. 127, 1967.

12. Katz, H. W. (ed.), *Solid State Magnetic and Dielectric Devices*, Wiley, New York, 1959.

13. Kittel, C., and J. K. Galt, "Ferromagnetic Domain Theory," *Solid State Phys.*, Vol. 3, Academic Press, New York, 1956.

14. Rodenback, G. W., *Thermomagnetic Generator*, Report NAA-SR-264, North American Aviation, 1958.

15. Sivertsen, J. M., "Resistivity of Ferromagnetic Alloys," *J. Appl. Phys.*, Vol. 35, No. 8, p. 2407, 1964.

16. Sparks, M., *Ferromagnetic Relaxation Theory*, McGraw-Hill, New York, 1964.

17. Stauss, H. E., "Efficiency of Thermomagnetic Generator," *J. Appl. Phys.*, Vol. 30, p. 1622, 1959.

18. Thomas, H., "Paramagnetic Behavior of Thin Ferromagnetic Films Above the Curie Point," *Zs. Angew. Physik*, Vol. 15, No. 3, p. 201, 1963.

19. Trombe, F., "Magnetic Properties of Rare Metals," *Ann. Physique (France)*, Vol. 7, p. 383, 1937.

20. Trombe, F., "Ferromagnetism and Paramagnetism of Metallic Dysprosium," *Compt. Rend.*, Vol. 221, p. 19, 1945 (French).

21. Weiss, P., "Hypothesis of the Molecular Field and Ferromagnetic Properties," *J. Physique*, Vol. 6, p. 661, 1907 (French).

22. Weiss, P., and R. Forrey, "Aimantation et Phénomène Magnétocalorique du Nickel," *Ann.* Physique *(France)*, Vol. 15, p. 153, 1920.

23. Williams, H. J., "Magnetic Properties of Single Crystals of Si-Fe," *Phys. Rev.*, Vol. 52, p. 747, 1937.

24. Williams, H. J., R. F. Bozorth, and W. Shockley, "Magnetic Domain Patterns on Single Crystal of Silicon Iron," *Phys. Rev.*, Vol. 75, No. 1, p. 155, 1949.

25. Williams, H. J., and W. Shockley, "A Simple Domain Structure in an Iron Crystal Showing a Direct Correlation with Magnetization," *Phys. Rev.*, Vol. 75, p. 178, 1949.

B. Nernst Power Generation

26. Angrist, S. W., "Galvanomagnetic and Thermomagnetic Effects," *Scientific American*, Vol. 205, No. 6, p. 124, 1961.

27. Angrist, S. W., "A Nernst Effect Power Generator," *J. Heat Transfer, ASME Trans.*, Vol. 85(2), p. 41, Feb. 1963.

28. Angrist, S. W., "On the Boundedness of the Dimensionless Index of Performance of a Nernst Effect Generator," *J. Appl. Mechanics, ASME Trans.*, Vol. 85(3), p. 291, June 1963.

29. Boer, A. C., *Galvanomagnetic Effects in Semiconductors*, Academic Press, New York, 1963.

30. Bridgeman, P. W., *Thermodynamics of Electrical Phenomena in Metals*, Macmillan, New York, 1934, p. 137.

31. Callen, H. B., "A Note on the Adiabatic Thermomagnetic Effects," *Phys. Rev.*, Vol. 85, No. 1, p. 16, 1952.

32. Colwell, J., G. Guthrie, and R. Palmer, *Static Energy Conversion Studies*, Technical Doc. Rep. No. ASD-TDR-63-372, April 1963.

33. Delves, R. T., "The Prospects for Ettinghausen and Peltier Cooling at Low Temperatures," *Brit. J. Appl. Phys.*, Vol. 13, No. 9, p. 440, 1962.

34. El-Saden, W. R., "Irreversible Thermodynamics and the Theoretical Bound on the Thermomagnetic Figure of Merit," *J. Appl. Phys.*, Vol. 33, No. 10, p. 3145, 1962.

35. Ettinghausen, A., and W. Nernst, *Wied. Ann.*, Vol. 29, p. 343, 1886.

36. Goldsmid, H. J., D. W. Hazelden, and H. E. Ertl, *Thermomagnetic Effects in Bismuth-Antimony Alloys*, Proceedings of International Conference on Physics of Semiconductors, A. C. Strickland (ed.), The Inst. of Physics and the Phys. Soc., London, 1962, p. 777.

37. Harman, T. C., and J. M. Honig, "Operating Characteristics of Transverse Anisotropic Galvano-Thermomagnetic Generators," *Appl. Phys. Letters*, Vol. 1, p. 31, Oct. 1962.

38. Harman, T. C., and J. M. Honig, "Theory of Galvano-Thermomagnetic Energy Conversion Devices," *J. Appl. Phys.*, Vol. 33, No. 8, p. 3170, 1962.

39. Harman, T. C., "Theory of the Infinite Stage Nernst-Ettinghausen Refrigerator," *Adv. Energy Conversion*, Vol. 3, p. 667, Oct. 1963.

40. Harman, T. C., J. M. Honig, and B. M. Tarmy, "Galvano-Thermomagnetic Phenomena and the Figure of Merit of Bismuth I," *Adv. Energy Conversion*, Vol. 5, p. 1, May 1965.

41. Harman, T. C., J. M. Honig, and L. Jones, "Galvano-Thermomagnetic Phenomena and Figure of Merit of Bismuth II," *Adv. Energy Conversion*, Vol. 5, p. 183, Nov. 1965.

42. Harman, T. C., and J. M. Honig, "A Note Concerning the Temperature Profile Within a Thermomagnetic Energy Converter," *Adv. Energy Conversion*, Vol. 6, p. 127, April 1966.

43. Harman, T. C., and J. M. Honig, *Thermoelectric and Thermomagnetic Effects and Applications*, McGraw-Hill, New York, 1967.

44. Honig, J. M., and T. C. Harman, "Galvano-Thermomagnetic Effects in Multi-Band Models," *Adv. Energy Conversion*, Vol. 3, p. 529, July 1963.

45. Honig, J. M., and T. C. Harman, "Structure of Equations Specifying Operating Characteristics of Energy Converters Constructed of Anisotropic Materials," *Adv. Energy Conversion*, Vol. 6, p. 149, July 1966.

46. Leduc, S., *Compt. Rend.*, Vol. 104, p. 1783, 1887 (French).

47. O'Brien, B. J., and C. S. Wallace, "Ettinghausen Effect and Thermomagnetic Cooling," *J. Appl. Phys.*, Vol. 29, No. 7, p. 1010, 1958.

48. Righi, A., *Compt. Rend.*, Vol. 105, p. 168, 1887 (French).

49. Simon, E., "Relationship between Nernst, Ettinghausen and Peltier-Seebeck Figures of Merit in Strong Magnetic Field," *Adv. Energy Conversion*, Vol. 4, No. 12, p. 237, 1964.

50. Tsidel'Kovskii, I. M., *Thermomagnetic Effects in Semiconductors*, Infosearch, London, 1962.

51. Ure, R. W., "Theory of Materials for Thermoelectric and Thermomagnetic Devices," *Proc. IEEE*, Vol. 51, No. 5, p. 633, 1963.

C. Thermophotovoltaic Power Generation

52. Bauduin, P., and P. Sibillot, "InAs Diodes Used for the Thermophotovoltaic Conversion of Energy," *Adv. Energy Conversion*, Vol. 6, p. 67, Jan. 1966.

53. Beck, R. W., *Study of Germanium Devices for Use in a Thermophotovoltaic Converter*, Progress Report No. 1, DA28-043-AMC-02543(E), General Motors Corp., Feb. 1967.

54. Bomel, R., "Detection of Nuclear Particles with Semiconductors," *Bull. Info. Sci. Tech. (Paris)*, No. 34, p. 2, 1959.

55. Braunstein, R., A. R. Moore, and F. Herman, "Intrinsic Optical Absorption in GeSi Alloys," *Phys. Rev.*, Vol. 109, No. 2, p. 695, 1958.

56. Crawford, J. H., and D. S. Billington, *Radiation Damage in Solids*, Princeton University Press, Princeton, N.J., 1961.

57. Crouch, D. P., and R. W. Beck, "Study of Germanium Devices for Use in a Thermophotovoltaic Converter," Progress Report No. 2, DA28-D43-AMC-01420(E) General Motors Corp., March 1966.

58. Dash, W. C., and R. Newman, "Intrinsic Optical Absorption in Single Crystal Germanium and Silicon at 77°K and 300°K," *Phys. Rev.*, Vol. 99, No. 8, p. 1151, 1955.

59. Eisenman, W. L., R. L. Bates, and J. D. Meriam, "Black Radiation Detector," *J. Opt. Soc. Am.*, Vol. 53, p. 729, 1963.

60. Fortini, A., P. Bauduin, and P. Sibillot, "Réalisation d'un Convertisseur Thermophotovoltaique," *Revue Générale de l'Electricité (Paris)*, Vol. 13, No. 9, p. 466, 1964.

61. Gritton, D. G., and R. C. Bourke, "Radioisotope-Photovoltaic Energy Conversion System," *Adv. Energy Conversion*, Vol. 5, No. 2, p. 119, 1965.

62. Gubareff, G. G., J. E. Janssen, and R. H. Forborg, *Thermal Radiation Properties Survey*, Honeywell Research Center, Honeywell Regulator Co., Minneapolis, 1962.

63. Kittl, E., *Thermophotovoltaic Energy Conversion*, 20th Power Sources Conference Proceedings, PSC Publications Committee, Red Bank, N.J., 1967, p. 780.

64. Liston, M. D., and J. U. White, "Amplification and Electrical Systems for a Double Beam Recording Infra-Red Spectrophotometer," *J. Opt. Soc. Am.*, Vol. 40, p. 36, 1950.

65. Nachtigall, D., "Isotope-Batteries," *Die Tech. (Germany)*, Vol. 13, p. 300, April 1958.

66. Ruth, R. P., and J. W. Moyer, "Power Efficiency for the Photovoltaic Effect in a Ge Grown Junction," *Phys. Rev.*, Vol. 95, p. 562, 1954.

67. Shapiro, S. J., *Thermophotovoltaic Spectral Analysis of a TPV System*, 21st Power Sources Conference Proceedings, PSC Publications Committee, Red Bank, N.J., 1967, p. 138.

68. Smith, A. H., and R. C. Bourke, *Radiant Energy Conversion*, General Motors Corp., Allison Res. and Eng., Vol. 5, p. 16, 1965.

69. Vul, B. M., et al., "The Conversion of the Energy of β-Particles to Electron Energy in Germanium Crystals with p-n Junction," *Kernenergie*, Vol. 1, p. 279, April 1958.

70. Watts, H. V., M. D. Oestreich, and R. J. Robinson, *A Nuclear Photon Energy Conversion Study*, Armour Research Foundation Quarterly Technical Report ARF 1214-TR-2, AD 286024, Oct. 1962.

71. Watts, H. V., and S. Nudelman, *A Nuclear Photon Energy Conversion Study*, Armour Research Foundation Quarterly Technical Report ARF 1214-TR-3, AD 286024, Jan. 1963.

72. Wedlock, B. D., "Thermophotovoltaic Energy Conversion," *Proc. IEEE*, Vol. 51, No. 5, p. 694, 1963.

73. Wedlock, B. D., A. Debs, and R. Siegel, *Investigation of Germanium Diodes for Thermophotovoltaic Converters*, Final Technical Report DA44-009-AMC-625(T), MIT, Cambridge, Mass., Sept. 1965.

74. Wedlock, B. D., and R. Siegel, *Investigation of P-I-N Germanium Diodes for TPV Conversion*, 20th Power Source Conference Proceedings, PSC Publications Committee, Red Bank, N.J., 1966, p. 182.

75. Werth, J., *Thermophotovoltaic Energy Conversion*, 17th Power Sources Conference Proceedings, PSC Publications Committee, Red Bank, N.J., May 1963.

76. Werth, J., *Design Study of a Thermophotovoltaic Converter*, 18th Power Sources Conference Proceedings, PSC Publications Committee, Red Bank, N.J., May 1964, p. 125.

77. White, D. C., B. D. Wedlock and J. Blair, *Recent Advances in Thermal Energy Conversion*, 15th Power Sources Conference Proceedings, PSC Publications Committee, Red Bank, N.J., May 1961, p. 125.

78. White, D. C., and R. J. Schwartz, *P-I-N Structure for Controlled Spectrum Photovoltaic Converter*, 6th AGARD Combustion and Propulsion Colloquium, Cannes, France, 1964.

D. Other Methods of Direct Energy Conversion

79. Anderson, J. H., and J. H. Anderson, Jr., "Thermal Energy from Sea-water," *Mechanical Engineering*, Vol. 88, No. 4, p. 41, 1966.

80. Bernstein, M. L., *Power Stations Using the Energy of the Tides and the Modern Need for Energy*, Moscow, 1961.

81. Birks, J. B., "Electrophoretic Deposition of Insulating Materials," in *Progress in Dielectrics*, J. B. Birks and J. H. Schulman (eds.), Wiley, New York, 1959.

82. Chester, M., "Thermodynamics of a Superconductivity Energy Converter," *J. Appl. Phys.*, Vol. 33, No. 2, p. 643, 1962.

83. Gardner, G., "Conversion of Sun Radiation to Electricity," *Sol. Energy*, Vol. 4, No. 1, p. 25, 1960.

84. Gibrat, R., *L'Energie des Marées*, Presses Universitaires de France, Paris, 1966.

85. Gross, P., and P. V. Murphy, "Currents from Make Detectors and Batteries," *Nucleonics*, Vol. 19, No. 3, p. 86, 1961.

86. Heindl, C. J., W. F. Krieve, and R. V. Meghreblian, "Fission Fragment Conversion Reactors for Space," *Nucleonics*, Vol. 21, No. 4, p. 80, 1963.

87. Helmholtz, H. L., *Wied. Ann.*, Vol. 7, p. 337, 1879.

88. Hoh, S. R., and W. J. Harries, *Power Conversion by Electron Convection*, Tech. Doc. Rept. No. ASD-TDR-62-693, AD 291683, 1962.

89. Jensen, A. S., and I. Limensky, *Photoemission Solar Energy Converter*, 14th Power Sources Conference Proceedings, PSC Publications Committee, Red Bank, N.J., 1960.

90. Kikoin, I. K., and M. M. Noskov, "A New Photoelectric Effect in Cuprous Oxide," *Phys. Zh. Sovjet*, Vol. 5, p. 586, 1934.

91. Leiby, C. C., Jr., and J. C. Fields, *Electrophoretic Power Generation in Thermally Ionized Plasmas*, Air Force Cambridge Research Laboratories, 1965.

92. Linder, E. G., and S. M. Christian, "The Use of Radioactive Material for the Generation of High Voltage," *J. Appl. Phys.*, Vol. 23, p. 1213, 1952.

93. Oldham, I. B., F. J. Young, and J. F. Osterle, "Streaming Potential in Small Capillaries," *J. Colloid Science*, Vol. 18, p. 328, 1963.

94. Osborn, J. A., "Magnetostriction Generators," in *Sources of Electric Energy*, AIEE, Jan. 1951.

95. Osterle, J. F., "Electrokinetic Energy Conversion," *J. Appl. Mechanics, Trans. ASME*, Vol. 31, Series E, p. 161, 1964.

96. Perdreaux, R., "Nuclear Applications," *Electro-Technology*, Vol. 69, No. 4, p. 152, 1962.

97. Plummer, A. M., and J. N. Anno, *Conversion of α-Particle Kinetic Energy into Electricity*, AMU-ANL Conference on Direct Energy Conversion, Nov. 4–5, 1963; published by Office of Technical Services, U.S. Dept. of Commerce, 1964, p. 170.

98. Raab, B., "Power-Producing Shield for Space Reactors," *Nucleonics*, Vol. 21, No. 2, p. 46, 1963.

99. Riddiford, A. W., and J. A. Krumhansl, "A Model for a Diffusive Magnetothermo-electric Generator," *J. Appl. Phys.*, Vol. 34, No. 12, p. 3572, 1963.

100. Rosensweig, R. E., J. W. Nestor, and R. S. Timmins, *Ferrohydrodynamic Fluids for Direct Conversion of Heat Energy*, Proceedings of the Symposium on Materials Associated with Direct Energy Conversion, Institution of Chemical Engineers, London, 1965.

101. Saucer, K. M., *Photo-Galvanic Cells*, Conference on Use of Solar Energy; the Scientific Basis, University of Arizona, Tuscon, Vol. 5, 1958, p. 43.

102. Simon, R. E., and W. E. Spicer, "Photoemission from Si Induced by an Internal Electric Field," *Phys. Rev.*, Vol. 119, p. 621, July 1960.

103. Taylor, J. M., *Semiconductor Particle Detectors*, Butterworths, Washington, D.C., 1963.

104. Van Roosbroeck, W., "Theory of the Photomagnetic Effect in Semiconductors," *Phys. Rev.*, Vol. 101, No. 6, p. 1713, 1956.

105. Wolfe, R., and G. E. Smith, *Effects of a Magnetic Field on the Thermoelectric Properties of Si-Sb Alloys*, International Conference on Physics of Semiconductors, Exeter, Institute of Physics and Physical Society, London, 1962, p. 771.

APPENDIX I

PHYSICAL CONSTANTS

a_g = standard gravitational acceleration = 9.806 m/sec^2

$\mathscr{B}_1$ = Stephan-Boltzmann constant = 5.67 × 10^{-8} W/m^2 (°K)^4

c = speed of light in vacuum = 2.99793 × 10^8 m/sec

e = charge of an electron = 1.6021 × 10^{-19} C

F_n = Faraday = 0.96487 × 10^5 C/mole

F'_n = Faraday's constant = 96,522 C

G = gravitational constant = 0.667 × 10^{-10} m^3/kg·sec^2

h = Planck's constant = 6.625 × 10^{-34} J·sec

k = Boltzmann's constant = 1.38054 × 10^{-23} J/°K

$\mathscr{L}$ = Lorentz' number = 0.37 × 10^8 °K^2/V^2

M_e = mass of an electron = 9.108 × 10^{-31} kg

M_n = mass of a proton = 1.674 × 10^{-27} kg

N_0 = Avogadro's number = 6.0225 × 10^{23} particles/mole

$\mathscr{R}$ = universal constant of gases = 8.3143 J/mole·°K

ε_0 = permittivity of free space = 8.854 × 10^{-12} F/m

μ_0 = permeability of free space = 1.257 × 10^{-6} H/m

APPENDIX II

CONVERSION FACTORS*

1 abampere = 10 amperes

1 abcoulomb = 10 coulombs

1 abfarad = 10^9 farads

1 abhenry = 10^{-9} henry

1 abohm = 10^{-9} ohm

1 abvolt = 10^{-8} volt

1 angstrom = 10^{-10} meter

1 atmosphere = 1.01325×10^5 newtons/meter^2

1 barn = 10^{-28} meter^2

1 Btu = 1055 joules

1 calorie = 4.18674 joules

1 degree Celsius = 1 degree Kelvin − 273.15

1 curie = 3.70×10^{10} desintegrations/second

1 degree = 0.01745 radians

1 dyne = 10^{-5} newton

1 electron-volt = 1.6021×10^{-19} joule

1 erg = 10^{-7} joule

1 degree Fahrenheit = $\frac{9}{5}$ degree Kelvin − 827.4

1 fermi = 10^{-15} meter

1 foot = 0.3048 meter

*   E. A. Mechtley, *The International System of Units, Physical Constants, and Conversion Factors*, NASA SP-7012, 1964.

1 gauss $= 10^{-4}$ weber/meter^2 (Tesla)

1 gilbert $= 0.7958$ ampere-turn

1 horsepower $= 746$ watts

1 inch $= 0.0254$ meter

1 kilogramforce $= 9.806$ newtons

1 langley $= 4.184 \times 10^4$ joule/meter^2

1 liter $= 10^{-3}$ meter^3

1 maxwell $= 10^{-8}$ weber

1 micron $= 10^{-6}$ meter

1 mile (U.S.) $= 1.6093 \times 10^3$ meters

1 millibar $= 100$ newton/meter^2

1 oersted $= 79.577$ ampere-turn/meter

1 ounce $= 0.02835$ kilogram

1 poise $= 0.1$ newton $\times$ second/meter^2

1 pound $= 0.4536$ kilogram

1 pound (wt) $= 4.448$ newton

1 roentgen $= 2.5798 \times 10^{-4}$ coulomb/kilogram

1 statampere $= 3.335 \times 10^{-10}$ ampere

1 statcoulomb $= 3.335 \times 10^{-10}$ coulomb

1 statfarad $= 1.1126 \times 10^{-12}$ farad

1 stathenry $= 8.988 \times 10^{11}$ henrys

1 statohm $= 8.988 \times 10^{11}$ ohms

1 statvolt $= 299.793$ volts

1 stoke $= 10^{-4}$ meter^2/second

1 torr (0°C) $= 133.322$ newton/meter^2

1 yard $= 0.9144$ meter

INDEX